Principles of Cartography

McGRAW-HILL SERIES IN GEOGRAPHY

John C. Weaver, *Consulting Editor*

Bennett—Soil Conservation

Cressey—Asia's Lands and Peoples

Cressey—Land of the 500 Million: A Geography of China

Finch, Trewartha, Robinson, and Hammond—Elements of Geography: Physical and Cultural

Finch, Trewartha, Robinson, and Hammond—Physical Elements of Geography
     (A republication of Part I of the above)

Pounds—Europe and the Mediterranean

Pounds—Political Geography

Raisz—General Cartography

Raisz—Principles of Cartography

Thoman—The Geography of Economic Activity

Trewartha—An Introduction to Climate

Trewartha, Robinson, and Hammond—Fundamentals of Physical Geography

Vernor C. Finch was Consulting Editor of this series from its inception in 1934 to 1951.

Principles of Cartography

Erwin Raisz

McGraw-Hill Book Company, Inc.

New York   San Francisco   Toronto   London

PRINCIPLES OF CARTOGRAPHY

Preface

Twenty years ago when I wrote "General Cartography" most of my thoughts concerned the subject matter. At that time cartography in colleges was relatively new, and the subject had to be defined, delimited, and organized to serve American demands. After teaching cartography for three decades, however, my orientation has changed somewhat as I have become more and more concerned with the student's reaction. The most logical organization of subject matter is of little use if the interest of the student is not aroused and if he cannot understand the subject and apply it to his purposes.

This book is smaller and somewhat simpler than "General Cartography." Its aim is to guide the student to understand the language of maps, to enable him to illustrate his own papers, and to give him a foundation if he chooses to become a cartographer. The book is written primarily for college students, but it is not too difficult for high school classes or for the general reader. No mathematics is required beyond the most elementary trigonometry.

The book is arranged so that the first exercises will take the students into the field to present the basic relationship between land and map. This will give him a foundation for further work, such as designing symbols, choosing colors, and understanding the principles of cartography. Field work will also bring student and instructor into closer contact.

The chapters are arranged primarily for a practical sequence of exercises. For instance, it may seem illogical to place the discussion of the earth's size and shape in the middle of the book, but the student does not need this for the first exercises. The chapters are handled as independent units, and their sequence can be altered by the instructor to adapt them to seasons and curriculum. It may also seem unusual to start the book with a chapter on tools. Yet, in the author's experience, this is necessary for starting the laboratory work. Students are eager to handle their new tools and their interest is awakened.

In 1920 there were only two universities giving courses in cartography. At present the number is well over 100. This progress is phenomenal, yet far from enough. Every college, high school, and normal school where geography is taught should have a course on maps, the most effective tool with which to express geographic relationships. To fulfill this growing demand, "Principles of Cartography" makes its hopeful appearance.

The author is indebted to Donald G. Bouma for his help on the chapter on photography. The chapter on modern tech-

v

niques was corrected by Samuel Sachs. Dr. L. Tisza prepared the table of chord distances. Prof. Philip W. Porter wrote the part on regressional equations. Mrs. Eileen Schell read the book critically and gave many helpful suggestions. Corrections in language were made by Mrs. Grace R. Smith and Marjorie Hurd. Mrs. Doretha Bouché typed the manuscript. Mrs. Clara E. LeGear helped in the selection of the atlas title page. The author wants to express his thanks for all the illustrations taken from various publications. The sources are acknowledged in the captions.

Erwin Raisz

Contents

Introduction

The Earth's Patterns. The nature of cartography can best be explained by the simile of the ant on a rug. The ant walking on the rug will notice the various colors and textures, but they will have no meaning for him. Let us imagine that some of the ants want to find out what these patterns mean. They assign parties to measure the whole rug and every patch on it; other groups will collect these measurements and will devise methods to draw them on such a small scale that the pattern of the whole rug with all its patches will be visible at a glance. They will be thrilled to see a beautiful design revealed which they did not know before. Immediately some of the wisest ants will propose various theories for the possible use and ultimate meaning of this pattern.

The ants' task is relatively easy, compared with that of man. The ant is a million times larger in proportion to the rug than man is in comparison with the earth. The richest oriental rug is much simpler than the carpet of the earth.

First the surveyors measure the land. Cartographers collect the measurements and render them on such small scale that the earth's wonderful pattern shows at a glance. The geographer analyzes this pattern, studies man's relation to it, and theorizes about its ultimate meaning. About this, we hope, he does better than the ant, who probably has only a vague idea, if any, of the existence of man, the creator of the rug.

The Scope of Cartography. The scope of cartography is shown in Fig. Int. 1. The aim of a map is to show some kind of geographic space relationship. In most cases we want to give the all-around picture, as on a topographic or atlas map. Often the purpose is to make special maps concerning geology, history, economics, roads, airways, etc.—almost anything that involves a large area. Maps are not the only means of cartographic expression; globes, models, and cartograms are others. (For definitions of unfamiliar terms see the Glossary and the Index at the end of this book.)

A map has an advantage over text in that it can be seen at a glance, while words must come in sequence. No verbal description of a region can rival the impact and retention possibilities of a map. Verbal description has its place, however, in regional analysis and interpretation of the map.

The primary sources of maps are explorers' accounts, surveying, and air photography. Secondary sources are other maps, books, and articles. The cartographer takes materials from these sources, adds generously from his own art and technique, and,

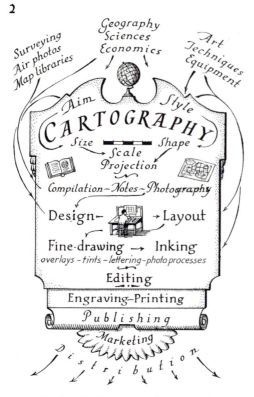

Fig. Int. 1. The scope of cartography.

sketch paper and plan our general design. We decide on the over-all style, where the title should be, what type of frame to use, etc. Next comes the pencil layout of the parallels and meridians, rivers, cities, boundaries, mountains—whatever is to go on the map. Fine drawing and inking follow. We may use such mechanical devices as cellotints, cellotype, negative scribing, etc. If colors are used, we prepare a fully colored layout and a color plate is prepared for each color. Careful editing will always disclose some mistakes, which must be corrected. Rarely does a cartographer do his own engraving and printing. Publishing and distribution are also usually done by specialists.

Who Will Make a Good Cartographer? His tasks may range from a battle plan of Marathon to a model of a tropical hurricane, so he should have wide background knowledge. He has to have a sense of proportion and some manual dexterity. Artistic ability helps decidedly, but, with the many mechanical aids available, almost anybody can produce satisfactory maps. But, most important of all, the cartographer must be intensely interested in land. The author is often asked by students whether they should go into cartography as a profession. The answer is: If, while riding on train or plane, he spends his time reading a magazine or even a textbook on cartography instead of looking out of the window studying the landscape, he will not make a happy cartographer. He may, however, push a graver for scribing in a large office quite successfully.

using his specialized equipment, forges them into a good, usable, and, we hope, beautiful map.

The cartographer is sometimes free to choose the size and shape of his map, but often he is limited by the page size of a book or atlas, the size of paper or printing press, etc. This enters into the choice of scale. This and a number of other factors enter into the choice of projection.

Next comes the compilation of map data. The maker of topographic maps (large-scale general maps showing water, land relief, and man-made features) will use air photos or the plane-table sheets of surveyors. However, geographers are more likely to use notes, sketches, and existing maps.

After compiling our material, we take

We may distinguish four types of cartographers. The *geocartographer* is trained in geography and he is interested more in small-scale and special maps. The *topocartographer* has an engineering background and is interested in surveying and producing large-scale topographic maps. The man transforming an air photo into a map is an *aerocartographer*. His work is closely related to that of the others, but he specializes in photogrammetry. The *cartotechnician* has his training in art, engraving, printing, and photographic processes. His work begins when he receives a layout or guide copy from a geo- or topocartographer. The larger the establishment, the more profound is the differentiation of functions. This book deals primarily with the function of the geocartographer, but it would resemble a house without foundations if it neglected the other functions.

Education. What courses should the cartographer take? First of all, a complete education in geography is necessary. A course in geomorphology will help him to understand the earth's basic patterns. Although this book discusses the fundamentals of surveying and photogrammetry, a cartographer can gain much from a special course in these subjects. A course in art, dealing with design, lines, colors, and techniques, is also recommended. The average geocartographer may be 40 per cent geographer, 20 per cent artist, 20 per cent technician, 10 per cent mathematician, and 10 per cent everything else.

History of Maps [1]

Today's cartography can be understood best from a study of its historical development. In Fig. Int. 2 we see a map—really a relief model—made by an Eskimo. He did not know how to read or write and never heard of cartography. Yet he built

[1] The history of cartography is treated in some detail in the author's "General Cartography." Here only a short outline is given.

Fig. Int. 2. Eskimo map of carved wood, pasted on sealskin. (*Library of Congress.*)

this model according to cartographic principles, to scale, and as seen from above. He never saw the region this way. He put together thousands of mental images of hundreds of voyages and, by complex mental processes, forged them into the concept this map represents. Hunters, nomads, and seafaring people can make very good maps, even if they draw them with sticks on the ground. The ability to make maps is an innate ability of mankind.

Antiquity. The oldest map known today is a small clay tablet showing the location of a man's estate in Mesopotamia dating from about 2800 b.c. (see Fig. Int. 3). The division of the circle into 360 degrees is also probably of Babylonian origin. The Egyptians staked out their precious land and measured and mapped it for taxation purposes; these were the ancestors of our "cadastral," or real estate maps. Doubtless the ancient Indians and Persians also made good maps, of which almost none survived. The Chinese, however, developed their cartography to a high degree in very early times, and we have many records of it and a few actual remnants. Maps have hard use and perish more easily than books or other records.

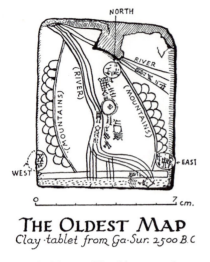

Fig. Int. 3. The world's oldest map is a small clay tablet preserved in the Semitic Museum of Harvard University.

Fig. Int. 5. The "Orbis Terrarum" of the Romans Note that almost the entire world is part of the Roman Empire.

Fig. Int. 4. Greek scientists recognized the spherical form of the earth, measured its size, and designed systems of parallels and meridians. The map of Eratosthenes is a reconstruction.

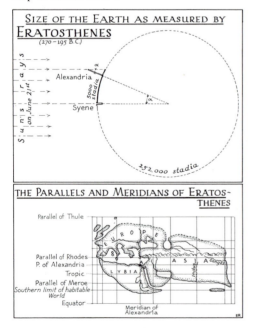

Scientific cartography was an achievement of the ancient Greeks. They discovered the spherical shape of the earth, measured its size, defined the poles, the equator, and the tropics. They designed the parallel-meridian system, divided into degrees as we use it today. The Romans, however, were more interested in a practical map for travel and war, and designed a disk-shaped map, the "Orbis Terrarum," which was widely imitated in the Middle Ages. (See Fig. Int. 5.)

The Ptolemy Atlas. The famous Ptolemy atlas as we know it today is probably of thirteenth-century Byzantine design, based on the works of Claudius Ptolemy of Alexandria, ca. 150 A.D. How much of the atlas is original with Ptolemy is not known. The world map, Fig. Int. 6, shows a projection system with converging meridians, and is divided into degrees. The latitudes are also divided into zones of *climata*, meaning the length of the longest day. The Ptolemy

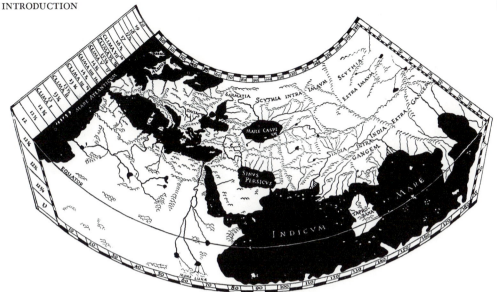

Fig. Int. 6. Ptolemy's map represents the summary of Greek geography. Note the conic projection and the system of climates (length of the longest day) on the left. This map is a part of an atlas, probably of thirteenth-century Byzantine design, based on Ptolemy's works.

maps had an enormous influence on posterity. Even some of their mistakes, like the exaggerated Mediterranean Sea, the mid-Saharan river, and others, were copied for many centuries.

Middle Ages. Medieval man was an introvert. He designed this world, not as it is, but to conform to his religious beliefs. The irregular outlines of the ancient maps did not agree with his idea of godly perfection, so he tried to change the world into a more harmonious shape. He went so far that, by manipulating the Roman map, he designed the cartogram shown in Fig. Int. 7a. He must have been delighted to notice that not only did God create this world in perfect symmetry, but also shaped according to its Latin initials. At the same time the Arabs also made a circular diagrammatic map, but with much better proportions, for use in school atlases (see Fig. Int. 7b).

A miraculous product of the late Middle Ages is the portolan chart. Figure Int. 8 is a later, somewhat improved, copy. Never before had such an accurate map been produced. No doubt it is the result of a systematic compass survey, most probably conducted by the Genoese admiralty in the second half of the thirteenth century, when the use of compass became universal.

Fig. Int. 7. a. The divine perfection and simplicity of the "T-in-O" ("Orbis Terrarum") appealed to the medieval mind. b. These cartograms were used in medieval Arabic school maps.

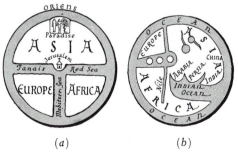

(a)   (b)

Fig. Int. 8. The portolan charts were the result of a compass survey. Note the difference between the accuracy of the Mediterranean and the crude design of the Baltic. The original is richly colored. (*Pseudo-Pilestrina chart after Wolfenbüttel in Imago Mundi.*)

Hundreds of copies survived because it was used with slight improvements for actual navigation for over three centuries. Most charts have 16 or 32 compass roses with radiating *rhumb lines* (lines showing compass direction), a motif often inserted on modern maps as decoration.

The Renaissance of Cartography. Starting with the portolan charts, this came to full bloom in the sixteenth century. A new interest in the outside world replaced the inward-looking attitude of medieval man. At this time, engraving and printing were invented. The revival of the ancient Greek and Roman culture resulted in over a hundred new editions of the Ptolemy atlas, with the addition of modern maps. The most far-reaching effects, however, came from the "Great Discoveries." Maps and

charts were needed to present the lay of the Americas and the routes to the Indies. The result was an outburst of cartographical activity unparalleled until very recent times.

The outlines of four maps shown in Fig. Int. 9 illustrate the confusion brought about by the discovery of America. La Cosa's map is the first to show America. The Contarini map shows North America as part of Asia. Waldseemüller's map is famous because it is the first with the name "America." It is a large, beautiful, wood-engraved map, based largely on Ptolemy. Diego Ribero's map, drawn after Magellan's voyage, shows the vast immensity of the Pacific Ocean and has an almost modern look.

The best maps of the late sixteenth and

the seventeenth centuries came from the Netherlands. Today, collectors pay hundreds or even thousands of dollars for atlases made by Mercator, Ortelius, Hondius, Janszoon, Blaeu, and others because of their beauty and the high quality of engraving. We are still learning balanced composition, expressive symbolism, and good lettering from these Dutch masters. Many of our artistic maps still reflect the style of this period.

The Reformation of Cartography. This period began with the longitude measurements of the French Academy at the end of the seventeenth century. We refer to the eighteenth century as the Age of Reason; the spirit of the period is reflected in its maps. They are far less decorative, but much more accurate, than their seventeenth-century Dutch predecessors (see Fig. Int. 10).

The greatest achievement of this age was the triangulation and topographic mapping of France, directed by members of the Cassini family.

This series set the pattern for the national surveys of the nineteenth century. Large-scale topographic maps and charts of a nation can be produced only by a large organization. The tasks of the independent cartographers were narrowed mostly to small-scale maps, a division which still persists.

The nineteenth century witnessed a great diversification of maps. Geologic, economic, educational, transportation, and other maps required new approaches—to the great enrichment of cartography. New engraving processes, lithography, wax engraving, photoengraving, and color printing made new techniques possible. Map making has become concentrated in large gov-

ernmental and private offices, producing maps by the millions, reaching great masses of people.

Airplane Photography. We are living in an age that has witnessed the most impor-

Fig. Int. 9. Geographical conceptions changed more rapidly in the first quarter of the sixteenth century than ever before or since. These are outlines of large detailed maps.

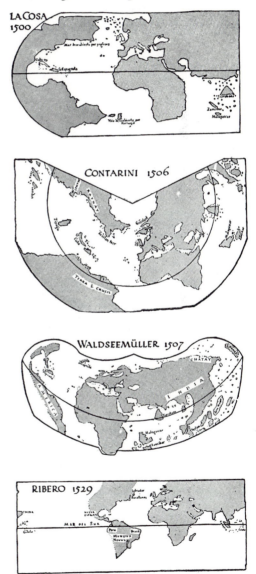

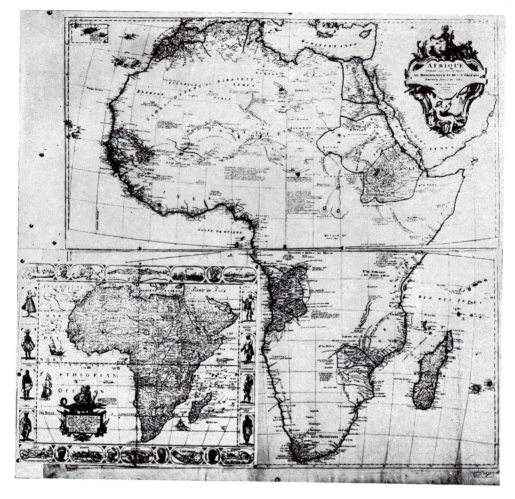

Fig. Int. 10. D'Anville's Africa illustrated the reformation of cartography. Its scientific accuracy and omission of all doubtful data are in contrast with the decorative map of Janszoon, in the lower left, drawn a century earlier.

tant event in the history of cartography since the "Great Discoveries." Aerial photography makes it possible to map even the most inaccessible areas of the world with great speed and accuracy. It enables us to see the earth from above, as the maps show it. Our emphasis on techniques often led us to loose touch with the land. Now, when we can look upon the face of Mother Earth, we cannot be quite satisfied with many of our products and we try to find a better symbolism to close the gap between land and map.

American Cartography. The map of this continent gradually emerged and by the eighteenth century a great number of fairly detailed maps were available. These maps, however, were American in subject only. They were made by English, French, Dutch, or Spanish cartographers.

Colonial cartography developed slowly in the eighteenth century (see Fig. 14-1). British and Spanish army officers produced many excellent maps and some of these men were employed later by Washington and Jefferson, both expert surveyors.

In the early nineteenth century, Army explorers added much to the knowledge of the newly acquired territories, and the older states were engaged in surveys that were more or less detailed. Private map makers were busy publishing maps and school atlases. Outstanding is Henry S. Tanner's "New American Atlas" of 1823, published in Philadelphia. In the midcentury, American cartography was influenced by the introduction of wax engraving and lithography (see Chap. 14). The lithographed county atlases are typical American products.

The systematic survey of our country had a late start. The U.S. Coast Survey was founded in 1807, but the first chart came out in 1844. The U.S. Geological Survey was organized in 1878, combining several individual surveys. Since then our cartography has become much diversified, and the present output, aided by airplane photography, runs into thousands of new maps yearly.

The somewhat parallel history of Canadian and Latin American cartography cannot be discussed in this short sketch.

Classification of Maps

Maps are of many kinds. Perhaps the most important difference is between serial and individual maps. Large-scale topographic maps and charts come in sets and are usually made in governmental offices with highly specialized equipment, and broken down to jobs with rather rigid standards. In the second class we have maps often on smaller scale which the individual can design and draw. In the first, the technical training is the more important; in the second, the knowledge of geography and certain ability in graphic expression.

Maps may be classified with reference to their scale and to their content, as follows:

1. *General maps*
   a. Topographic maps drawn on a large scale and medium scale, and pre-

Fig. Int. 11. Arab surveyors from a sixteenth-century Venetian book reading a quadrant (right), a diopter (center), and placing a sun dial (left). (*Courtesy of The Arab World.*)

senting general information, including relief.

b. Planographic maps, the same as before, but without relief.

c. Maps representing large regions, countries, continents, or the whole world on a medium or small scale. (Atlas maps belong to this class.)

d. World maps of general content.

2. *Special maps* emphasizing certain factors

a. Charts for navigation in the sea or in air.

b. Thematic or single-factor maps.

(1) Qualitative, such as geologic or vegetation maps.

(2) Statistical maps.

(3) Cartograms (or diagrammatic maps).

c. Land-use maps.

d. City maps.

e. Transportation maps, showing railroads, airways, auto roads.

f. Political and historical maps, general maps with emphasis on political units.[2]

g. Maps of the various sciences (often single-factor maps).

h. Maps for illustration and advertising.

i. Cadastral maps, drawn on a large scale to show land ownership.

3. *Globes and models*

Although graphs or diagrams cannot be called maps, they are used on, or combined with, maps and they are also discussed in this book.

[2] Map publishers differentiate between *political maps*, where the political units are colored, and *physical maps*, which have altitude tints.

1

Tools and Equipment

Proper tools at the proper place in the proper condition will save time and nerves in map drawing, and can make all the difference in the quality of the product. The best eraser is worth nothing if it hides under the table, and the best pen does not work if clogged with dried ink.

The following items will satisfy the needs of this course and should be bought by each student. All these materials come in various makes and grades, as explained below. It is best to buy the highest quality so that the item will last, but if one cannot afford that at present, it is better to buy the cheapest than none at all. Borrowing from one's neighbor is not conducive to popularity.

Pencils, HB, H. Nail file for sharpener
Gillott pens Nos. 303 and 404
Round-writing pens, Nos. 3, 4, 5

Esterbrook Drawlet pens, Nos. 11, 12, 13, or equivalents
Round-nibbed pens in three sizes
Double-nibbed pen
Six penholders
Compass with detachable tips for pencil and ruling pen, and with lengthening bar
Protractor
Scale, 12 in. engineer's type (optional)
X-acto knife (optional)
India ink, black, waterproof
Set of colored pencils, wax base
Set of water colors
Sable brush, No. 4 (or Nos. 2 and 6)
Three sheets of one-ply Strathmore board, kid finish, large size or equivalent
Erasers, soft and hard, knead gum
Triangle, 12 in. long, either 30°–60° or 45°
T square, 24 in. long
Tracing paper, one block (12 by 18 in.)
Scotch (masking) tape
Slide rule

Some materials can be bought more economically by groups of about six students:

Rubber cement
White paint, opaque
Pounce (pumice powder)
Zip-a-tone of various patterns
Cellotype alphabets
Acetate ink (Artone or Craftint)
Cross-section paper
Ross board, No. 1
Craftint Doubletone sheets
Vinylite or Diryte sheets, or equivalents
Glass-laminated sheets

NOTE TO THE INSTRUCTOR: If all the necessary equipment is bought at the beginning of the course it will save much confusion later. As the acquisition of instruments takes some time, it helps if they are explained and demonstrated at the first meeting so that the student may know how to buy intelligently and economically.

Coated film for negative scribing
Carbon pencils (Wolff crayons)

Some tools will be used for many years and can best be bought as permanent equipment by the department, one for every five to ten students:

Pivot (contour) pen
Road pen
Drop-bow compass
Beam compass
Steel straightedge
Large T square
Flexible curve
French curves
Proportional divider
Honing stone
Pencil sharpener
Lettering set
Pantograph
Planimeter
Stereoscope
Stapler
Electric eraser and shield
Tools for negative scribing

(This list does not include material used for relief-model making, which will be discussed in Chap. 25. Nor does the list include photographic and contact-printing equipment, which are discussed in Chap. 26. Duplicating machines are likely to be available in the departmental office and are described in Chap. 14. Additional specialized tools will be described in their respective chapters.)

Pencils are made from finely powdered graphite mixed with clay. The more clay, the harder the pencil. They are graded from 9H (hardest) through HB (medium) to 6B (softest). Hard pencils may wedge down into the paper, and their marks are difficult to erase. It is better to use a softer pencil and keep it sharp. A special long-point sharpener is helpful. Some prefer me-

chanical pencils in which the leads can easily be changed to the desired type. Emery-board nail files are cheap and very good for sharpening. Graphite photographs poorly; lines and tones reproduce better if drawn with carbon crayons (Wolff crayon or equivalent). These also come in various degrees of hardness.

Erasers. Soft erasers are preferable because hard ones often have emery in them and spoil the surface of the paper for inking or painting. Hard erasers are used on inked lines only. An electric eraser, used with a shield, is a popular piece of equipment.

Pen Points. Pen making is a high-precision industry, and American makers, such as Hunt and Esterbrook, are now competing with Gillott of England. The Gillott No. 290 is a very fine and very flexible pen. Some prefer the stiffer "crow quill" pens for fine lines. Gillott's Nos. 303 and 404 are suitable for lines of average width. Wide lines are made with round-nibbed pens, but if one can afford it, it is better to buy a set of Payzant, Wrico, or Leroy pens. The latter two can also be used in lettering devices. Fountain pens filled with india ink, such as Graphos, are becoming increasingly popular.

The round-writing (flat-nibbed or stub) pens will be useful for fast and informal lettering on maps (see Chap. 5). The double-nibbed pen is used for roads. Good penholders should hold the pen firmly. Penholders should be marked with different colored tapes for type, and the distance of the tape from the end of the penholder may indicate the weight of the line. Pens are best kept in a tall glass with shreds of linen in the bottom to keep the nibs dry.

Ruling Pen. A ruling pen is expensive

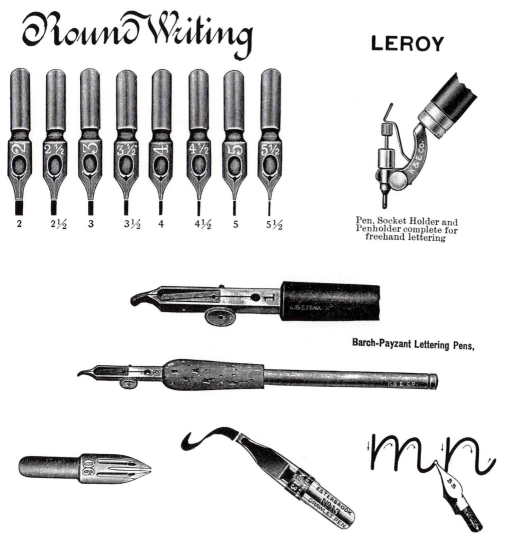

Fig. 1.1. Some special pen points used by map makers. Leroy pens are numbered from 000 to 8; the 000 is very fine. Payzant pens are numbered the opposite way, No. 8 being the finest. Wrico pen is shown in Fig. 6.4. Double-nib, flat-nib, and round-nib pens (bottom) come in different makes and sizes. Note the difference between round-writing and round-nib pens. (*Courtesy of Keuffel & Esser Company.*)

but indispensable for inking straight lines (see Fig. 1.2). It can be kept sharp with a honing stone. Ruling pens must be used along a rule, triangle, french curve, or T square, and this may cause a little difficulty at first. We fill the pen from inside with a quill or pen; no ink should adhere to the outside. We hold it vertically or slightly inclined in the direction of the line. We never tilt it toward or away from the rule. The pen has to be cleaned every few minutes.

Pen pressed against T-square too hard

Pen sloped away from T-square

Pen too close to edge, ink ran under

Ink on outside of blade, ran under

Pen blades not kept parallel to T-square

T-square (or triangle) slipped into wet line

Not enough ink to finish line

Fig. 1.2. Beginners' common faults in the use of ruling pen. (*From French, "Engineering Drawing."*)

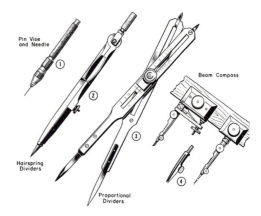

Fig. 1.4. Needle is used for pricking and cutting cellotones. The proportional divider can be set to reduce or enlarge ten times. Beam compass is used for flat arcs. (*Courtesy of the U.S. Army.*)

Compass. The compass is used to draw circles. As arcs may be flat, the larger the compass the better, and it should have a lengthening bar. Both a pencil point and a ruling-pen point for inking are needed. The latter may save buying a separate ruling pen. For very flat arcs, we use a beam compass. If one is not available, we can improvise one with a rod, a pin, and a pencil or pen held by elastics.

Divider. The divider is used for laying out distances. For classwork it is not indispensable, as the compass may be used instead. For precision work, however, it is

Fig. 1.3. Compass with changeable points and lengthening bar. Holder at bottom can be attached to ruling pen. (*Courtesy of Keuffel & Esser Company.*)

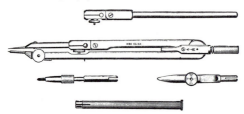

useful. A proportional divider is used for enlarging or reducing.

Triangles are usually cut out of sheets of plastic. They come with 45° and with 30–60° sides. It is better if not all students in the class have the same kind and the same sizes.

T squares come in various lengths. For classwork a 24-in. square is large enough, but a few of greater length should be available. For map work a simple T square is best; adjustable heads are not needed. We often reverse the T square and use it for ruling.

Scale. An engineer's scale should be used with the inch divided into tenths, rather than an architect's scale which is calibrated into sixteenths. A wooden or plastic scale is handy, but not indispensable, as a strip of cross-section paper serves almost as well. A centimeter scale may also be useful.

Protractor. A protractor is necessary for laying out angles. The best is a full circle printed on plastic or transparent paper.

Knives are used for cutting, sharpening,

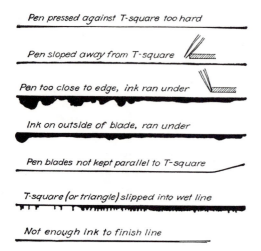

and even for removing ink. X-acto knives, or similar makes, have changeable blades of many designs. Razor blades in a holder also serve well.

Inks. Maps are inked with india ink, which consists of fine carbon particles suspended in a liquid of the same specific gravity. It has to be very opaque black, yet penetrating, waterproof when dry, and relatively quick-drying. Many good inks are on the market. The ink bottle should be closed when not in use. Colored waterproof inks are also available. They tend to coagulate, and have to be stirred from time to time.

Colored pencils are of two kinds: waterproof and water-soluble. The map maker uses them for quick layouts and rough coloring. Shading with a soluble pencil can be spread evenly with water and brush, and the wax-based crayons can also be made to spread with a cotton tip dipped in gasoline.

Water Colors. An inexpensive set of school paints is as good as or better than expensive artist's colors in tubes. Artists want the paint to adhere to one place; thus they prefer paint with much glue in it. The map maker usually wants an even spread of color. For this purpose some excellent ready-made liquid paints can be bought in art-supply stores. Dyes are not recommended as they cannot be washed off.

Opaque white is used for painting out imperfections after inking. None of the commercial products seem to be opaque enough, but the painted area does not photograph even if it looks slightly gray.

Brushes. Sable brushes are expensive, but they are resilient and make a fine point. Camel's-hair brushes are cheaper and fairly good. Brushes are used by map makers to color in certain areas on the map and also to paint out the small irregularities of ink-ing with opaque white. For the first, a No. 6 or 8 brush is best; for the second, a No. 1 or 2. If only one brush is bought, a No. 4 makes a good compromise.

Papers. A good paper for map making has to be strong, ink-absorbent, stable in size for ordinary changes of weather, erasure-proof, and should have a surface not too smooth and not too rough. Only the best rag papers fulfill these demands. Papers come in various thicknesses and surfaces. Thin papers have the advantage of being more translucent over the light table. A very smooth surface does not let the ink penetrate deeply, and fine lines may rub off when the map is erased after inking. A great variety of special papers may be used in map making, and students should learn how to use them. Each student needs a pad of transparent tracing paper for sketching and layouts.

Much professional map work is done now on transparent plastics, such as Vinylite, Diryte, or Mylar sheets, which are very stable in size. They are excellent for colored maps, where each color has to be drawn on separate sheets which must fit exactly. They are not as pleasant to work on as paper because ink cannot penetrate these plastic sheets deeply, and lines may rub off in erasing. Some special inks like Artone or Craftint acetate inks are especially made for plastics. They etch the surface slightly for better penetration. Vellums are midway between plastics and tracing paper and can be used for small color overlays. Tracing cloth is not quite as shape-proof; it clogs the pen and is used much less now than years ago. Plastic-coated cotton cloth, however, is a promising new material. Paper mounted on aluminum is a favorite with surveyors.

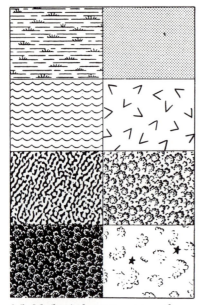

Fig. 1.5. Mechanical patterns are used on maps for differentiating type areas. (*Courtesy of Para-Tone Company.*)

In recent years various glass laminates have appeared on the market (MM Glass Laminate, Stabilene, Permascale). These consist of a layer of glass fiber sandwiched between two thin perforated sheets of plastic. These combine the stability of plastics with the absorbency of drawing papers and, except for price, make an ideal combination.

Fig. 1.6. A map drawn with ink and black crayon on stippled (Ross) board. Orthographic projection. (*By Richard Edes Harrison.*)

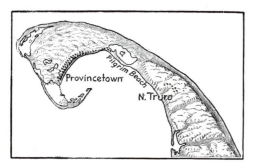

Tints and patterns are very often used by map makers and there are several media to choose from; they will be discussed in Chap. 13.

Cellotones, such as Zip-a-tone, Craftone, and others, are cellophane sheets printed on the underside and coated with adhesive. They come in hundreds of patterns and colors (see Fig. 1.5). Application is easy. We cover the area with a sheet of the desired pattern, tap it down lightly, and cut it around with a needle. Then we pull off the excess and rub down the cellotone with a burnisher, a rounded piece of ivory, plastic, or glass. We often use red cellotone instead of blackening large areas with ink, as red cellotone photographs as black. Ink swells the paper and hides any design underneath, and it is difficult to remove. Do not place cellotone on a hot radiator; the waxy adhesive melts and the pieces fall off. There are, however, heatproof cellotones on the market.

Craftint Singletone and Doubletone are drawing papers with invisible patterns printed on them. A developer applied to the desired area will bring out the pattern in black. Doubletone has a lighter and a darker pattern brought out by two developers. Numbers 206, 214, and 267 are recommended for map work. They are often used to show mountains in a subdued tone.

Ross boards and Coquille boards are papers covered by an enamel into which minute knobs are pressed, about 50 to 100 in an inch. If we draw on them with black crayon, only the top of the knobs is darkened. The stronger we press, the more the knob is blackened. Thus the penciled area is automatically cut into minute dots which, when printed without using a halftone screen, will look darker or lighter gray (see Chap. 14). Ross board comes in various patterns also. Number 1 is recommended for showing mountains on maps. Ross board can be blackened with ink, and white lines can be cut into it, so that it acts as a scratch board. Artistic effects can be obtained by inking certain areas and cutting white lines into the enamel with a knife.

Much of the fine map work of the government and of large companies is done now by negative scribing (see Chap. 27). A Mylar or Vinylite sheet is covered with an emulsion and the lines are cut into it with steel tools called *gravers*. This makes a direct negative for reproduction. Although the average student rarely will have an opportunity to use scribing, he should know how it is done. Coated plastic sheets can be obtained from all major engineering and art-supply companies. Professional scribing tools are expensive; yet each cartographic laboratory should have at least one rigid graver. For simpler work, phonograph needles or discarded dentists' tools can be honed to variously shaped points.

Cross-section (graph) paper is used for drawing profiles, diagrams, etc. For us the most useful type is printed with nonphotographic light blue lines, divided into tenths of inches and millimeters. They come in dozens of designs, and are available in college bookstores.

Adhesives. Thumb tacks ruin the drawing table, so paper should be fastened with small pieces of Scotch tape. Both the transparent and opaque types are used, with preference for the opaque. A special drafting tape is less likely to tear the paper when removed. Rubber cement is used for mounting because it is waterproof, does

not swell the paper, and can be rubbed off. In about ten years rubber cement deteriorates and makes brownish grease spots. Permanent methods of mounting will be discussed later (see Chap. 11).

Furniture. Most departments use large horizontal tables which are designed for architects and engineers and are not the best for cartographers, who work seated and work on unattached, freely sliding paper. The design in Fig. 1.7 has a comfortably inclined top, a shelf for tools and books, and a frame for maps. The paper can be freely moved and can hang down under

Fig. 1.7. Side view of a cartographer's drawing table. A short tube of fluorescent light can be attached to the upper left corner of the back frame. Some light can be used under the glass. Length of table is 5 ft.

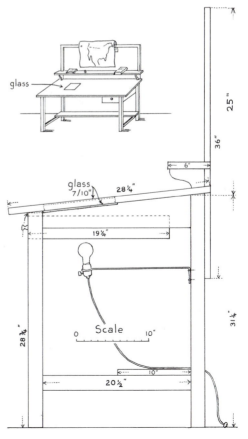

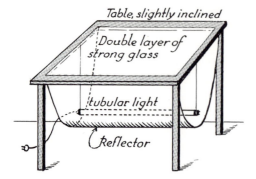

Fig. 1.8. Copying maps and drawing color overlays are best done over the light table.

the shelf. A drawer is useful for tools. Some tables have a 7- by 10-in. glass set flush into the table, with a lamp underneath for copying. A short fluorescent light tube in

the upper left corner of the frame provides sufficient lighting. The same lamp can be used underneath the glass for copying. The short T square is usually used with the head in front of the table rather than on the side.

A light table for copying large maps is shown in Fig. 1.8. It can be made easily at home. A sheet of white drawing paper or thin metal plate makes a good reflector.

Cases for maps and materials will be discussed later. Soft wallboards 4 by 8 ft are used to display maps for instruction. All students' work should be pinned on them and exhibited immediately. There never seem to be enough of these display boards.

Fig. 1.9. Exercises in the use of drawing instruments.

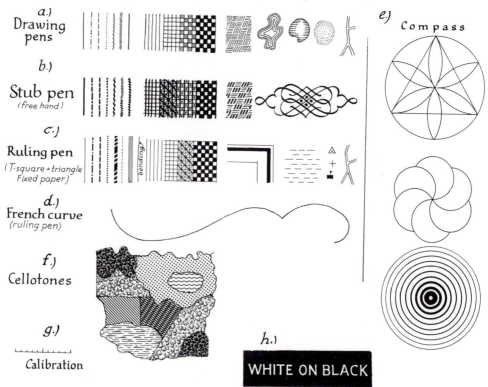

Exercise 1.1 *

a. Draw sequence A free hand with a fine pen, with a medium pen, and with a Leroy or similar pen on loose paper.
b. Draw B freehand with a stub pen. Do not turn the pen.
c. Draw C with a ruling pen using T square and triangle on fixed paper.
d. Draw a smooth curve with pencil and ink it in exactly, using a ruling pen and french curve.
e. Draw some geometric patterns with compass and ink. Use your imagination.

f. Apply various cellotones over a layout of your own. You may experiment with two cellotones, one over another.
g. Calibrate a line 1.1 in. long freehand, dividing it into 10 equal parts.
h. Paint your name in white on black, using a fine brush.

Figure 1.9 is just a guide; it need not be copied exactly. Your drawing should be larger. Practice first on a separate sheet, then make a layout lightly in pencil on good drawing paper, and ink it in.

* NOTE TO INSTRUCTOR: This exercise cannot be made at the first meeting, as students need some time to acquire equipment. The first laboratory period may be used for air-photo reading (see Chap 2). This exercise may be given during one of the later laboratory periods.

2

Air-photo Reading

Looking down from a plane upon field and forest, houses and rivers, one cannot fail to note how much the scene looks like a map. Indeed, there is no better way to understand maps and map problems than to study the landscape below and its record on air photographs.

A vertical air photo looks strange to us. Even from an airplane, we rarely look directly down upon the land below us. We also miss the natural coloring. With some reason and imagination, however, we have little difficulty in reading the photograph.

The technical aspects of airplane photography will be discussed in Chap. 16. Here it will be sufficient to mention that air photos are taken either vertically down or obliquely. Most photographs for mapping are verticals; obliques are used more for reconnaissance. The photographs shown here are verticals.

Farming Country. Some features are obvious. Nobody would fail to recognize in Fig. 2.1 the even curve of a double-track railway flanked by a black road and by a white road. In the middle of the picture we see a driveway branching off the black road, leading to a farm. The largest building is likely to be the barn. The farmer's house will probably be in the area shaded by the large trees. A road leads from the farm to a small creek, which can be recognized more easily by the trees growing along it than by the white stream bed meandering underneath the trees. The way the little streams converge shows that the creek flows to the right, and it passes underneath the black road, the railway, and the dirt road. The dark coarsely mottled area on the upper left is forest. We cannot see the creeks and trails through the dark canopy of the trees; thus the picture must have been taken in the summer or the fall. How many farmyards are visible in the picture?

We may even make some guesses about what grows in the fields. All the land between the forest and the farmhouse is uniformly smooth gray, crisscrossed by light-colored trails—apparently cow paths. Several single trees are left standing. All this suggests a hayfield which is also used for pasture. Next to the farm buildings are smaller, light-colored, slightly streaked

NOTE TO THE INSTRUCTOR: Instead of using the halftone reproductions of photographs in this text, try to obtain local pictures. City planning boards, state highway departments, etc., usually have them. If not, write to the U.S. Map Information Office. If only a few pictures are available, they can be projected.

fields, typical of wheat or other small grain. At the bottom of the picture are two large fields, finely mottled and streaked, characteristic of corn. The fields on the right side of the railway are smooth, dark, and undivided, indicating a tall growth of grass.

We notice that almost all fields and roads are laid out at right angles parallel to each other. This indicates that the region is west of the Appalachian Mountains, where the "township and section" system prevails. The shadows are short; thus the picture was taken near noontime when the shadows point north, thereby indicating that the direction North is toward the bottom left.

Downtown Boston. This photograph certainly presents a contrast to the previous one. No fields and forests are here. Every space is occupied by buildings, streets, railways, and harbor facilities. It is somewhat

surprising that the water appears black. This is because water reflects very little light straight up. Ordinarily we see water as light-colored because we look obliquely at the reflection of the sky. Where the sun reflects directly into the camera, as on the right of the picture, the water is light. Note the large waves and the ferryboat plowing through them.

Pier after pier juts into the water, harboring seagoing steamers. Some ships at the piers at top center are more slender. What kind of ships are these? At the oblique pier is the historic frigate *Constitution*. Note the ship in dry dock.

All bridges can be opened to allow small boats to go upstream to the Charles River. The bridge in the center with the adjacent park (where Science Park is now built) is really a dam separating the salt water from the fresh water of the Charles River Basin.

Fig. 2.1. Vertical air photo of farm land.

Fig. 2.2. Part of Boston Harbor. This picture was taken before the Mystic River Bridge and approaches were built.

Over the next bridge the subway crosses, plunging under Beacon Hill on the right. Otherwise the presence of the hill is hardly perceptible. The hill is crowned by the State House, with its gilded dome, looking down on historic Boston Common. On the right center are the big buildings of the business district. The elongated building is the famous market, Faneuil Hall. The large buildings are in strong contrast to the small old brick houses between the business district and the harbor.

Bunker Hill Monument, casting a long shadow, is in the center of Charleston on the upper left. Note the Boston and Maine railway yards with roundhouses and sheds, the large gas tanks and factories. Fourteen tracks of the Boston and Maine Railroad lead into North Station, the terminal, across the river. The small houses of the workers at the lower left contrast with large apartment houses laid out in a rectangular grid on the lower right. This is the fashionable section called "Back Bay" because it is built on fill. Before filling, tidewater reached the Common.

Figure 2.3 shows a rural landscape in New England. Try to identify all numbered features before turning to the explanation of air-photo analysis at the end of the chapter.

Clues for Air-photo Reading

From the reading of these pictures and of air photos of your own region, you will gain certain clues as to what to look for in identifying different features.

Size. First of all, we have to know how large an area is shown in the photograph. After some practice, the sizes of houses, barns, roads, etc., will give a fair estimate. People are keen in estimating distances,

and this is even easier from the air. Often the photo itself is marked, giving the altitude from which it was taken and the focal length of the camera. From this, the scale of the photo can be calculated according to Fig. 2.4.

Sometimes we have other clues. West of the Appalachian Mountains most of the land is divided according to the public-land system, comprising 1-mile-square sections. Farmers often halve and quarter the sections but, from the sizes of the houses and barns, it is not very hard to judge whether one is looking at half-, quarter-, or eighth-of-a-mile farms. In Fig. 2.1 note the trace of a prerailway section line near the creek, continued by the road in the upper right. From this trace to the forest on the left, the distance seems to be ½ mile, judging from the size of the barns, the width of the roads, and the layout of the fields.

A baseball diamond is 90 ft square. Tennis courts are 80 ft from base to base, but with runways they will be about 120 ft long. Railway freight cars average 45 ft in length; passenger cars are twice as long. In good apple orchards, the trees are 35 to 40 ft apart. Fire lanes or cuts for high-tension lines are, in most cases, 100 ft wide. City blocks are usually laid out 1,000 ft long, but the other dimension varies. All of these proportions apply only to the United States; dimensions overseas are strikingly different.

Shape. The general rule is that natural features have irregular shapes, and man-made features have regular outlines. There are many exceptions; for instance, a gravel pit or dump is artificial yet irregular. Natural features rarely have rectangular shapes, but circular lakes are common.

Fig. 2.3. Parts of Framingham and Southboro, Massachusetts. 1. Railway; 2. Cut (shadow inside); 3. Fill and overpass (shadow outside); 4. Deciduous forest (coarsely mottled); 5. Coniferous forest (much darker than deciduous forest, indicating that picture was taken in spring); 6. Bush (finely mottled); 7. Mixed forest; 8. Rough pasture (smooth with dots); 9. Hayfield (smooth); 10. Plowed land (streaked); 11. Orchard (trees 40 ft apart); 12. Nursery (trees 20 ft apart); 13. Farm buildings; 14. Small farm; 15. Residence with tennis court; 16. Concrete four-lane highway; 17. Asphalt two-lane highway; 18. Dirt road; 19. Highway intersection with red light; 20. Aqueduct; 21. Dam with spillway; 22. Reservoir; 23. Creek; 24. Swamp; 25. Drainage ditch; 26. Stone fence; 27. Power line; 28. Abandoned gravel pit; N. North; D_1 and D_2 Drumline.

Tone. Tone means the variation of the gray. This can often be deceiving. Tone depends entirely upon how much light is reflected directly back to the plane. We have seen that the tone of water is black unless it reflects the sun. Muddy water is lighter than clear water. Grass is medium gray; wherever trampled on, however, it is very light. A newly cut hayfield is much lighter than where the grass is tall, as in Fig. 2.1. Sand is usually white in photos. From the whiteness of the dirt road on the right in Fig. 2.1, we may conclude that the soil is sandy. The black road to the left of the railway indicates that it is tarred. Airplane photos bring out strikingly the small differences in the wetness of the soil. Lower, more humid portions are darker. We often see a river pattern from the air which would not be visible from the ground. Archeologists have often found the sites of ancient cities by these signs.

Texture. The pattern of the land surface in a photograph indicates its texture. This is the most important clue in recognizing the use of the land. In general, the following key may be helpful:

Forest—Coarsely mottled
Bush—Finely mottled
Meadow—Smooth
Rough pasture—Smooth with spots
Plowed land—Streaked
Orchard—Closely and evenly set large dots

A deciduous forest in winter sunshine shows a peculiar streakiness as we see the shadows of the trees. If there is a very great difference in tone between parts of the forest, we may assume that the picture was taken in winter or early spring and that the trees are mixed soft and hard woods, as in Fig. 2.3.

An abandoned field often grows up in summer and fall with tall herbaceous plants and resembles bush. In winter and spring, however, there will be quite a difference between this and the bush, which is really a young forest.

Rough pasture often has cow paths that are easily visible, as in Fig. 2.1. Meadows

Fig. 2.4. The scale of a photograph equals the focal length of the camera lens divided by the height over the land. Displacement depends on the height or depth of the object and on its distance from the plumb point.

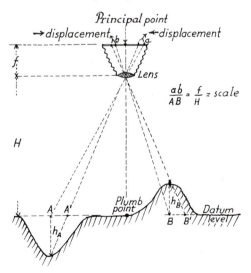

$$\frac{ab}{AB} = \frac{f}{H} = scale$$

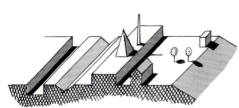

Fig. 2.5. The length of shadow depends on the height of the object, the angle of the light, and the slant of the walls and ground.

cut with a machine may show fine streaks. On the other hand, a field of grown wheat may not show any streakiness at all. After harvest, stacks of wheat and corn are easily recognized. Freshly plowed fields may show a peculiar coffin shape made by the turn of the plow. Part of this is visible in Fig. 2.1 on the lower left. Contour plowing and terracing is a good indication of relief.

Shadows. Often these are the best, and sometimes the only, clue to the height of an object. The general rule is this: If the shadow is outside, the object is raised; if the shadow is inside, the feature is below the general level. First we ascertain the

angle of the sun's rays from some familiar objects, such as houses, barns, trees, or roofs. From this we can figure the height of other objects if the ground is not sloping. Note the long shadow of Bunker Hill Monument in Fig. 2.2. Pictures used for mapping in the middle latitudes are usually taken between 10 A.M. and 2 P.M. because long shadows would be disturbing in making maps from the photos. Hills are often hardly visible in such pictures. Bunker Hill and Beacon Hill rise steeply, yet they are not visible on the picture. Late afternoon pictures are often preferable to bring out the nature of the relief. Height, however, can most accurately be seen on stereoscopic pairs, which will be explained later.

Shadows have a peculiar effect on air photos. If the shadows fall away from you, the pictures look inverse, the hills look like valleys. We have to turn the picture around so that the shadows fall toward us in order to see the relief properly (see Fig. 2.6). In the Northern Hemisphere, this means having South on top—opposite to

Fig. 2.6. Craters of the moon look like disks. To see them as craters, hold the book upside down.

maps. None of the photos shown here have North on top.

Approaches. Many times houses are discovered only by driveways leading to them. Roads and paths are particularly visible in air photos and so are often very revealing. Factories normally have railroad sidings. In Fig. 2.3 the aqueducts can be distinguished from roads because they have no approaches from the houses. Converging cow paths indicate either a barn or a water hole. In war, dugouts and gun emplacements were more often discovered by paths leading to them than by their actual well-camouflaged position.

Relationships. If we see a large building with white, trampled ground around it and a baseball diamond nearby, we think of a school. In Fig. 2.3 we see the roadlike feature 2 starting at a dam, first above ground and then in a cut unconnected with any road, and then suddenly disappearing under a hill. Obviously this is a gravity aqueduct. At 28 we see a quarry overgrown with bush and connected to a main road by a gray, unused road (an active road would be white). This was a gravel pit which was used to build the big four-lane road and has since been abandoned. As it is not completely overgrown, the highway is probably less than twenty years old.

Common sense, combined with a keen eye for observation when traveling, will be most helpful when reading air photos. An essential part of a cartographer's education is to visit the areas seen in local photographs and make a mental connection between the land and its image. The best way to do this is to mark the photographs in field with letters and symbols. A method how to do this is described in Chap. 23 under Geostenography.

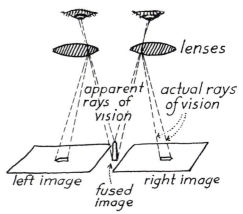

Fig. 2.7. In stereoscopic vision, we see a fused image standing in high relief.

Stereovision. Let us look at a nearby landscape and close one eye. The view flattens out as if looking at a picture. Our eyes are only 2¼ in. apart, yet this tiny difference in the angle of view enables us to see three-dimensionally. Viewed with grandmother's stereoscope, two pictures taken from slightly different angles come alive. Some modern stereoscopic cameras are reviving this idea.

When the photographic plane flies along a flight line at great height the photographer takes successive pictures which overlap each other more than half. Thus every point appears at least on two pictures. Although the two adjacent pictures are taken hundreds of feet apart, their effect is the same as that of the two inborn "cameras" on the two sides of our nose because of the great height from which the pictures were taken.

Most of us have good stereovision unless one eye is weak. Looking at Fig. 2.8 we can fuse the two pictures into a three-dimensional image without difficulty. The same thing can be done after some practice with two adjacent photographs. Most

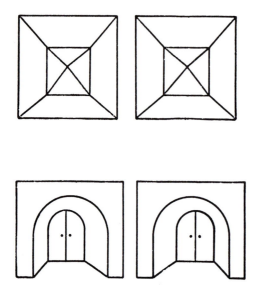

Fig. 2.8. Looking through the book, as if it were transparent, fuses the two pictures into a stereoscopic image. Placing a card perpendicularly between the two images also helps.

of us, however, will prefer to use a stereoscope—an instrument with the help of which mirrors or lenses will fuse the two pictures without straining our eyes. We can make the simple lens stereoscope of Fig. 2.9 by ourselves, but it has a rather

Fig. 2.9. Stereoscope made of two lenses eye distance apart and held at focus length over the photographs. To fuse the two pictures, place the finger tips over identical features and pull them eye distance apart. Only the overlapping parts of the pictures can fuse. (*From E. Raisz, "Mapping the World," Abelard-Schuman, Inc., Publishers.*)

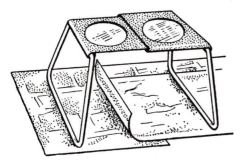

restricted field. Large mirror stereoscopes, often equipped with binoculars, are more expensive, but provide larger view area.

Looking with the stereoscope at pairs of pictures will aid us greatly in air-photo reading and in preparation of the following exercises.

Parallactic Displacement. From Fig. 2.4 it is obvious that only the central area is photographed truly vertically. All other points are seen somewhat obliquely. The farther they are from the plumb point, the point directly underneath the camera, the more obliquely they are seen. If an object is above the general (*datum*) level, it will be displaced outward on the photo; if it is below the datum level, it will be displaced inward. This displacement is called *parallax*. Thus parallax is increasing with the height of the object and its distance from the center along a line radial to the center. The center point of a picture, where the camera axis pierces the photograph, is called the *principal point*. This can be found on most photos from four marks on the sides. If the camera axis is truly vertical, the picture of the plumb point will appear at the principal point. The parallactic displacement is the basis of all stereoplotting instruments discussed in Chap. 16.

Air-photo Analysis

The simple reading of air photographs is relatively easy, and everybody with some practice and field experience can do it. A trained geographer, a geologist, hydrographer, botanist, agronomist, city planner, or military scientist can go way beyond the simple reading of photos by detailed analysis. He will not use merely a single photo or a pair of photos but will study the whole area. He can actually measure the darkness of tones, the length of shadows, and compare patterns with other regions. The geologist will study the various rock types which can be differentiated by their response to erosion and the vegetation they support. The color, shape, and texture of rocks will help him to map the various formations by which he can read the structures and make inferences for their economic value. Similarly, the soils will be studied by an expert. The hydrographer will study soil moisture, types of drainage, river patterns, and flow conditions as they vary through the seasons. Coastal features show up particularly well on air photos.

The biologist will study the vegetation cover from photographs taken at various seasons. The forester can tell the type, age, and value of timber. The agronomist can report the amount and type of cultivation and projection of crops.[1]

[1] Kirk H. Stone, Aerial Photo Interpretation of Natural Vegetation in the Anchorage Area, Alaska, *Geog. Rev.*, vol. 38, 1948, pp. 365–374; and C. F. Kohn, The Use of Aerial Photographs in the Geographic Analysis of Rural Settlements, *Photogr. Eng.*, vol. 17, 1951, pp. 716–725. See also works listed in the Bibliography, Appendix 3.

The city planner will clarify the types of urban land use, streets, traffic patterns, and available public places.

All these experts can prepare *guide keys*, classifying and describing the features as they appear on the photographs taken at various seasons and provide a kind of pictorial glossary, with reproductions of the various features and their interpretation added.

The geographer's task is to integrate the various studies for any given region and interrelate natural features with human occupation; the cartographer makes a comprehensive and balanced map of the whole. To this end our first exercises are dedicated.

Exercise 2.1

For a study of air-photo marking and transportation, each student should be supplied with a copy of an air photo of a rural area in his own vicinity. He should drive back and forth on every road in the area, making frequent stops, and mark the air photo with letters and numbers describing the area. The student may choose his own index letters or use the ones described under Geostenography in Chap. 23.

3

The Principles
of Map Making

SCALE
SELECTION
SYMBOLIZATION
GENERALIZATION

Differences between Map and Photograph

The map in Fig. 3.1 shows the same region which we saw in the air photo in Fig. 2.3. The area shown on the map is larger, the map is on a smaller scale and is oriented differently than the air photo, but there is no difficulty in recognizing all the major features. The most apparent difference is that the photo shows the region in various shades of gray, while the map is almost all line work. The original edition of this map is in color and could not be reproduced here.

There are, however, some more essential differences. Features such as roads and buildings are emphasized on the map, while the camera records everything indiscriminately. The patterns of fields, forests, orchards, and pastures, so prominent on the air photo, are absent on the map. On the other hand, the county boundary does not appear on the photograph. Roads and houses appear in various forms on the photo, but they are shown by uniform symbols on the map. Hills, almost invisible on the photo, stand out boldly on the map through the use of contour lines. The intricate curves of shore lines and rivers are simplified on the map. Most features are named or numbered on the map, in contrast to the photo.

Principles of Cartography

Analyzing the differences between photograph and map, we may deduce some fundamental principles of cartography.

1. Maps are drawn in a predetermined *scale*. Each feature is placed exactly in the proper direction from other points at a horizontal

NOTE TO THE INSTRUCTOR: Obtain topographic sheets of the areas shown by the local air photos, one for each student, and use these rather than the samples in the book. Besides these, sets of the following U.S. Geological Survey maps are recommended: Boston South, Mass., 1:24,000; Framingham, Mass., 1:24,000; Cedar Creek, Del., 1:62,500; O'Fallon, Mo. and Ill., 1:125,000; Everett, Pa., 1:62,500; Palo Alto, Calif., 1:62,500; Mt. Rainier, Wash., 1:125,000; Awawatz Mts., Calif., 1:250,000. Sets of 25 and 100 interesting maps have been compiled for teaching purposes and can be obtained at reduced price from the U.S. Geological Survey.

distance proportionate to the scale of the map. (This principle will be modified for small-scale maps in the various projections.)

2. Maps are *selective*. Only those features are shown which are important for the purpose of the map.

3. Maps *emphasize* certain of the selected features.

4. Maps are *symbolized*. All features are shown by standardized symbols.

5. Maps are *generalized*. Intricate detail is simplified, particularly on small-scale maps.

Fig. 3.1. Map of the same region shown in air photo, Fig. 2.3. (*After the Framingham and Marlboro, Massachusetts, colored topographic sheets of the U.S. Geological Survey.*)

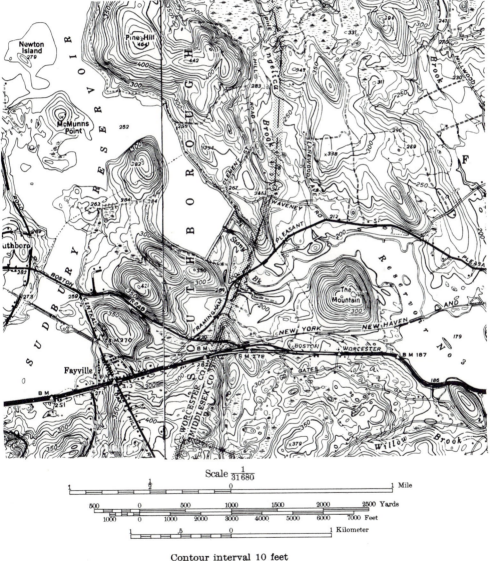

Scale $\frac{1}{31680}$

Contour interval 10 feet
Datum is mean sea level

There are two other principles which, however, we will discuss later.

6. Maps are usually *lettered, titled,* and *labeled.*
7. Maps are usually related to a *system of parallels and meridians.*

Definition of a Map. From these principles we may try to define a map as a selective, symbolized, and generalized picture of some spatial distribution of a large area, usually the earth's surface, as seen from above at a much reduced scale. Most maps are lettered and related to a coordinate system.

As we shall see later, the idea of a picture is sometimes very obscure as, for instance, in a rainfall map. Maps can be made of the stars or moon—not only of the earth's surface. There is a great difference between a map and a picture in the ordinary sense. We can see the object of a picture at a glance. The earth is so large, however, that we have to compile its picture from air photos and measurements and reduce it to small scale. We are indeed like an ant on a rug in some respects.

Scale. The proportion between land and its map is its scale. It can be expressed in different ways. At the bottom of the map we see the scale $\frac{1}{31,680}$ on Fig. 3.1. This we call the *natural scale* or *representative fraction.* It simply means that every inch or any other unit of length on the map is 31,680 times larger in nature. This seems to be an odd number, but a mile is 5,280 ft, or 63,360 in. Thus every inch on this map is exactly ½ mile on the ground. A square inch on the map is ¼ sq mile on the ground.

Underneath the natural scale are graphic scales. These are most convenient for visualizing the scale and for measuring distances. We may take any length on the map and read it off on the graphic scales in miles, in yards, in feet, or in kilometers. Of course, if we measure along a curving road we use a divider or an instrument called a *cartometer.*

Another way to express the scale is by saying how many inches represent how many miles. Thus the above map is 2 in. to 1 mile or 2 in./mile (not 2 in. = 1 mile, which is absurd). If we speak of half-inch or quarter-inch maps, we mean maps 1:126,720 or 1:253,440, respectively.

The first feature we look for on a map is its scale. If the scale is larger than 4 miles to an inch, or roughly 1:250,000, we call it a large-scale map. Topographic sheets, the basic maps which the government surveys publish, and from which all smaller-scale maps are reduced, are a good example. City maps, real estate (cadastral) maps, and construction plans are also of large scale and also serve as basic maps.

Medium-scale maps are 1:250,000 to 1:1,000,000—4 to 16 miles/in. Of course if a mile is shown only as $\frac{1}{16}$ in., one cannot draw houses or fields. The map has to be strongly selective, symbolized, and generalized. For instance, in Fig. 3.2 most of the names from Fig. 3.1 are omitted, and lines in general are simpler. Auto road maps, airway maps, and many special maps are of medium scales.

If the map shows more than 16 miles/in., it cannot show much besides major rivers, mountains, cities, and boundaries. These are small-scale maps, and we find them mostly in atlases, books, and magazines. A one-page world map in a geography book may be on the scale of 1:150,000,000 or

2,360 miles to an inch, and not much detail can be shown. The given scale of a small-scale map is true only along certain lines. Elsewhere there may be considerable distortion due to the projection used (see Chaps. 17 and 18).

Exercise 3.1

What is the numerical scale if 1 in. on the map represents 11 miles? Draw a graphic scale showing 50 miles divided into 10-mile parts.

Understanding Scale. This comes most readily from flying or from looking at air photos. If we fly over Lake Michigan and see the enormous sweep of sand dunes on the south, the smokestacks of Gary, and the skyscrapers of Chicago in the great distance, we will be impressed by what immense territory is included in that little curve on our small-scale map. Every time we fly, we should carry along a map and try to locate cities, roads, rivers, and lakes. Only then will maps have real meaning, and it will improve cartographic work.

Changing the Scales. In map compilation we often have to use maps which are too large or too small. The simplest way to change their scale is to photograph them to size. Photostats are cheaper, but they often have some distortion in one direction. Often a *camera lucida* or an *optical rig* of lenses, mirrors, and prisms is used in larger establishments. An enlarging-reduc-

Fig. 3.2. Medium- and small-scale maps centered on the region of map shown in Fig. 3.1. (*After colored U.S. Air Force Aeronautical chart.*)

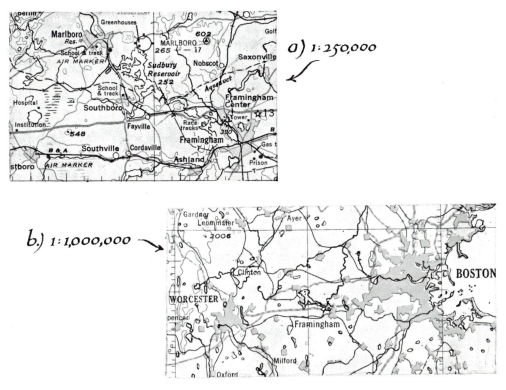

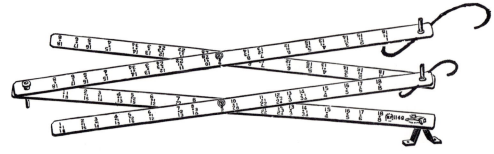

Fig. 3.3. Pantograph for enlarging and reducing. (*Courtesy of F. Weber Company.*)

ing camera is very useful in a cartographic laboratory (see Chap. 26).

If the map is simple, we can superimpose on it a half-inch *grid of squares*. Then we make our own grid larger or smaller, and draw the lines by hand. Instead of squares, parallels and meridians can be used, and thus we can change even the projection system. But this comes later in this study. There are special instruments to enlarge or reduce a map, such as a *pantograph* (see Fig. 3.3).

Exercise 3.2

Enlarge a portion of the map shown in Fig. 3.1 (or any other map) twice, by the square-grid method.

Selection. The maker of the large-scale map in Fig. 3.1 has already done a great deal of selecting from the air photo. He does not show fields, orchards, or pastureland. He omitted them because, with all of these included and lettering added, the map would be overcrowded.

The quarter-inch airway map, Fig. 3.2a, omitted a great deal more. Gone are all smaller roads, houses, creeks. Only the major roads, rivers, villages, and airports are shown. Contour interval is 250 ft, while on the 2-in. map it was 10 ft. But lakes, airports, race tracks, and green-

houses are very carefully drawn because these can be easily recognized from the air, and the main purpose of the map is to guide the pilot of a plane. The 1:1,000,000 aeronautical chart left out even the smaller cities and most of the landmarks. The lowest contour line is 1,000 ft.

The rule of selection is to *omit all that is unnecessary for the purpose of the map, and retain as many of the helpful features as can be shown without crowding.*

Emphasis. Among those features retained in a map, we emphasize those which are important for the purpose of the map. In Fig. 3.2 airports and air markers are emphasized because they are important for air navigation. On a marine chart the lights, buoys, and harbor facilities are emphasized. Large symbols, heavy lines, prominent colors, inscriptions, pointing arrows are the methods of emphasis.

Symbolization. On the air photo a railway looks very much like a road; careful analysis is needed to tell them apart. On the map the railway has a distinctive symbol—a line with short crossbars. On the photo the transmission line is hardly visible; on the map it is shown by a special line. Similarly a house symbol with a flag means a school, and with a cross it means a church. The medium- and small-scale maps are symbolized even more. Villages are shown on them by a small circle; an airport is a large circle with spokes. While both tracks of the Boston and Albany Railroad are shown on the large-scale map, on smaller scale it is a single line with double cross bars. Thus small-scale maps have more abstracted symbols than do large-scale ones.

The rule of symbolization is that a symbol should be *simple*, yet *distinctive*, *small*, and *easy to draw*. A good symbol can be recognized without a legend.

Symbols are usually divided into five categories:

1. Hydrography, or water features (usually blue)

2. Culture, as in man-made features (black or red)
3. Hypsography, or relief features, such as hills or mountains (brown)
4. Vegetation and cultivation (green)
5. Special symbols, as on aeronautical charts, etc. (purple, orange)

Colors add greatly to the distinctiveness of symbols. The use of colors will be discussed later.

Water Symbols. The conventional color is blue, but it is not difficult to render water symbols also in black. Shorelines and rivers are usually the first features to be drawn on maps because by them we locate cities, mountains, etc. In drawing a whole river system, the thickness of lines should gradually increase slightly down-

Fig. 3.4. Examples of hydrographic symbols.

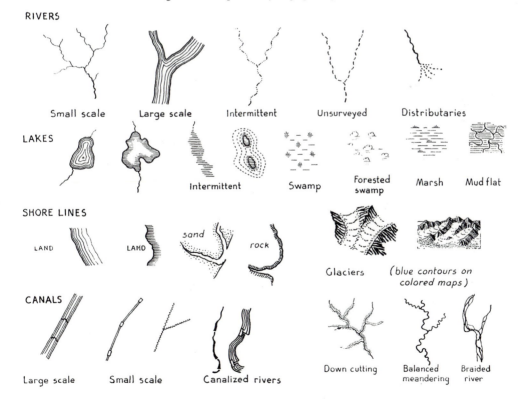

RIVERS

Small scale   Large scale   Intermittent   Unsurveyed   Distributaries

LAKES

Intermittent   Swamp   Forested swamp   Marsh   Mud flat

SHORE LINES

LAND   LAND   sand   rock   Glaciers   (blue contours on colored maps)

CANALS

Large scale   Small scale   Canalized rivers   Down cutting   Balanced meandering   Braided river

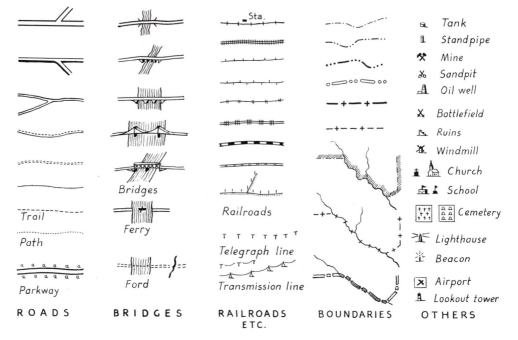

Fig. 3.5. Symbols of man-made features are often more easily recognized from side views than from strictly vertical views.

stream, somewhat like the branches of a tree. This symbolization is not always true, however; the mouth of the Colorado River is often dry and is better shown by the symbol of an intermittent river. Even on small-scale maps it is possible to show whether a river is braided or meandering, though we may not be able to show every twist and turn exactly in place. Canalized rivers, such as the Ohio, Tennessee, and upper Mississippi, should be shown with a series of dams. It is regrettable that this important feature is often omitted from maps.

The conventional cultural symbols on topographic sheets are usually designed from the shape of these objects as seen from directly above. The house symbol is a square; a lighthouse is a radiating point, etc. Although such symbols are carto-

graphically correct, they are hard to recognize. Symbols which are side views, such as the swamp symbol or the lighthouse, are much more familiar (see Fig. 3.5). In a test, the author's students on the average were able to name only 60 per cent of air-chart symbols derived from vertical views, but recognized 90 per cent of side views.

Vegetation Symbols. Vegetation and cultivation are usually shown by colors, either in flat patches or crystallized into patterns. They are possible on black-and-white maps, too, but care must be taken not to overburden the map. The patterns are shown in Fig. 8.1, derived from side views of plants.

Symbolism of *relief features* include contour lines, hachure lines, plastic shading, and landform symbols. These will be discussed in Chap. 7.

Standardization of Symbols. It is important in war and peace that an American soldier or airman be able to read a French map, or that a Canadian mountaineer should be able to climb the Andes using a Chilean map. Although each survey publishes a key of symbols, this is often not at hand. The tendency is now to explain the more unusual symbols on every map. Efforts have been made on a national and international scale to standardize symbols. The Millionth Map of the World, the International Civil Aviation Organization (ICAO), and many surveys work toward unified symbolization. No standard set of symbols has gained universal popularity; this is due chiefly to the fact that they are often poorly designed, difficult to recognize, and do not adapt themselves to different scales. Different regions often need different symbols. For instance, a swamp symbol in a desert may often mean a salt-crusted depression rarely filled with water,

while the same symbol in Siberia means a forested taiga. A cartographer should be able to design good symbols. In the following exercises, do not try to follow precedents, but use the symbols which appeal to you most.

Keys or *legends* are the dictionaries of the language of cartography. There is no need to explain such an obvious symbol as that for a river, but the symbol for an intermittent river should be listed. Legends also call our attention to unusual features on the map, such as shrines, ruins, caves, etc.

Exercise 3.3

Transform the air photo used in your previous study into a map. Copy all lines and shapes through a light table, but try to make a map instead of an exact copy of the photo. Use the symbols and colors which you think best express the nature of the land. Show vegetation and cultivation. Add a scale, a North arrow, and a key. Although this exer-

Fig. 3.6. City symbols vary with the scale and style of the map. They should always indicate the size of the city.

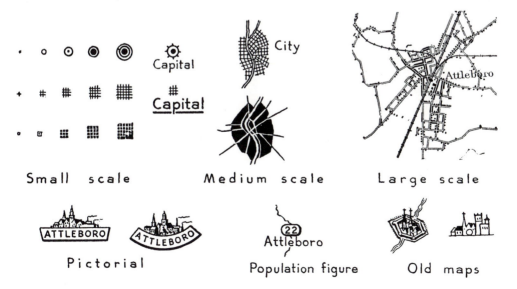

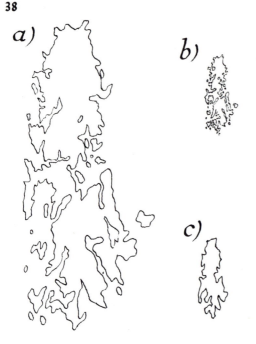

Fig. 3.7. Contour lines need to be generalized on smaller-scale maps. Compare the 1,000-ft contour west of Worcester, Massachusetts, 1:250,000, with its reduction to one-fourth by pantograph and the line actually used on the 1:1,000,000 map.

cise is not for lettering, try to add a title and some names.

Generalization. Even a large-scale map does not show, at every farm compound, the house, barn, shed, etc., separately; instead it shows only a single small black square. To show all the buildings would require symbols so minute as to be unreadable. The medium-scale map does not show all the buildings of a city but shows an outline only, crossed with the main roads and railways. The small-scale map omits even these roads and railways and shows smaller cities by a square or a circle (see Fig. 3.2). If all the curves of the Sudbury River were photographed down from a

2-in. map to a 1:1,000,000 map thirty-two times smaller, the eye could not follow them. On the small-scale map the river is drawn with a curving line to show the nature of it, although only the larger curves are shown.

To generalize contour lines one must know the nature of the country. Figure 3.7 shows the hills west of Worcester, Massachusetts, on a medium-scale map, its photographic reduction to one-fourth, and the actual line used on the small-scale map. If the small-scale map showed the photographic reduction, the contours of these gentle hills would give the impression of great ruggedness. The maker smoothed the lines to show hills rather than rugged ridges, but he carefully preserved their general North-South trend. Intelligent generalization demands a good knowledge of geography and a sense of proportion.

No rules can be given for generalization. We *combine* small features, such as houses in a city or the little isolated hills on the western side of Fig. 3.7. We *omit* some others as we did in the same figure. We *simplify* the lines, but still retain their character. For instance, we would not miss the characteristic sharp point of Cape Hatteras on a map at any small scale.

We have to keep in mind the fact that people's first glance at a map will be at their own region, which they know intimately. They may judge the entire map by it. As Gösta Lundquist remarked in his excellent paper on generalization: [1] "Other people's maps are extremely good—except for their treatment of Sweden."

[1] Paper given at the Second International Cartographic Conference, Evanston, Ill., 1958.

4

Field Methods

COMPASS TRAVERSE

PLANE TABLING

FIELD SKETCHING

At the present time the work of the cartographer is usually separated from that of the surveyor. Yet some generations ago they were one and the same person. Even now, a cartographer-geographer often has to perform simple surveys. It is quite essential that the cartograther and the surveyor should each understand the job of the other. Only a few simple field methods are discussed in this chapter; more thorough treatment is given in Chap. 16.

Compass Traverse

Select a piece of land about 20 acres—an estate or park with some variety. It should be open, relatively flat country, with some trees and a few buildings on it. Avoid the college campus because of buried iron pipes. The best choice would be the local area used in air-photo reading. The student should take along a sighting compass, a dozen sheets of notebook paper, a writing pad, pencil, red crayon, eraser, inch scale, and protractor. A ready-made sketch book can be used.

Before starting out, measure off 100 ft. Each student walks the 100 ft with his usual walking steps, counting the number of steps back and forth. The 100 ft will require somewhere between 36 and 42 steps. Each student should prepare a special scale for feet and steps. Assuming that an inch represents 100 ft and 38 steps are required to walk that distance, 1 step will equal 100/38 or 2.63 ft. Take a strip of profile paper on which the inches are divided into tenths. Make a red mark at every 0.263 in. for every 10 steps (26.3 ft), marking first 2.63 in. and dividing it in 10 parts to avoid cumulative errors. Now the scale shows not only feet but steps as well.

A *traverse* is a connected series of straight lines on the earth's surface (called *legs*), the length and direction of which have been determined. Before starting the traverse, walk around the estate to make a

NOTE TO THE INSTRUCTOR: The chief purpose of simple surveying exercises is to train the student to recognize the basic relationship between land and map, and to practice orientation and judging of distance. Coming early in the course, they also help to establish closer contact among the members of the class. Instruments can usually be obtained from military science or engineering departments.

plan. A "closed" traverse is made by returning to the starting point. Stations will be 200 to 300 ft apart in order that every part of the estate be visible. At each selected station, stick a peg into the ground. It can be made more easily visible by attaching to it a white rag or piece of paper.

Procedure. Figure 4.1 will explain the procedure. Stand on A and read the compass to B. Some compasses show *degrees*, 360 in a circle; others show *mils*, of which there are 6,400 in a circle. It does not matter which is used, as long as it is used consistently. Face the target squarely, hold the compass with both hands, and wait until the needle stops swinging. It can be slowed down by locking the compass for a second at the halfway point. Directions measured clockwise from North are called *azimuths*. Mark A near the bottom center of the sheet; draw a center line straight up and another line for magnetic north, using the protractor. Aiming at B, read the compass and mark the reading on the center line (in this case, 86°). On sighting the left shore of the lake, draw a line in the direction of the reading (55°) and mark the line. The right shoreline need not be sighted; merely note that it is about 10 ft from the center line. Sight both edges of the house, the fountain, the garden, the gate, the large tree, all other important points, and mark the azimuths.

Now walking to B, count the steps (in this case 118), and mark off B with the step scale. Stand at B and sight back to A. The reading should be 180° + 86° = 266°. (If the reading had been 210°, we would have subtracted 180°, resulting in 30°.) If the difference is more than 2°, check

both readings; if less, divide the error. Sight all visible points which were aimed at before. Locating a point by sighting it from two points of known distance is called *intersection*. Sketch in freehand the surrounding landmarks. It is good to take a reading to a very distant point from every station to discover hidden magnetic errors.

At first, use the protractor every time an angle is laid out; later the work can be speeded up by estimating the direction of the line. Later, in home compilation, all measurements will be laid out exactly.

Placing a new sheet on the pad, start from B, looking toward C. This direction will be the center line for this leg. Mark magnetic north, then sight to new points and also back to A and to some other points of the previous leg. Count steps to C; mark it off; intersect angles from C and sketch as before. On the next sheet, sight from C to D, and so on. Repeat the procedure going around leg after leg until you return to A.

Compilation. This is done in the laboratory, using the field notes. First list the distance and azimuth of each leg. Then convert steps into feet and magnetic north to true north. The map should be oriented to true north. Magnetic declination can be obtained from an airway map or a topographical sheet. If magnetic north is west of true north, subtract the declination to get true azimuths; if it is to the east, add it.

It is now time to draw the map. Fasten a large piece of paper to the drawing board. Put A approximately in the proper location, and with the T square draw a vertical line through it. From this, lay off the true azimuth of A to B, draw the line, and

Compass traverse. McGraw Estate, Sheet A–B

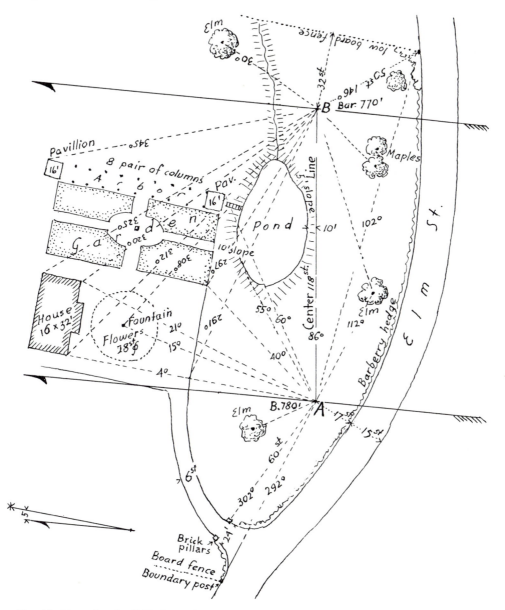

Fig. 4.1. Example of a field sheet of a compass traverse. Several such sheets will complete the traverse.

locate B at proper direction and distance. Draw a vertical line through B′ and lay out C′ similarly. Proceed in this way, leg after leg, and the last leg should come out at A —but it never does.

If the difference between A′ and A is

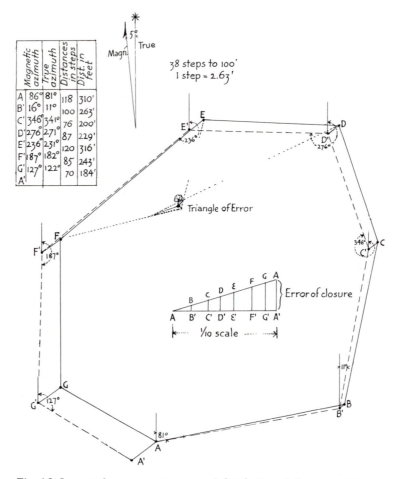

Magnetic azimuth	True azimuth	Distances in steps	Dist. in feet
A	86°	81°	118	310'
B'	16°	11°	100	263'
C'	346°	341°	76	200'
D'	276°	271°	87	229'
E'	236°	231°	120	316'
F'	187°	182°	85	243'
G'	127°	122°	70	184'
A'				

Fig. 4.2. Layout of a compass traverse and distribution of the error of closure. Dash lines show the initial layout; solid lines show the traverse after distribution of the error of closure. Note the compromise on the "triangle of error."

larger than one-twentieth of the total distance, the whole traverse must be repeated until the mistake is found. If the "error of closure" is small, the error may be distributed. To do so, lay out the length of one leg after another on a horizontal line. The result is the same if a fraction of the lengths, one-quarter or one-tenth, is used, as in Fig. 4.2. On the A' end, measure off vertically the error of closure and draw a

triangle. Draw parallels from each point, and thus obtain lengths BB', CC', etc. Now push A' to A. Then push G' to G along a line parallel to A'A, but only to the distance GG' taken from the triangle. After we have pushed F', E', D', C', and B' similarly, we have a traverse which starts at A and ends at A, with which we can now draw a map.

Lay out all the azimuths from every

point with a protractor, and sketch in the rest. Do not be distressed if the azimuths of a tree laid out from three stations do not come to an exact point. If the triangle of error is small, we halve the angles and use the resulting center, as shown in Fig. 4.2. Not more than 5 per cent exactness can be expected from a compass traverse.

After the layout is ready, walk over the terrain once more to add missing detail. Add lettering, title, scale, North arrow, and

Fig. 4.3. After major features are laid out from the field sheets, detail is added while one walks over the area with the layout in hand.

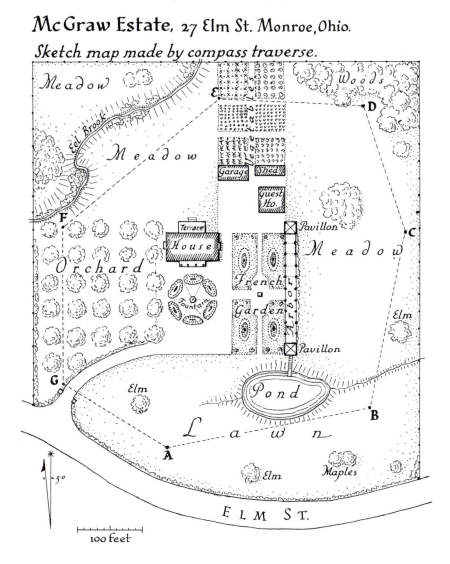

McGraw Estate, 27 Elm St. Monroe, Ohio.

Sketch map made by compass traverse.

Meadow

Woods

Eel Brook

Meadow

Meadow

Vegetables

Garage Shed

Guest Ho.

Pavillon

Meadow

F

Terrace

House

Orchard

French

Fountain

Garden

Arbor

Elm

Pavillon

G

Elm

Pond

B

L   a   w   n

A

Elm

Maples

ELM ST.

-5°

100 feet

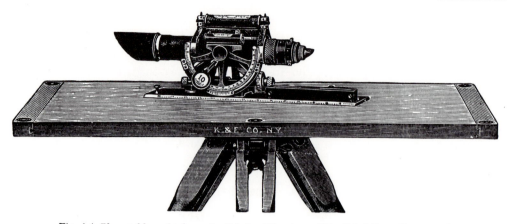

Fig. 4.4. Plane table and telescopic alidade. (*Courtesy of Keuffel & Esser Company.*)

legend. Before inking, it is good to rub the paper with eraser powder or pounce to take out some of the pencil and dirt. Ink the lettering first. Add some light tones of water color—and tack your achievement proudly on the exhibit board.

Plane Tabling *

Plane tabling is much more accurate than compass traverse, but needs a great deal of equipment. The plane table is a drawing board set up on a tripod and covered by drawing paper. On this we draw with a sighting rule called an *alidade*. It is quite possible to make a map with only this equipment. For efficient work, however, a telescopic alidade which has stadia wires and a graduated rod is generally used.

If no telescopic alidade is available, simple equipment can be made in a shop. For the plane table, a small drawing board

about 12 by 15 in. is used with a weight hung from the bottom center. A tripod can be made according to Fig. 4.5. The rod is a one-by-two lath 10 ft long, painted as in the figure, with each foot divided into 10 parts. The alidade is a 12-in. rule; on each end is fastened upright a strip of tin. One tin has pinholes; the other is cut on one side like a saw with teeth exactly ⅛ in. apart. The pinholes and the edge of the saw teeth must be aligned exactly parallel with the rule's edge. A round level bubble vial can be bought in photographic stores.

Stadia. The picture of a tree 1,000 ft away will be half as high as one of the same size at 500 ft distance. If it is X times as far, it will be X times smaller. On this is based the *stadia* measurement. When we look through the pinhole, our eye is about ½ in. from it and 12½ in. from the saw

---

* NOTE TO THE INSTRUCTOR: Divide your class into parties of three, one rodman and two plane-table men who alternate. Secure plane table, alidade, tripod, and rod for each party. If this is not possible, parties may alternate, some making compass traverse, others field-sketching. The plane-table map can be left in pencil as a cooperative effort of the party. There is no need for each student to make an individual map, as the finished map will be very much like the compass-traverse map.

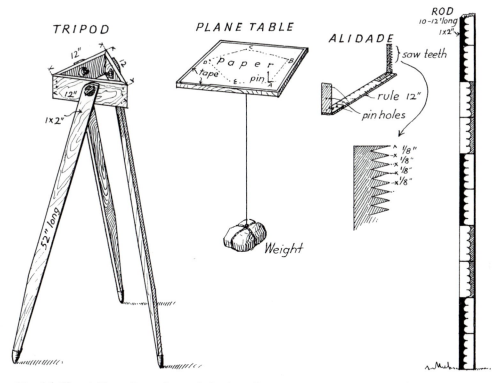

Fig. 4.5. Plane-table outfit can be made in shop, if no great accuracy is expected. Much depends on the exact cutting of the saw teeth.

teeth. This is 100 times larger than the saw-tooth interval of ⅛ in. If, for instance, we see 2.3 ft on the rod between two saw teeth (rod intercept) then we must be 230 ft away from it. We repeat the reading between several saw teeth and take the average.

If a telescopic alidade is available, look at the rod between the upper and lower cross hairs, as shown in Fig. 4.6. In this way, distances up to 1,000 ft can be measured with fair accuracy.

Plane tabling is a rather obvious process. First look over the area and select stations. Stations should be intervisible, if possible, and should command good views. They can be farther apart than in a compass traverse—300 to 1,000 ft, depending on the density of forest and the type of terrain. Each station is marked with a stick or cairn.

Procedure. Set up at A, as in Fig. 4.3, and make the plane table horizontal, using the round level bubble. Make a line showing magnetic north as a major control. Use a

Fig. 4.6. Principle of stadia reading.

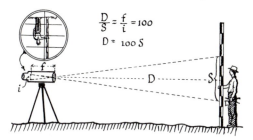

$$\frac{D}{S} = \frac{f}{i} = 100$$
$$D = 100\,S$$

pin to mark A at its approximate position on the paper. For this, carefully consider the shape of the land and the scale, as it would be most troublesome if some of the map should be off the paper.[1]

From A, radiate. Advise the rodman at what points to stop. When the rodman takes his first position, sight to the rod with the alidade touching the pin. Draw a line lightly from the pin, read stadia, and mark the distance according to scale. Read point after point, seen from A, until the rodman has made his circuit. Label each point, with tiny letters, elm, fountain, rock, E corner, etc. Make a particularly careful reading to station B. It is helpful to check this distance by tape.

The most common source of trouble comes from jarring the plane table. Never rest an arm on it, even when drawing the lines, and do not kick the tripod. Check the AB line frequently. The rodman keeps the rod exactly vertical, marks each of his stops with a stick or stone, and watches for hand signals. Sometimes he may have to lift the rod so that foliage will not interfere with the reading. If it is not possible to read between the upper and lower cross hairs, read to the middle cross hair and double the reading.

After marking all readings from A, pick up the plane table and set it up at station B. Stick the pin into the paper at B and lay the alidade along the BA line; turn the plane table until the aim is back exactly to station A. (Using a Johnson tripod, un-

lock the lower screw.) Now the plane table is exactly parallel to its previous orientation at A. The slightest mistake in this will throw off all later work. Check also the magnetic north. Send out the rodman and radiate as before. Repeat this whole process at each station.

From the last station, we sight and read to A. We should come out very close to our first pinhole for A. If the mistake is less than 2 per cent of the whole length of the traverse, we may distribute the error as in a compass traverse; otherwise part of the survey has to be repeated. After we have obtained a satisfactory set of points, we walk around the terrain with the map and a compass and sketch in the detail. For doing this the sticks or stones left by the rodman are very helpful.

In the foregoing we assumed that the terrain was more or less level. If the slopes are not more than 5 per cent, their effect may be ignored. At present we do not plan to draw contour lines, but cut banks, bluffs, etc., can be shown by simple hachures.

If the reading is on a slope, we have to make a correction for the horizontal distance (HD) and also obtain the difference in vertical elevation (VE). To get the true slope we aim the telescope so that the center hair is at plane-table height, about 5 ft on the rod. The following tables give us the corrections. For HD multiply the stadia intercept by the value in Table 4.7 and subtract the product from the horizontal distance. For vertical height, multiply the stadia intercept by the value in Table 4.8.

Example: Vertical angle 4°20′, stadia intercept 3.58 ft.

---

[1] If Johnson tripod is available, clamp the upper screw and turn the plane table until it is in the desired orientation. Then clamp the lower screw and the plane table is locked.

TABLE 4.7 *

Vert. angle	Hor. cor. for 1.00 ft	Vert. angle	Hor. cor. for 1.00 ft
0°00′		7°09′	
	0.0 ft		1.6 ft
1°17′		7°23′	
	0.1 ft		1.7 ft
2°13′		7°36′	
	0.2 ft		1.8 ft
2°52′		7°49′	
	0.3 ft		1.9 ft
3°23′		8°02′	
	0.4 ft		2.0 ft
3°51′		8°14′	
	0.5 ft		2.1 ft
4°15′		8°26′	
	0.6 ft		2.2 ft
4°37′		8°38′	
	0.7 ft		2.3 ft
4°58′		8°49′	
	0.8 ft		2.4 ft
5°17′		9°00′	
	0.9 ft		2.5 ft
5°36′		9°11′	
	1.0 ft		2.6 ft
5°53′		9°22′	
	1.1 ft		2.7 ft
6°09′		9°33′	
	1.2 ft		2.8 ft
6°25′		9°43′	
	1.3 ft		2.9 ft
6°40′		9°53′	
	1.4 ft		3.0 ft
6°55′		10°03′	
	1.5 ft		
7°09′			

* Horizontal distance is always shorter than inclined reading. For every 100-ft distance (1-ft stadia intercept) we subtract the length given in table. (*From Keuffel & Esser Company, Solar Ephemeris.*)

Fig. 4.9. Beaman arc is used for inclined stadia readings. It gives horizontal distance on the left as percentage of inclined distance. Vertical difference is obtained by multiplying the stadia intercept by the number of units left or right of 50 to the index arrow at VERT. The central scale is for degrees which can be used with Tables 4.7 and 4.8.

$$HD = 358 - (3.58 \times 0.6) = 356 \text{ ft}$$
$$VE = 3.58 \times 7.53 = 27 \text{ ft}$$

Many telescopic alidades have a "Beaman arc," as in Fig. 4.9. Horizontal distance is obtained from the H scale. It gives the percentage by which the reading has to be reduced. For instance, the oblique distance is 640 ft and the H arrow points to 97.

$$HD = 640 \times 0.97 = 621 \text{ ft}$$

If the telescope is horizontal, the arrow for the vertical elevation is set for 50 to avoid negative values. Let us suppose the stadia

TABLE 4.8 *

Minutes	0°	1°	2°	3°	4°	5°	6°	7°	8°	9°
0	0.00	1.74	3.49	5.23	6.96	8.68	10.40	12.10	13.78	15.45
10	0.29	2.04	3.78	5.52	7.25	8.97	10.68	12.38	14.06	15.73
20	0.58	2.33	4.07	5.80	7.53	9.25	10.96	12.66	14.34	16.00
30	0.87	2.62	4.36	6.09	7.82	9.54	11.25	12.94	14.62	16.28
40	1.16	2.91	4.65	6.38	8.11	9.83	11.53	13.22	14.90	16.55
50	1.45	3.20	4.94	6.67	8.40	10.11	11.81	13.50	15.17	16.83

* Difference in height in feet can be obtained if we multiply the value given in the table by the stadia intercept. (*From Keuffel & Esser Company, Solar Ephemeris.*)

Fig. 4.10. In field sketching we establish a horizontal and a vertical reference line to locate a number of key points exactly before drawing detail.

intercept is 6.4 ft. The slope is downward and we read opposite the V arrow 33.

$$50 - 33 = 17$$

With this we multiply the stadia intercept:

$$VE = 17 \times 6.4 = 108.8 \text{ ft}$$

Field Sketching

One need not be an artist to make satisfactory field sketches. Old-time geographers, before the age of the camera, all drew sketches, usually very good ones. The camera is fast, accurate, and effortless, yet sketching has many advantages.

First of all, we can be selective. A great deal of unnecessary foliage can be omitted; we may concentrate on important fea-

tures, such as land use, industries, etc. Photographs usually have too much foreground; geographers are often more interested in the more distant landscape, which shows up poorly on photographs. Notes, labels, place names, angular scale and compass direction may be easily added to our sketches. Sketches can be reproduced by line cuts; photographs need halftones, which look dull and are expensive. But the chief merit of sketching is that it stimulates observation in the student, who discovers a great deal which he might have missed with a hurried snapshot. There is something personal and prepossessing in a sketch—it attracts people.

One may use a ready-made sketch book, or cut 8 by 12-in. sheets of drawing paper

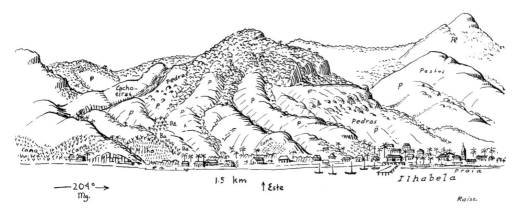

Fig. 4.11. Geographic field sketch made from shipboard in Brazil. Note a certain amount of symbolization and annotations. Actual sketching time was about 7 min. Inking took much longer. (*P*—pastures; *Ba*—bananas; *Fl*—forest; *Milho*—corn; *Caña*—sugar; *Cachoeiras*—rapids.)

and fasten them to a writing pad. Pencil, eraser, an inch scale, sunglasses, and compass are all the equipment needed. The process is simple. Select a suitable spot with a good view and decide how much to include in the sketch. Choose a center point and halve the paper with a vertical line. Then draw a horizontal reference line —the horizon, or a distant lake shore, or road. Select six to ten key points and measure their distance from the center with the inch scale held at arm's length. Lay the points out on paper at half of the measured distances, for if the full distance is used, the area may be too narrow.

In drawing detail, *proceed from large features to smaller ones.* By adding detail to detail we are bound to come out wrong. A certain symbolization may be used— forest with "tree-top" lines, small rectangular lines for distant villages, straight ruling for fields, etc. A slight vertical exaggeration of distant features near the horizon is not only permissible, but desirable.

The coulisse effect (like the side panels of a stage) must be brought out very clearly; there should be no doubt which hill obscures the other. This will bring depth to the picture. Lines depicting distant features should be much lighter than those of nearer features. Forested-hill profiles should be coarse nearby, and finer and finer with distance. If the sketch is painted, warm red, brown, and yellow colors can be used in the foreground and cold light blue-grays in the distance.

Fig. 4.12. Compass traverse by the sixteenth-century Nuremberg surveyor Pfinzing. (*After E. Gage.*)

Do not hesitate to add a great deal of lettering to your picture. Names can go above or below, with arrows pointing to the features. The purpose of a field sketch is to record rather than to produce a work of art.

Inking and coloring can be done in the laboratory. Coloring should be light and done with thin water colors. Heavy coloring may spoil the picture, but it can always be washed off with a sponge. Coloring with crayons is far less effective.

5

The Principles

of Lettering

The maps of primitive people, such as the Eskimos, obviously had no writing and had to be explained orally (see Fig. Int. 2). If the map has to be used by a person not present, the features have to be identified by written names. Even the earliest maps which have survived were heavily lettered. Whether inscriptions were in Babylonian cuneiform, Chinese characters, Greek letters, or Arabic script, they became a dominant feature on maps. It often happened, indeed, that the names were so bulky that they crowded out the topographic symbols. Occasionally the symbol was completely omitted and replaced by lettering. On the map in Fig. 5.1 the name "Anti-Lebanon" is curved along the crestline of the range without any mountain symbol at all. Thus the lettering itself became the symbol.

The problem of lettering is more difficult with small-scale maps. On a small world map the name "Copenhagen" in the smallest legible type may easily reach beyond Moscow, and the name "Switzer-

land" or "The Netherlands" may not fit inside the respective countries. On a standard double-page map of the United States, such as in the "Goode's World Atlas," the name "Philadelphia" is 170 miles long.

Yet certain places and features can be best described by their names, such as political units like Syria, natural regions like Jazira, seas and bays, etc. (see Fig. 5.1). Here the lettering indicates the extent and location of the area. Lettering attracts attention. Map publishers have found by experience that maps without names do not sell. An unnamed feature will not be remembered easily. There was a time, not too long ago, when a map publisher's chief claim to superiority was that he could press more names into his maps than the competitors. Nowadays we recognize that a map should be more than a multitude of place names, and we omit all names not necessary for the purpose of the map. An overlettered map is not only unattractive, but it does not allow us to tell much about the land.

The Style of Letters

Our letters are derived from ancient Rome. The Roman lettering which survived was chiseled into stone, as it was on Trajan's Column in Rome, which is regarded as the standard. This is beautiful, well-proportioned lettering and is the basic type for our capital letters. The chisel, as the Romans used it, produced wide vertical and narrow horizontal lines. To give a

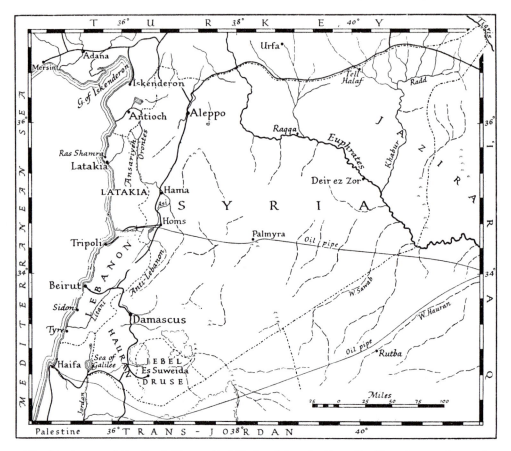

Fig. 5.1. A map of the Royal Geographical Society lettered with quills. Note the prevalence of lettering over symbols in the West and also how names are used to indicate areas and mountain ranges.

clean-cut form to a letter, they added some extra chisel strokes at the end of each line, and this is the origin of serifs. Later the medieval monks used a reed pen or quill with chopped-off nib. They simplified the roman letters and produced our lower-case lettering.

In the fifteenth century a more cursive style of letter developed. The easiest motion of a human hand is not vertical or horizontal, but slightly oblique. Slanted letters can be drawn faster, and the motion comes more naturally than does that for vertical

lettering. This slanting was first applied to the roman letters in Italy. At present our printers call slanted letters *italic* (also *"inclined"* or *"oblique"*). The usual angle of the originally vertical lines is about 70° to 75° with the horizontal.

The simplest style of lettering has no serifs and no variation in thickness. This is variouslly called sans-serif, block, or single-stroke lettering. The printers call it gothic. The recent tendency is toward sans-serif letters. Several foreign surveys, our own Army maps, and most scientific maps use almost exclusively gothic types. Maps lettered with the ordinary Leroy or Wrico sets are obviously sans-serif. The variety of styles, however, adds to the expressiveness of the map. The styles of letters used on maps are shown here.

Various styles of letters can express the various features of the map. Most of our government maps use roman lettering for political divisions, italic for hydrography, gothic for relief features, and inclined gothic for buildings and engineering features like roads, dams, etc. Most atlases and road maps, however, use gothic for cities as this saves space.

Light-face types are used to deemphasize large letters. For instance, auto road maps use them for county names, which are large but less important. Lydian style combines the elegance of roman with the economy of gothic, and is becoming popular both in upright and inclined form. Phantom (open-face) letters are good for very large names, to reduce their weight.

Shape and Weight of Letters

If we look at the roman letters in Fig. 5.2, it is apparent that letters vary greatly in width. The letters MW and CDGOQ are the widest. AHNTUVXYZ are medium in width, while the rest of the letters are narrow. This variation adds greatly to

Fig. 5.2. Roman and italic lettering designed for the Ordnance Survey of Great Britain. Note the ascenders and descenders on the numbers. This is more legible than numbers of uniform height.

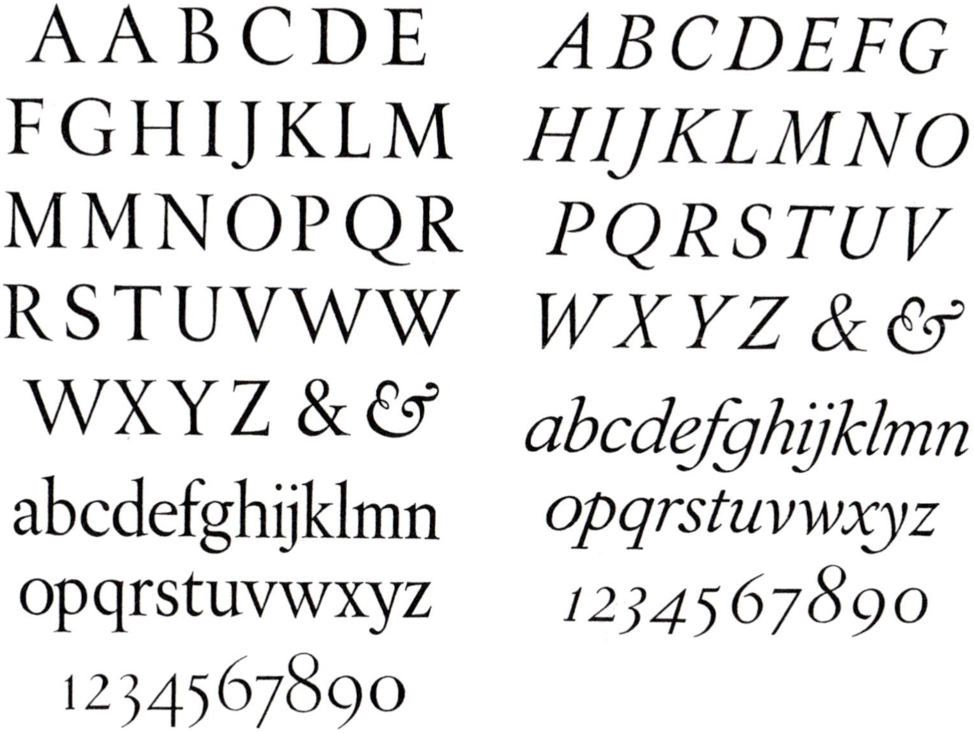

AABCDE
FGHIJKLM
MMNOPQR
RSTUVWW
WXYZ & &
abcdefghijklmn
opqrstuvwxyz
1234567890

ABCDEFG
HIJKLMNO
PQRSTUV
WXYZ & &
abcdefghijklmn
opqrstuvwxyz
1234567890

LETTERING ON MAPS
Withycombe

Roman of variable thickness and serifs

Italic of variable thickness and serifs, inclined

Gothic of even thickness, no serifs

Lightface Roman of even thickness with serifs

Lightface Italic of even thickness with serifs, inclined

Inclined gothic or gothic italic of even thickness, no serifs, inclined

Variable thickness, no serifs, as **Lydian**

Sans serif more modern gothic types

Cursive Manuscript a kind of roman produced by stub pen

Phantom produced by double lines

Fig. 5.3. Styles of letters used on maps.

beauty and legibility. There is, however, little variation in the width of lower-case letters.

As for the whole alphabet, the letters can be *wide* (extended), *normal,* or *condensed.* A normal, medium-width letter, like an H, is about seven-tenths as wide as it is high. The M type letters fit almost into a square, while the width of narrow letters is about half of their height. We rarely use wide letters on maps, but narrow CONDENSED letters are common, since in crowded maps they save space. They are often used for "spread" lettering. Condensed letters can

Fig. 5.4. Width and height of letters.

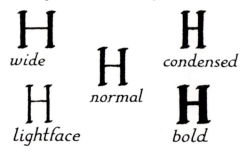

be made higher without adding too much weight, and the larger space between them allows for more topography. It is perfectly permissible to have small names in the space between spread large letters.

The width of line in a normal gothic letter is about one-tenth the height. If the letter is thinner, we call it lightface; if it is heavier, the printers called it **bold.** In roman and italic types the wide strokes are about one-seventh of the height, while the narrow parts are about half as thick. The cartographer can use the weight of letters for emphasis and deemphasis. Bold letters are good for mountain ranges; lightface letters are often used between inner and outer margins to complete the lettering inside (see Fig. 5.1).

Capitals and Lower-case Letters

When shall we use capitals and when the lower-case letters? For a simple rule: *Names inside the feature are capitals; outside the feature, names are lower case with initial capitals.* For instance, on a small-scale map where Boston is a dot, we use lower case. But on a topographic map where the city covers several square inches, and the lettering is inside the area, spread the name out in capitals. The Ohio River, if shown by a line, will have the name in lower case; but on a large-scale map the name will be inside the river in capitals. There are exceptions for uniformity. Names of countries are usually inside in capitals; on a small-scale map, however, names like "Guatemala" or "Netherlands" will often be outside, yet in capitals. In general, point and line symbols will be named by lower-case letters, while large areas will have capitals.

Lettering should express by its style the

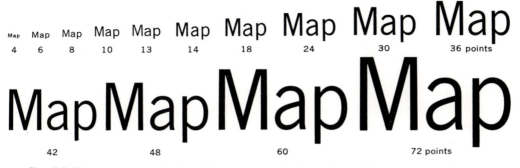

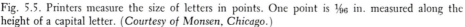

Fig. 5.5. Printers measure the size of letters in points. One point is 1/96 in. measured along the height of a capital letter. (*Courtesy of Monsen, Chicago.*)

nature of the feature, by its size the importance of the feature, by its placement the location, and by its spacing the extent of the feature.

Size of Letters

The general rule is: The larger the feature, the larger the lettering. A large city should be shown by larger letters than a small one, and the name "Asia" should be larger than "Iran." This does not mean that the size of lettering is proportionate to the size of the feature. If we start by indicating a small city of a thousand people with the smallest readable type, a city of a million cannot have lettering more than two or three times larger. Rarely are more than four to six gradations used, each just slightly larger than the other, but heavier letters often indicate the larger feature.

Printers express the height of letters in points, a point being 1/72 in. measured along the whole metal body of the letter, including descenders (as in a y) and margin. Thus the height of capital letters will be only about 1/96 in. for each point. The smallest legible type is about 4 point, with about 1/24-in.- high capitals. Printers call 12 points a *pica*, which is 1/6 in.

Much depends on the use to be made of maps. A reference atlas will have smaller letters than a map in a periodical. The largest letters will be used on wall maps, which are seen from a distance. Not all lettering on any wall map can be made legible from a great distance, since to be readable from 40 ft the smallest letters would have to be 1 in. high (see Fig. 5.5). Figure 5.6 shows the relative readability with distance. This is true only of black let-

Fig. 5.6. The smallest readable type at various distances for readers of average eyesight and under good light.

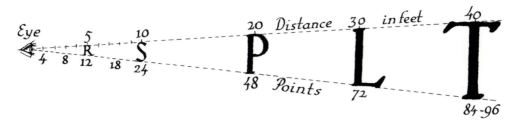

ters on white or yellow. White letters chalked on the blackboard have to be about 50 per cent larger.

Lettering for Lantern Slides. We can use Fig. 5.6 for figuring out how large the lettering should be on maps made for lantern slides. We must know the size of the room and the size of the screen. Let us assume that the rear seats are 60 ft from the screen, which is 10 ft square, and we made the map 1 ft wide (or high). The question is: How far has the map to be held from our eyes to cover the screen?

$$X{:}w = D{:}W \text{ and } X = \frac{wD}{W}$$

In our case

$$X = \frac{1 \times 60}{10} = 6 \text{ ft}$$

where X is the distance from map to eye, W is the width of map, D the distance from eye to screen. If we place the map in front of the man in the rear seat, at $60{:}10 = 6$ ft distance, the map would just cover the width of the screen. Thus our smallest lettering should be visible from 6 ft. According to Fig. 5.6, from 6 ft away the minimum size is 14 point. We assume that the projector uses the full 10-ft width of the screen. It is true that the map will be photographed to a small-sized slide, but

Fig. 5.7. In the upper name, the distance between letters is equal yet the lettering appears uneven. In the lower name, the area between letters is nearly equal.

MILWAUKEE

MILWAUKEE

if it is projected so that the full screen is filled, the visibility remains the same.

As one frequently does not know exactly the size of the room and the projector beforehand, it is good to be on the safe side. Fourteen-point lettering on a 12-in.-wide map is usually adequate, but in smaller rooms 12 or even 10 point letters can be used. Lower-case letters read almost as well as capitals of the same point size.

The Placing of Names

With point symbols, such as cities or buildings on small-scale maps, the name is usually placed to the right of the symbol. This makes it clear to which symbol the name belongs. It is better not to place the name exactly in line with the symbol, but slightly above or below. If we want to indicate that the city is on the west side of a river, we place the name west of the symbol. The names of harbors should be preferably toward the sea. We may have to choose other arrangements in congested areas. Any placing is permissible as long as there can be no mistake about the application of the name. Sometimes short arrows leading from the name to the symbol help.

Names of line symbols, such as roads, rivers, etc., curve along the line and preferably above it. We do not spread the letters all along a river, but keep them close together. River names may have to be repeated to indicate which of the branches carries the name of the main river. Names of lakes and islands can be either outside or inside, but only in an emergency is the shoreline cut with the name.

Spacing of Letters

In Fig. 5.7 the letters in the top name have the same space between them, yet the lettering looks uneven. In the lower name the letters are spaced so that the space between the open part of the letters is about equal. The space between two letters, even at their widest spacing, should be considerably less than the width of normal letters—usually about half of that unless the letters are spread as in Fig. 5.6. The larger the letter, the smaller (relatively) the space between individual letters. The most common mistake of beginners is to space their letters too unevenly and too widely. Two words are separated by the width of one letter.

Spreading of Letters

The skill of a cartographer is best demonstrated in the placing of names of areas, such as countries, mountains, regions, seas, etc. Here the lettering has to be spread to indicate the size of the area. If the region is nearly round, it should be lettered horizontally—that is, parallel to the parallels. If the region is elongated, the lines are curved along the axis of the area. The procedure is shown in Fig. 5.8. First the letters are counted, and the first and last

Fig. 5.8. Names of elongated areas and mountain ranges are spread along the whole extent of the area.

letters are placed near the two ends of the area. We draw the bottom guide line for the letters and space the letters evenly, with due regard to their width. After drawing short upper guide lines, one for each letter, we are ready for lettering. The names of mountain ranges are spread along the crestline of the range and should follow the main trend very exactly.

6

The Practice of Lettering

GEOGRAPHIC NAMES

Lettering takes a large portion of the cartographer's time, whether it is done by hand, by machine, or by pasting printed names. Very few large establishments use hand lettering in their published maps but even these lay out their lettering in pencil. People are more critical of lettering and spelling than of the actual content of maps.

Hand Lettering

With the availability of mechanical devices, the art of hand lettering is somewhat neglected. Yet no machine lettering can compete with the beauty, flexibility, and speed of good hand lettering. There is no better way to learn good spacing and placing. To be able to letter well is a lifelong asset.

It takes thousands of nerve impulses just to stay still. To make such fine and complex motions as lettering needs the channeling of thousands more. No need to be discouraged if the first attempts are far from perfect. To establish these channels takes concentration and practice. The following directives may help:

1. Lay out your lettering very lightly in pencil. Do not try for perfection in form, but concentrate on the proper spacing of the letters.
2. Draw fine, sharp guide lines with pencil. Do not press the pencil down into the paper. For lower-case letters the guide lines are at "waist" height, but a third line for tall and capital letters may be added. You may use the device in Fig. 6.1 or ready-made perforated triangles. There are also special dividers with two pencil arms. The guide lines will control the size and placing of names.
3. Rub the pencil work of the map with eraser powder or pounce to take out the unnecessary graphite and dirt and to make the paper more absorbent. Dust off the paper very thoroughly.
4. Select your pen for proper width or shape. It will be necessary to clean it often with water. Special solvents for india ink are on the market. An ink retainer made from copper wire is helpful (see Fig. 6.2). After a few weeks' use, a pen should be sharpened on an oil stone under a strong magnifier. The same pen may be used for months; this is better than breaking in new pens all the time. The pen should not wobble in the holder.
5. Sit erect, rest body on left arm, and place the paper in the most comfortable position. Avoid, if possible, lettering on immovable paper. Rest your right lower arm on the table; it will move sufficiently over its padding of muscles. Your little finger gives support and should slide easily over the paper.
6. Every stroke, however short, should be

58

clearly defined. The strokes in general are toward you, and preferably left to right. Make them with full-arm motion, not with wrist and finger. Dip your pen and mark the end of the stroke with a dot. Stretch your arm a little beyond the beginning of the stroke, aim, and bring down your pen. Keep your eye on the end dot and not on your pen point. Move slowly and with even pressure. After you have reached the end dot, follow through, lifting the pen up in the direction of the stroke. Curved letters are made up from two to four segments. Do not "paint" your letters, going over the same line several times.

7. Keep up an even, slow rhythm in lettering. Relax, but maintain the utmost concentration. If somehow the lettering does not go well, check your sitting position and the angle of the paper. Dust off the paper and clean the pen. Beginners usually try to letter too fast, and miss out on placing their end dots. It is also hard to change over from finger motion to arm motion. This needs a great deal of self-discipline, but is a must for a letterist. The same holds true for following through, yet success depends on this too. It is better to go beyond the guide lines than short of them; the surplus can easily be painted out with white. And remember: Even motion brings even results.

Stub-pen lettering is the best way to learn good lettering with full-arm motion. It is fast and attractive looking, but it cannot be done well with finger motion. For practice take an Esterbrook Drawlet pen No. 13 and try first lower-case italics, ½ in. high at waist, as in Fig. 6.3. If you can make some good swash lines, you have mastered the motion. It helps to practice on the blackboard with an inch-long chalk held down flat with four fingers.

Serifs are made by getting contact with the paper in the beginning, and by giving a little jerk at the end of the stroke to finish

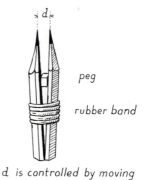

d is controlled by moving the peg up or down

Fig. 6.1. Device for drawing guide lines for lettering.

it off neatly. It is possible, however, to omit serifs altogether, as in the Lydian type. The Royal Geographical Society uses almost exclusively stub-pen lettering made with hand-cut quills (see lettering in Fig. 5.1).

Left-handed people have little handicap in full-arm lettering. It comes to them much more easily than writing, and many of them become fully proficient.

Lettering Devices

The great majority of papers presented at geographical meetings include maps, tables, and diagrams lettered by Leroy,

Fig. 6.2. An ink retainer provides an even flow of ink.

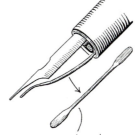

copper wire with ends hammered flat

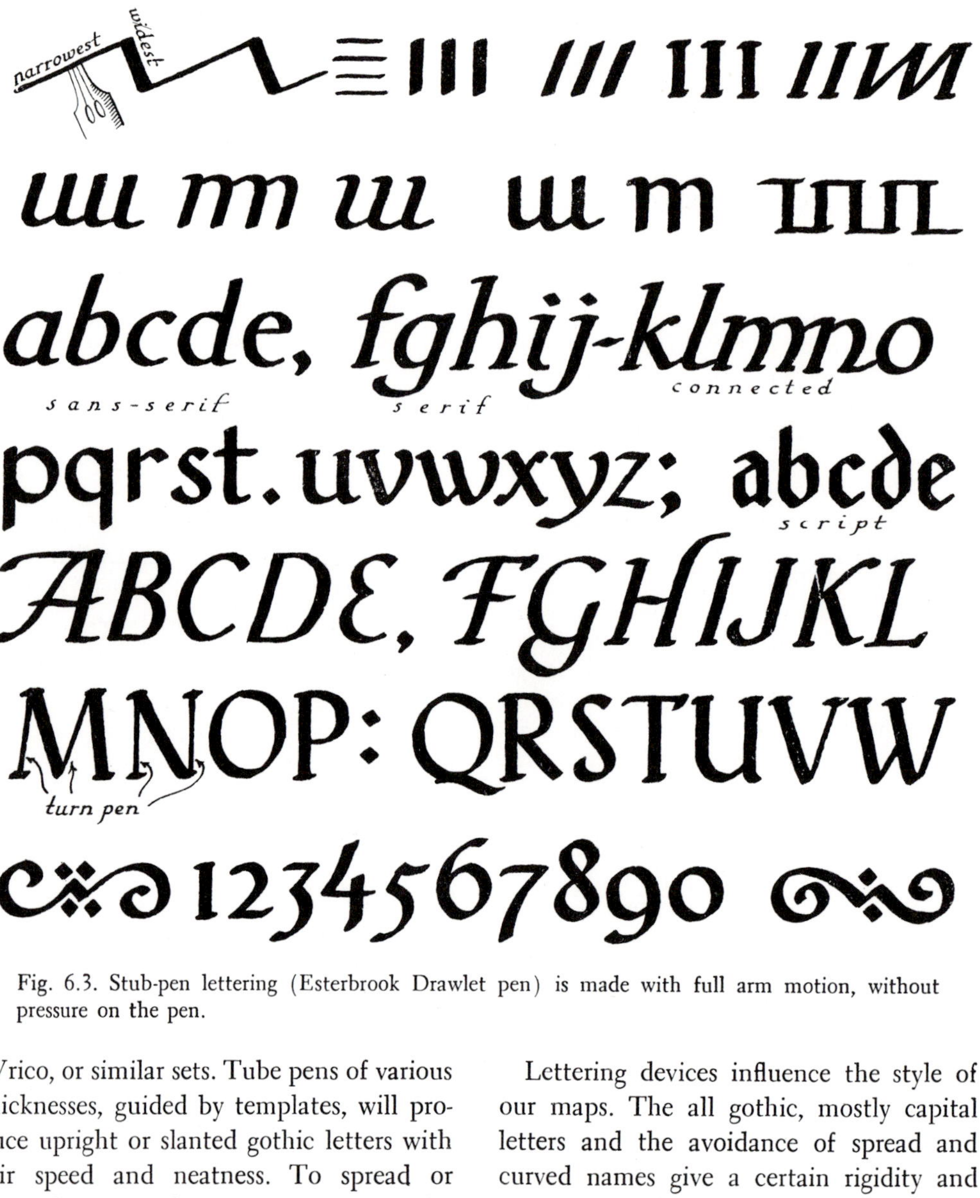

sans-serif   *serif*   *connected*

script

turn pen

Fig. 6.3. Stub-pen lettering (Esterbrook Drawlet pen) is made with full arm motion, without pressure on the pen.

Wrico, or similar sets. Tube pens of various thicknesses, guided by templates, will produce upright or slanted gothic letters with fair speed and neatness. To spread or curve the names, however, is not easy. The chief problem is the proper selection of size and thickness, and keeping the paper and pen clean. Even spacing of letters needs attention. Too often the letters are spaced unevenly and too far from each other.

Lettering devices influence the style of our maps. The all gothic, mostly capital letters and the avoidance of spread and curved names give a certain rigidity and mechanical appearance to the maps. As small-sized lettering is difficult to do, maps are prepared two to three times publication size, rendering them even more unnatural when reduced. But the names are neat, clear, and legible, and the maps are uniform. Thus the use of these devices is

Fig. 6.4. Normograph and Wrico lettering sets use perforated templates.

Fig. 6.5. Leroy lettering set uses the same embossed guide for vertical or slanted letters.

often justified. Their danger lies in preventing the practice of hand lettering, to the detriment of the profession.

Varigraph and similar devices can make the letters larger or smaller, wider or narrower, by shifting dials, using the same template. Over a hundred different templates are available, in roman, italic, gothic. The acquisition of such a machine is justified in large departments.

"Stick-up." For most of our formal maps, the lettering is set up in type and pasted on the map. The letters can be printed on gummed or plain paper, or they can be printed on thin rice paper and "floated," with a brush that has been dipped into a sticky liquid. The most popular is *cellotype*, with the names printed on transparent cellophane that is backed with sticky paraffin or other paste. Several companies produce cellotype and gladly send a type specimen book upon request.

The usual procedure is to make a list of names specifying the style and size and send it to the company. For this, a rich assortment is offered by each company; Fig. 6.7 shows some from Monsen in Chicago. It is economical to write the names continuously and not on separate lines. The company will print the names on the un-

Fig. 6.6. All letters are made from the same Varigraph template by shifting dials.

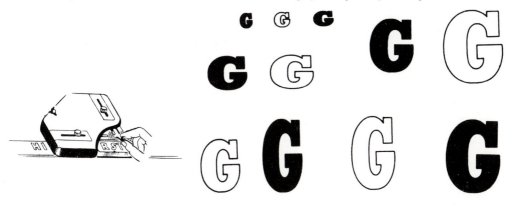

Fig. 6.7. Types used on maps. (*Courtesy Monsen, Chicago.*)

QUICK REFERENCE TO MAP TYPE STYLES

The specimens shown below cover most of the type styles used for map making. In each style we show as much of the alphabet as possible. The numbers preceding the letters indicate the point size. In the Copperplate Gothics where there are several type sizes of each point size the foundry number follows each point size. The name of the type with the sizes that are machine set or hand set is shown under each specimen.

5 ABCDEabcde   7 FGHIJfghij   8 KLMNklmn   9 OPQRopqr   10 STUstu   11 VWXvwx   13 YZyz
MONSEN LIGHT GOTHIC NO. 621J—5 TO 13 PT. MACHINE SET

5 ABCDEabcde   7 FGHIJfghij   8 KLMNklmn   9 OPQRopqr   10 STUstu   11 VWXvwx   13 YZyz
MONSEN MEDIUM GOTHIC NO. 512J—5 TO 13 PT. MACHINE SET

5 ABCDEabcde   7 FGHIJfghij   8 KLMNklmn   9 OPQRopqr   10 STUstu   11 VWXvwx   13 YZyz
MONSEN BLACK GOTHIC NO. 612J—5 TO 13 PT. MACHINE SET

7 ACEace   7 ACEace   7 ABCDEabcde   9 FGHIfghi   11 JKLMjklm   7 NOPQRnopqr   9 STUVstuv   11 WXYZwxyz
NO. 472J   NO. 813J   MONSEN LIGHT GOTHIC ITALIC NO. 621K   MONSEN MEDIUM GOTHIC ITALIC NO. 512K
ALL STYLES AND SIZES SHOWN ABOVE ARE MACHINE SET

5 ABCDEF   6 ABCDEabcde   7 FGHIJ   8 KLMNklmn   9 OPQRopqr   12 STUVstuv   14 WXYZwxyz
DRAFTSMAN'S ITALIC NO. 124K—5 TO 14 PT. (5 & 7 PT. CAPS ONLY) HAND SET

5 on 6 ABCDEFGabcdefg   6 ABCDEFGabcdefg   8 HIJKLMNhijklmn   10 OPQRSTopqrst   12 UVWXYZuvwxyz
DEGREE GOTHIC NO. 2—5 ON 6 TO 12 PT. HAND SET

6-1 ABCDEF   6-2 GHIJKL   6-3 ABCabc   6-4 DEFdef   8 GHIghi   10 JKLjkl   12 MNOmno   14 PQpq   18 RSrs
NO. 452J—6 TO 12 PT. MACHINE SET—14 TO 36 PT. HAND SET

5 ABCDabcd   6 ABCDabcd   7 EFGHefgh   8 IJKLijkl   10 MNOPmnop   12 QRSTqrst   14 UVWuvw   18 XYZxyz
NEWS GOTHIC CONDENSED NO. 204J—5 TO 72 PT. MACHINE SET—14 TO 72 PT. HAND SET

6 ABCDEFabcdef   8 GHIJKghijk   10 LMNOlmno   12 PQRSpqrs   14 TUVtuv   18 WXwx
NEWS GOTHIC NO. 206J—6 TO 12 PT. MACHINE SET—14 TO 72 PT. HAND SET

5 ABCabc   6 DEFdef   8 GHIghi   10 JKLjkl   12 MNOmno   14 PQpq   18 RSrs
MONOTONE GOTHIC—5 TO 24 PT. HAND SET

6-5 ABCD   6-4 EFGH   6-3 IJK   6-2 LMN   6-1 OPQ   12-3 RST   12-2 UVW   12-1 XYZ
LIGHT GOTHIC ITALIC (ALSO CALLED INCLINED OR SLOPE GOTHIC)—6 AND 12 PT. HAND SET

6 ACEace 8GKNgkn 10 OSTost 12 UVYuvy   6 ACEace 8 GKNgkn 10 OSTost 12 UVYuvy   6 ACEace 8 GKNgkn 10 OSTost 12 UVYuvy
ALTERNATE GOTHIC NO. 51J   ALTERNATE GOTHIC NO. 77J   ALTERNATE GOTHIC NO. 3
6 TO 12 PT. MACHINE SET—14 TO 84 PT. HAND SET   6 TO 12 PT. MACHINE SET—14 TO 72 PT. HAND SET   6 TO 72 PT. HAND SET

5½ Aa 6Cc 8GKgk 10NOno 12STst   6 ACac 8 GKgk 10 NOno 12STst   5 ACac 6 EFef 8GKgk 10NOno 12STst
GOTHIC NO. 149J—5½ TO 12 PT. MACHINE SET   GOTHIC NO. 249J—6 TO 12 PT. MACHINE SET   GOTHIC COND. NO. 49J—5 TO 12 PT. MACHINE SET

14 ABab   18 CDcd   24 Ee   **14 FGfg**   **18 HIhi**   **24 St**
MONSEN MEDIUM GOTHIC NO. 512J—14 TO 24 PT. MACHINE SET   MONSEN BLACK GOTHIC NO. 612J—14 TO 24 PT. MACHINE SET

6-31 ABCD   6-32 EFGH   6-33 IJK   6-34 LMN   12-35 OPQ   12-36 RST   12-37 UVW   12-38 XYZ
LIGHT COPPERPLATE GOTHIC CONDENSED NO. 341J—6 AND 12 PT. MACHINE SET—18 AND 24 PT. HAND SET

6-11 ABCD   6-12 EFGH   6-13 IJK   6-14 LMN   12-15 OPQ   12-16 RST   12-17 UVW   12-18 XYZ
HEAVY COPPERPLATE GOTHIC CONDENSED NO. 343J—6 AND 12 PT. MACHINE SET—18 AND 24 PT. HAND SET

6-1 ABCD   6-2 EFGH   6-3 IJK   6-4 LMN   12-5 OPQ   12-6 RST   12-7 UVW   12-8 XYZ
LIGHT COPPERPLATE GOTHIC NO. 340J—6 AND 12 PT. MACHINE SET—18 AND 24 PT. HAND SET

6-21 ABCD   6-22 EFGH   6-23 IJK   6-24 LMN   12-25 OPQ   12-26 RST   12-27 UV   12-28 WX
HEAVY COPPERPLATE GOTHIC NO. 342J—6 AND 12 PT. MACHINE SET—18 AND 24 PT. HAND SET

6-51 ABC   6-52 DEF   6-53 GHI   6-54 JKL   12-55 MN   12-56 OP   12-57 QR   12-58 ST
COPPERPLATE GOTHIC ITALIC NO. 346K—6 AND 12 PT. ALSO 12 PT. 55, 56 AND 57 MACHINE SET—18 PT. 58 UP TO 24 PT. HAND SET

8-41 ABCD 6-42 EFGH 6-43 IJK 6-44 LMN 12-45 OPQ 12-46 RST 12-47 UV 12-48 WX
COPPERPLATE GOTHIC BOLD NO. 345J—6 AND 12 PT. MACHINE SET—18 AND 24 PT. HAND SET

8 Pt. 606J ABCDEabcde   8-605J FGHIJfghij   8-604J KLMNOklmno   8-603J PQRSTpqrst   8-607J UVWXYZuvwxyz
SPARTAN LIGHT—6 TO 36 PT.   SPARTAN MEDIUM—6 TO 72 PT. SPARTAN DEMIBOLD—6 TO 72 PT.   SPARTAN BOLD—6 TO 84 PT.   SPARTAN BOLD COND.—8 TO 84 PT.

8 Pt. 606K ABCDEabcde 8-605K FGHIJfghij   8-604K KLMNOklmno   8-603K PQRSTpqrst   8 607K UVWXYZuvwxyz
SPARTAN LIGHT ITAL.—6 TO 36 PT. SPARTAN MED.ITAL.—6 TO 72 PT. SPARTAN DEMIBOLD ITAL.—6 TO 72 PT. SPARTAN BOLD ITAL.—6 TO 72 PT. SPARTAN BOLD COND.ITAL.—8 TO 72 PT.

6 Pt. 605K ABCDEfabcdef   8 GHIJKghijk   10 LMNOlmno   12 PQRSpqrs   14 TUVWtuvw   18 XYZxyz
SPARTAN MEDIUM ITAL. — 6 TO 72 PT. — THE ABOVE THREE SPECIMEN LINES SHOW THE SPARTAN SERIES WHICH IS ALSO CALLED FUTURA AND 20TH CENTURY. EVERY POINT SIZE IS A LITTLE SMALLER THAN MOST OTHER TYPE STYLES. ALL 6 TO 12 PT. SIZES ARE MACHINE SET AND LARGER SIZES ARE HAND SET.

6 ABab   8 CDcd   10 EFef   12 GHgh   14 IJK ijk   6 ABab 8 CDcd 10 EFef 12 GHgh 14 IJKijk
PARAGON—6 TO 24 PT. HAND SET   PARAGON ITALIC—6 TO 14 PT. HAND SET

5 ABCDEFGHIJKLM 6 NOPQRSTUVWXYZ 8 ABCDEFGHIJKLM 10 NOPQRSTUVWXYZ
VICTORIA ITALIC NO. 224K—HAND SET

8 Pt. 190J ABCDEFabcdef   8-290J GHIJKLghijkl   8-790J MNOPQRmnopqr   8-390J STUVWXstuvwx
STYMIE LIGHT—6 TO 72 PT.   STYMIE MEDIUM—6 TO 72 PT.   STYMIE BOLD—6 TO 36 PT.   STYMIE EXTRA BOLD—6 TO 72 PT.

8 Pt. 190K ABCDEFabcdef   8-290K GHIJKLghijkl   8-790K MNOPQRmnopqr   8-390K STUVWstuvw
STYMIE LIGHT ITALIC—6 TO 24 PT. STYMIE MEDIUM ITALIC—6 TO 72 PT. STYMIE BOLD ITALIC—8 TO 72 PT. STYMIE EXTRA BOLD ITALIC—8 TO 48 PT.
THE ABOVE TWO SPECIMEN LINES SHOW THE STYMIE SERIES. ALL 6 TO 12 PT. SIZES ARE MACHINE SET AND LARGER SIZES ARE HAND SET.

6 ABCDEFabcdef   7 GHIJKghijk   8 LMNOlmno   6 ABCDEFabcdef   7 GHIJKghijk   8 LMNOlmno
MODERN ROMAN NO. 13A—MACHINE SET   MODERN ROMAN ITALIC NO. 13C—MACHINE SET

8 ABCDEabcde   10 FGHIfghi   12 JKLMjklm   8 ABCDEabcde   10 FGHIfghi   12 JKLMjklm
WALBAUM—MACHINE SET   WALBAUM ITALIC—MACHINE SET

8 ABCDEabcde   10 FGHIfghi   12 JKLMjklm   8 ABCDEabcde   10 FGHIfghi   12 JKLMjklm
WALBAUM MEDIUM—MACHINE SET   WALBAUM MEDIUM ITALIC—MACHINE SET

6 Aa 7 Bb 8 Cc 9 Dd 10 Ee 11 Ff 12 Gg 14 Hh   6 Aa 7 Bb 8 Cc 9 Dd 10 Ee 11 Ff 12 Gg 14 Hh
TIMES ROMAN—6 TO 14 PT. MACHINE SET—18 TO 48 PT. HAND SET   TIMES ROMAN ITALIC—6 TO 14 PT. MACHINE SET—18 AND 24 PT. HAND SET

6 ABCDEabcde   7 FGHIJfghij   8 KLMNklmn   9 OPQRopqr   10 STUstu   11 VWXvwx   12 YZyz
CENTURY EXPANDED NO. 20A—6 TO 12 PT. MACHINE SET—14 TO 36 PT. HAND SET

6 ABCDEabcde   7 FGHIJfghij   8 KLMNklmn   9 OPQRopqr   10 STUstu   11 VWXvwx   12 YZyz
CENTURY EXPANDED ITALIC NO. 20C—6 TO 12 PT. MACHINE SET—14 TO 36 PT. HAND SET

6 ABCDEabcde   7 FGHIJfghij   8 KLMNklmn   9 OPQRopqr   10 STUstu   11 VWXvwx   12 YZyz
CENTURY SCHOOLBOOK NO. 420A—6 TO 18 PT. MACHINE SET—24 TO 48 PT. HAND SET

Reproduction Size = 25% Reduction of Original Type Size (4 to 3 reduction)

6 ABCDEabcde 7 FGHIJfghij 8 KLMNklmn 9 OPQRopqr 10 STUstu 11 VWXvwx 12 YZyz
CENTURY SCHOOLBOOK NO. 420A—6 TO 18 PT.

6 ABCDEabcde 7 FGHIJfghij 8 KLMNklmn 9 OPQRopqr 10 STUstu 11 VWXvwx 12 YZyz
CENTURY SCHOOLBOOK ITALIC NO. 420C—

5 ABCabc 6 DEFdef 8 GHIghi 10 JKLjkl 12 MNOmno 5 ABCabc 6 DEFdef 7 GHIghi 8 JKjk 10 LMlm 12 NOno
CHELTENHAM BOLD CONDENSED NO. 88J   CHELTENHAM BOLD NO. 86J

6 Aa 7 Bb 8 Cc 9 Dd 10 Ee 11 Ff 12 Gg 14 Hh   *6 Aa 7 Bb 8 Cc 9 Dd 10 Ee 11 Ff 12 Gg 14 Hh*
TIMES ROMAN—6 TO 14 PT.   TIMES ROMAN ITALIC—6 TO 14 PT. MACHINE SET—18 AND 24 PT. HAND SET

5 ABCDabcd 6 EFGHefgh 8 IJKLijkl 10 MNOmno 12 PQRpqr 14 STUstu 18 VWvw
CLEARFACE ITALIC NO. 289K—6 TO 12 PT.

WEISS ITALIC 11 12 14
ABCDEFGHIJKLMNOPQRSTUVWXYZ&
abcdefghijklmnopqrstuvwxyz

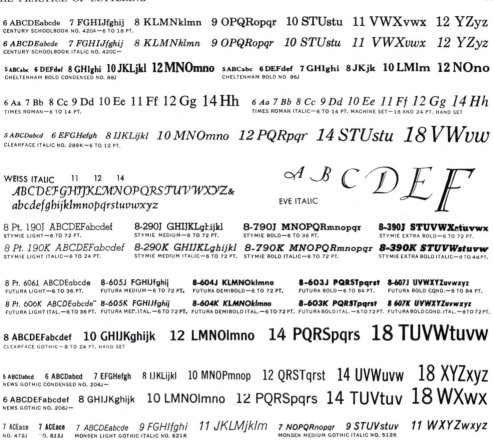

EVE ITALIC

8 Pt. 190J ABCDEFabcdef   8-290J GHIJKLghijkl   8-790J MNOPQRmnopqr   8-390J STUVWXstuvwx
STYMIE LIGHT—6 TO 72 PT.   STYMIE MEDIUM—6 TO 72 PT. STYMIE BOLD—6 TO 36 PT.   STYMIE EXTRA BOLD—6 TO 72 PT.

8 Pt. 190K ABCDEFabcdef   *8-290K GHIJKLghijkl*   *8-790K MNOPQRmnopqr* ***8-390K STUVWstuvw***
STYMIE LIGHT ITALIC—6 TO 24 PT. STYMIE MEDIUM ITALIC—6 TO 72 PT. STYMIE BOLD ITALIC—8 TO 72 PT. STYMIE EXTRA BOLD ITALIC—8 TO 48 PT.

8 Pt. 606J ABCDEabcde   8-605J FGHIJfghij   **8-604J KLMNOklmno**   **8-603J PQRSTpqrst**   8-607J UVWXYZuvwxyz
FUTURA LIGHT—6 TO 36 PT.   FUTURA MEDIUM—6 TO 72 PT. FUTURA DEMIBOLD—6 TO 72 PT. FUTURA BOLD—6 TO 84 PT. FUTURA BOLD CQND.—8 TO 84 PT.

8 Pt. 606K ABCDEabcde *8-605K FGHIJfghij* ***8-604K KLMNOklmno*** ***8-603K PQRSTpqrst*** *8 607K UVWXYZuvwxyz*
FUTURA LIGHT ITAL.—6 TO 36 PT. FUTURA MED. ITAL.—6 TO 72 PT. FUTURA DEMIBOLD ITAL.—6 TO 72 PT. FUTURA BOLD ITAL.—6 TO 72 PT. FUTURA BOLD COND. ITAL.—8 TO 72 PT.

8 ABCDEFabcdef   10 GHIJKghijk   12 LMNOlmno   14 PQRSpqrs   18 TUVWtuvw
CLEARFACE GOTHIC—8 TO 24 PT. HAND SET

5 ABCDabcd   6 ABCDabcd   7 EFGHefgh   8 IJKLijkl   10 MNOPmnop   12 QRSTqrst   14 UVWuvw   18 XYZxyz
NEWS GOTHIC CONDENSED NO. 204J—

6 ABCDEFabcdef   8 GHIJKghijk   10 LMNOlmno   12 PQRSpqrs   14 TUVtuv   18 WXwx
NEWS GOTHIC NO. 206J—

7 ACEace   **7 ACEace**   *7 ABCDEabcde*   *9 FGHIfghi*   *11 JKLMjklm*   *7 NOPQRnopqr*   *9 STUVstuv*   *11 WXYZwxyz*
NO. 472J   NO. 813J   MONSEN LIGHT GOTHIC ITALIC NO. 621K   MONSEN MEDIUM GOTHIC ITALIC NO. 512K

5 ABCDEabcde   7 FGHIJfghij   8 KLMNklmn   9 OPQRopqr   10 STUstu   11 VWXvwx   13 YZyz
MONSEN MEDIUM GOTHIC NO. 512J—

14 ABab 18 CDcd **24 Ee**   FRANKLIN GOTHIC CONDENSED
MONSEN MEDIUM GOTHIC NO. 512J—14 TO 24 PT. MACHINE SET   **ABCDEFGHIJKLMNOPQRSTU**
                                                          abcdefghijklmnopqrstuvwxyz

Lydian   *Lydian Italic*
14 ABCabc 18 DEFdef 24 G   *14 ABCabc 18 DEFdef 24 GH*

BODONI OPEN ABCDEFGHIJKLMNO   FUTURA INLINE
abcdefghijklmnopqrstu   ABCDEFGHIJKLMNOPQRSTUV

SHADOW   ORPLID
ABCDEFGHIJKLMNOPQRSTUVWXYZ&   ABCDEFGHIJKLMNOPQRSTUVW

TUDOR No. 5 ABCDEFGHIJKLMNOPQ   *Goudy Handtooled Italic*
abcdefghijklmnopqrstuvwxyz   *14 ABab 18 CDcd 24 EF*

derside of cellophane, so that they cannot rub off.

We consider carefully the amount of reduction. Maps are usually drawn larger than publication size, and the letters will be reduced photographically with the rest of the map. A 12 point letter, if reduced to half, is weaker than a 6 point letter. The type-specimen books also show how type will look after various amounts of reduction. It is better not to draw the map more than 50 per cent larger than publication size if cellotype is to be used.

On the type-specimen list, Fig. 6.7, certain sizes of certain types are marked "Machine Set." These are cheaper than the hand-set types if many names are ordered. Most types can be ordered either black or white. In addition, standard map symbols can be obtained, and the company may also print symbols from given design.

Selection of type for stick-up needs experience. As it takes some time to receive the names—rush orders cost more—they have to be ordered ahead of time. It is a common experience that by the time the names arrive, the emphasis has been changed and certain names are too large or too small. It is most difficult to match the printed letters by hand if something needs to be added. Stick-up is better for formal maps or for series maps with set standards than for the individual efforts of a geographer-cartographer.

Application. The map maker erases and cleans the paper and draws pale blue lines for guidance. Then he cuts out the names, places them properly with needle and pincers, finally rubs them down hard with a burnisher. If a mistake is made, the name can be pried off. Names can be curved by cutting every letter almost,

but not completely, apart on one side.

Spread letters can be ordered and cut apart, but it is more economical to order ready alphabets—not real alphabets, but letters which are repeated according to their frequency: eeeeeettttaaaaiiiissssoooo nnnnhhhhrrrddddllluuucccmmmmffggbbppw wkkjjqxyz1234567890. This is modified from the standard frequency of lower-case letters in the English language for a cartographer's use, with the addition of more k's and j's, as these are common in foreign languages. The same frequency can be used for names written in all caps, names for countries, seas, and mountains. For initial caps for city names, the sequence SSSSSSS SSCCCCCCCMMMMMMMBBBBBBBPPP PPTTTTTTTAAAAALLLLLKKKKGGGG NNNNHHHDDDWWWRRRREEEFF OOVVIIJJUYZQ is indicated by the index of the "Goode's World Atlas."

Phototypography. Some cartographic establishments, such as the National Geographic Society, design their own letters on a large scale and photograph them to smaller size. A special machine sets up the names from their letters, which again are photographed to size, and these names are pasted on (see Phototypography in Chap. 27).

Exercise 6.1

For practice in muscular coordination, draw the routine of Fig. 6.8 on a piece of notebook paper with a fountain pen. Use full-arm motion and try to follow through. Begin the motion in the air, bring your pen down on the paper, and when you lift it off at the end of the line, continue the motion in the air. This brief routine may be repeated before each exercise in lettering. It is interesting to repeat these lines with eyes closed. A man with good muscular control will do almost as well with his eyes closed as with his eyes open.

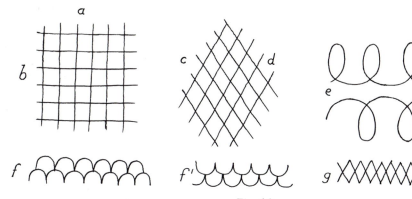

Fig. 6.8.

Exercise 6.2 *

a. Read the rules of lettering carefully.

b. Take a sheet of ruled notebook paper; with a fountain pen draw vertical block (gothic) capitals. Draw each letter five times, following the rules of lettering strictly. Be careful that the letters are vertical and that their standard width is maintained.

I	H	L	F	E	T				
V	W	M	N	Z	A	K	X	Y	
O	Q	C	G	J	S	D	P	B	R
I	2	3	4	5	6	7	8	9	0

c. Draw the vertical gothic lower-case letters in the same order. Use the letters in Fig. 6.7 for models.

d. Draw the inclined gothic capitals in the order of the alphabet. Try to space the letters evenly.

e. Draw an alphabet of the inclined gothic lower-case letters.

f. Draw a row of numerals in inclined gothic.

g. Draw your own name in vertical and inclined gothic.

Geographical Names

One can produce excellent topography—mountains, water, vegetation, all in proper place—but should a name be misspelled, the map will be rejected. More heated arguments are heard over the spelling of a name than on the location of the feature. Yet if we hear a Chinese pronounce the name "Yangtse Kiang," we will find out what a slender amount of reality underlies some of our conventional spelling. A cartographer, however, must be familiar with the principles of toponymy, or the names of places.

There is little difficulty with English names. They have an official spelling, decided upon by the U.S. Board on Geographical Names (BGN) of the Department of the Interior, or by the Permanent Committee on Geographical Names of the Royal Geographic Society. A list of post offices also helps.

In countries using the Latin alphabet, we follow the official spelling. If there is a conventional English name, that is used. Vienna is easier for the reader than Wien, which most people would not know how to pronounce. The official name, however, should be given in parentheses, as this has

---

* NOTE TO THE INSTRUCTOR: Every student should practice lettering with mechanical lettering sets and cellotype. As there are usually not enough sets to go around, this could be done on the various map exercises. Spacing and placing are also best practiced on the maps.

#					#			
1	А	а	a	a	18	Р	р	r
2	Б	б	b		19	С	с	s
3	В	в	v		20	Т	т	t
4	Г	г	g		21	У	у	u
5	Д	д	d		22	Ф	ф	f
6	Е	е	e		23	Х	х	kh
	Ё	ё	e		24	Ц	ц	ts
7	Ж	ж	zh		25	Ч	ч	ch
8	З	з	z		26	Ш	ш	sh
9	И	и	i		27	Щ	щ	shch
10	I	ᵏ	i		28	Ъ	ᵏ ъ	omit
11	Й	й	j	Omit after ы, и, or i, at the end of a word.	29	Ы	ы	y
					30	Ь	ь	omit
					31	Ѣ	ᵏ ѣ	e
12	К	к	k		32	Э	э	e
13	Л	л	l		33	Ю	ю	yu
14	М	м	m		34	Я	я	ya
15	Н	н	n		35	Ѳ	ᵏ ѳ	f
16	О	о	o		36	V	ᵏ v	i
17	П	п	p	ᵏ *obsolete*				

Fig. 6.9. Russian writing developed from the Greek. Obsolete letters are commonly found in older maps. (*Courtesy of U.S. Board of Geographical Names.*)

to be used in sending a letter. Names of countries and regions are usually in English. Diacritical marks, so common in foreign languages, must not be omitted. A mark changes the *s* sound into an *sh* sound in "Iaşy" or "Niš." A cartographer often needs to pronounce foreign names, and a reading knowledge of them is an asset.

For languages written in other than the Latin alphabet, such as Russian, a letter-by-letter transliteration system is made (see Fig. 6.9). The same is true for Greek and Arabic, Indian and Siamese. Names in colonies and dependencies are spelled according to the official list of the colonizing power. Many former colonies are now independent but, until a new set of names is published, Libyan names are still spelled in Italian, former French-African names in French, etc. India and Indonesia now have new official spelling. Japanese and Korean names are spelled according to the list of the BGN.

The spelling of Chinese names is somewhat controversial. The Army and the BGN have accepted a respelling of the conventional Chinese names according to the Wade-Giles system, with the syllables hyphenated. Thus the former capital of China, Nanking, is now spelled Nan-ch'ing. Our newspapers and magazines, which usually follow the National Geographic Society's spelling, have not yet accepted this.

In transliteration of names which have no official spelling, the general rule is to spell the consonants with their nearest equivalent in English and the vowels as in Italian. Map makers find good guidance in the lists of the BGN, the Columbia-Lippincott Gazetteer, Webster's "Geographical Dictionary," and the maps of the National Geographic Society and the American Geographical Society. The indexes of up-to-date atlases are also reliable.

We are not supposed to repeat foreign designations in English. Such names as Rio Grande River, Tien Shan Mountains, Tüz Gölü Lake, Kara Kum Desert are repetitions. This, however, presupposes more knowledge of foreign languages than many map makers possess. Strictly speaking, Mississippi River is also repetitive.

Abbreviations. These help to reduce the bulk of names and should be used freely. Good abbreviations can be read without a key and cannot be mistaken for something else. Unusual abbreviations need a key. Unfortunately, some of our usual abbreviations have double meanings. For instance, Pt. often stands (incorrectly) for Port or Point; St. for Saint or Street; Str. for Stream or Strait. Br. can be Branch, Bridge, Brook, or British.

The following abbreviations are approved by the U.S. Geological Survey:

Br.—Branch
Bk.—Brook
C.—Cape

Co.—County
Cr.—Creek
E.—East
Fd.—Ford
For.—Forest
Ft.—Fort
Fy.—Ferry
Gl.—Glacier
Hb.—Harbor
I.—Island
Is.—Islands
L.—Lake
Mon.—Monument
Mt.—Mountain
Mts.—Mountains
N.—North
Pd.—Pond
Pen.—Peninsula
Pk.—Peak
Pt.—Point (Port is written out)
R.—River
Ra.—Range
Rd.—Road
Ry.—Railroad
S.—South
Sch.—School
Spr.—Spring
St.—Saint or Street
   (St. Martin, or Martin St.)
Sta.—Station

Str.—Strait
W.—West

Standard abbreviations of the states of the United States need not be repeated here. The provinces of Canada from east to west are N.F., N.S., P.E.I., N.B., P.Q., or Que., Ont., Man., Sask., Alta., B.C., N.W.T. American dependencies are C.Z., P.R., V.I. Philippine Islands is P.I. and Long Island is L.I. In the British Commonwealth the following abbreviations are common: G.B., W.I. (West Indies), N.S.W. (New South Wales), N.Z., U.S.-Afr. (Union of South Africa). Colonies and islands are often designated as Br., Fr., Sp., Port., Du., Ger., Jap., Mex., etc. U.S.S.R. is better known than the full name "Union of Soviet Socialist Republics," but R.S.F.S.R., Russian Soviet Federated Socialist Republic, is less familiar. At the time of this writing, no decision has been made on the abbreviation designating the Commonwealth for members of the former British Empire. The same is true of the Communauté of the French.

As in the reading of printed words or a musical score, precision and speed in the reading of maps can pretty rapidly be carried further and further. Soon the map is read, as it were, not word by word, but phrase by phrase, the meaning of whole passages leap out; you see with something like a summary grasp. . . .

C. E. Montague

7

Relief Methods

Relief is the vertical dimension of the configuration of the land with its hills, valleys, mountains, plateaus, and other forms. Relief has four major elements: (1) elevation above sea level; (2) relative relief, the height of mountains over the adjacent plains or valleys; (3) average slope, the steepness of the land in a certain area; and (4) texture, which means the spacing of the small rivers and also the minor features of the surface caused by rock outcrops, moraines, sand, etc.

Contour Lines

The most common way of showing relief is by contour lines. The simplest way to understand contour lines is to take a dish and build up in it some imaginary mountains from plastiline. Then we place a vertical scale in the dish, assuming that the bottom of the container should represent the sea level, as in Fig. 7.1. We add water until it reaches the first mark on the scale and drop a bit of white oil paint. This will leave a trace, but if it is not distinct enough,

we trace the shore with a pencil, marking the first contour line. Then we add water until it fills up to the second mark, and trace the second contour line, and so on, until the whole mountain is submerged. We drain out the water and examine the contour lines. We photograph the model from far above, and obtain a contour map, as shown in Fig. 7.1. From this we find the following:

1. Contour lines are always horizontal and perpendicular to the dip of the land, the direction in which water would run at that place.
2. All contour lines are closed lines, unless cut off by the margin of the map.
3. The steeper the slope, the closer the contours.
4. On rivers, contour lines point upstream, except on some alluvial fans.
5. If contour interval is too large, minor hills may not be recorded.

Contour Interval. Contour lines are usually reckoned from mean sea level, which is called datum or base level. The contour interval depends on the scale of the map. A 1:24,000 map usually has a 10-ft interval; the 1:62,500 maps have 20-ft intervals. The average formula is

Contour interval $= 20-25 \times$ miles per inch

Thus in a 4-miles/in. map the contour interval would be 100 ft. We use, however, larger intervals for mountainous areas and smaller ones for flat land. The interval depends also on the use of the map. Irrigated land, for instance, is usually contoured with 5 ft interval. On 1:1,000,000 or smaller-

68

scale maps, the contour interval is not even. It is larger as we go higher up: 500, 1,000, 2,000, 3,000, 5,000, 7,000, 10,000, 14,000 ft, which is also a usual scale for atlas maps. This increase in interval has its dangers. We may miss in Nevada a 2,500-ft range sticking out from a 7,000-ft high basin.

The first contour map, made by Cruquius in 1729, was not of land but showed the depth of water, the mouth of the Merwede River in the Netherlands. To prevent running the ship aground we must know the depth of water; therefore, contoured sea charts were made much before land maps. Contour maps for land came into general use only in the last century.

How to Obtain Contour Lines. In the sea they are obtained by sounding. No such simple method is possible on land. Using a barometer we can tell our altitude, but only within 50 to 100 ft. Triangulation and leveling are common methods and will be explained in Chap. 16. Sometimes plane-table parties, using telescopic alidade, measure the spot elevation of all important points and draw the contour lines right in the field. Most modern topographic maps, however, are contoured with great accuracy from air photos by complex stereovisual machines (see Chap. 16).

Exercise 7.1

Draw 100-ft contours to conform with the spot elevation on the map in Fig. 7.2.

How to Draw a Contour Map. After the contours are laid out in pencil, they are inked with a fine pen, preferably a mechanical pen (Leroy, Graphos, etc.). The pivot pen makes a clean even line, but it has a tendency to produce rounded lines, and sharp cliffs are often lost in the process. Every fourth or fifth contour is drawn

with heavier line. This helps to visualize the map, but it gives undue importance to arbitrary lines.

The contour lines should be frequently numbered. Contours cannot bifurcate, cross each other, or go in spirals. In case of an overhanging cliff, the contours would cross each other, but at very steep crags the contours are usually replaced by the cliff symbol. Depressions are shown by barbed contours. Elsewhere the river system will indicate the direction of slopes.

Profiles. It takes some practice to visualize the form of the land from contour maps. It is not difficult at all, however, to visualize a profile. It shows the slopes as we are accustomed to see them. The best

Fig. 7.1. A plastiline model of a small island is placed in a dish. Pouring water into the dish and covering it to heights indicated on the scale will effectively illustrate the nature of contour lines.

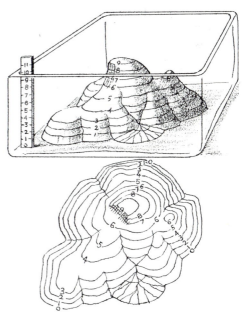

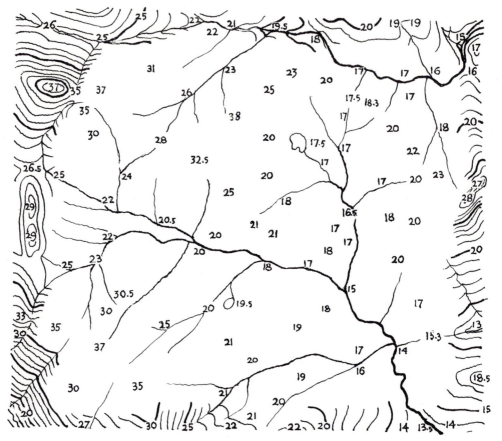

Fig. 7.2. Mountainous region in New England, 1:62,500. Numbers indicate 100 ft. Complete the 100-ft contour lines. Start with every fifth line, working alternately from crests and from rivers. Assuming even slopes between spot heights, the twenty-fifth contour will be five-eighths the distance from 20 to 28. It helps to draw first the divides (crestlines) faintly.

way to visualize contour maps is to draw a few characteristic profiles. We draw profiles for many other purposes too: to see the steepness of a road or the gradient of a river or to draw a geological section, etc. Profiles are usually drawn along a straight line. If we want to show the gradient of a river or the steepness of a road, we draw the profile by "pulling out straight" a line which is curved on the map. How to draw a profile along a straight line is shown by Fig. 7.3. We can use profile paper but this is by no means necessary.

Vertical Exaggeration. It is a part of human nature that we are far more sensitive to the up and down of the land than to its horizontal measure. We usually overestimate the steepness of slopes, and artists draw mountains steeper than they really are. If you show a profile of land drawn to the same scale vertically as horizontally, it looks incredibly flat, particularly if a long section is shown (see Fig. 7.4).

To bring out the ups and downs to correspond better with our mental concept, we exaggerate the vertical component. The de-

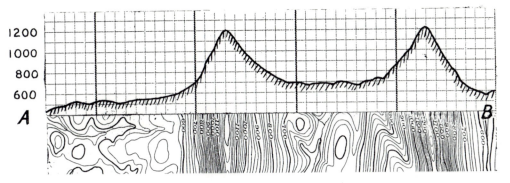

Fig. 7.3. Drawing of profile along line AB on a 1:62,500 map. Horizontal scale is 5,280 ft to the inch, and vertical scale is 1,000 ft to the inch, resulting in considerable vertical exaggeration.

gree of exaggeration depends on the scale, on the nature of the country, and on the purpose of the profile. In a hilly country a rough rule is

$$VE = 2\sqrt{\text{miles per inch}}$$

Thus we exaggerate on a map of the scale

1 mile	1 in.	2 times
4		4
16		8
64		16

In mountainous regions we exaggerate less; in plain regions, considerably more. We would use greater exaggeration in planning an aqueduct than, for instance, in a geologic section.

Exercise 7.2

Draw profiles on the topographic sheet of a region which you know. This will help to visualize vertical exaggeration. Choose various intervals.

Slope. Slope can be expressed in several ways. From a contour map the easiest way is to read the steepness in *feet per mile*. For instance, on a 1-mile/in. map with 20-ft contour interval a 200-ft/mile slope

will have the contours 0.1 in. apart, as shown on the slope indicator (see Fig. 7.5).

Engineers like to express slope in *percentage*. Thus, with a 200-ft/mile slope

$$\frac{X}{100} = \frac{200}{5,280} = 3.78 \text{ per cent}$$

For computations, the simple fraction called the *gradient* is used:

$$\frac{\text{Vertical distance}}{\text{Horizontal distance}}$$

In this case

$$\frac{200}{5,280} = \frac{1}{26.4} \quad \text{or} \quad 0.0378$$

the same as one one-hundredth of the percentage. The surveyor measures the slope in *degrees*. The gradient of 1° is 1/57.3. A 100-ft/mile slope is about 1°05′. This figure can be used as long as the angle is not steep. Thus a 200-mile/in. slope is 2°10′ but, at a slope of over 20°, the gradient has to be calculated by the function $\tan \alpha = \frac{h}{D}$, where α is the angle of the slope, h is the height, and D is the horizontal distance. To find α, we can use Table A.4 in the Appendix.

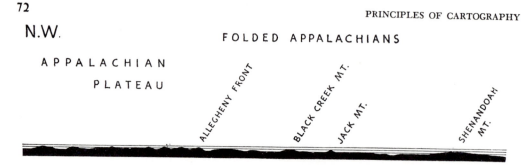

Fig. 7.4. A profile across the Appalachian Mountains without vertical exaggeration looks too flat.

Reading of Contour Maps.* It is regrettable that many of our educated people, including even geography students, cannot visualize the land from contour maps. Part of the difficulty is that we are not accustomed to seeing land from directly above. Even vertical air photos look unfamiliar to us. Besides, there are very few lines in nature which correspond to contour lines—unless we are in an area of horizontally bedded rocks. Sometimes the contour interval is too large and the contours are badly generalized. Some practice, how-

Fig. 7.5. Contour density indicator.

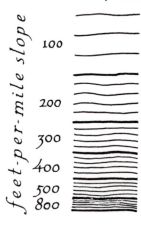

TABLE 7.6. COMPARISON OF SLOPES (EXPRESSED IN PERCENTAGES, DEGREES, AND FEET PER MILE) *

Per cent	Degrees	Feet per mile
1	0°35′	52.8
2	1°09′	105.6
3	1°42′	158.4
4	2°18′	211.2
5	2°52′	264
6	3°28′	316.8
7	4°01′	370
10	5°04′	528
15	8°32′	792
20	11°19′	1,056
30	16°42′	1,584
50	26°34′	2,640
100	45°	5,280
200	63°27′	10,560

* For a similar table comparing even degrees with other methods, see Appendix 1.

* NOTE TO THE INSTRUCTOR: First, study in the classroom the topographic sheets of the home area; then take the class to a lookout point and identify the hills and valleys, etc. Practice to estimate distances, also the relative height of hills and the steepness of slopes, noting the corresponding use of land.

The U.S. Geological survey has one set of 25 and another of 100 maps of regions of various types, which can be bought at reduced price. Foreign maps may be obtained from the captured material in the Library of Congress or directly from the various national surveys (see Chap. 9). It is helpful to read some 1:250,000 to 1:1,000,000 maps and air charts too. Several meetings should be devoted to map reading not only for land forms, but also for human geography.

S.E.

GREAT   VALLEY   BLUE RIDGE   Charlottesville Va

ever, will enable everybody to read contour maps.

First let us try the simple geometrical figures. Figure 7.7 shows their side view and their expression by contour lines. It is very much easier to visualize their shape from their side views than from their contour map. To transform mentally a contour map into an image of land relief requires imagination and much practice. Figure 7.8 illustrates this well. No one has any difficulty in visualizing the oblique view, but to visualize the contour map below requires scrutinizing the elevation figures and the density of contour lines. We have to draw mental profiles, particularly along crestlines.

The best way to learn map reading is to practice on maps of diverse lands and of different scales, starting with the home area, where we can directly correlate the hills and valleys with their expression by contour lines. We count contours at various places and try to visualize the slope. This can be done only after field practice in estimating slopes.

Layer Tints (Altitude Tints, Hypsometric Coloring). For small-scale maps, contour lines have to be generalized. In Fig. 3.7 we have seen that simply photographing 1:250,000 contours to 1:1,000,000 would result in an unreadable jumble. Even if we left out all but every tenth line, the

reduction still would give the impression of jagged rocky land in places where there were rounded, gentle hills. Thus the rounded character of the contours can be preserved only by omission, selection, and combination, as discussed under Generali-

Fig. 7.7. Geometric forms expressed in contour lines. The dashed lines are hypsographic curves showing the area of land between the contour lines.

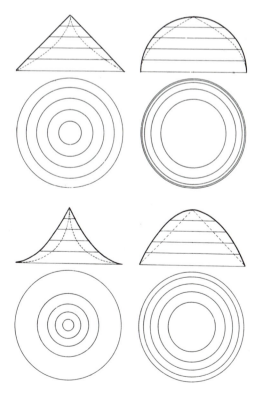

Fig. 7.8. Contour-line maps are difficult to visualize.

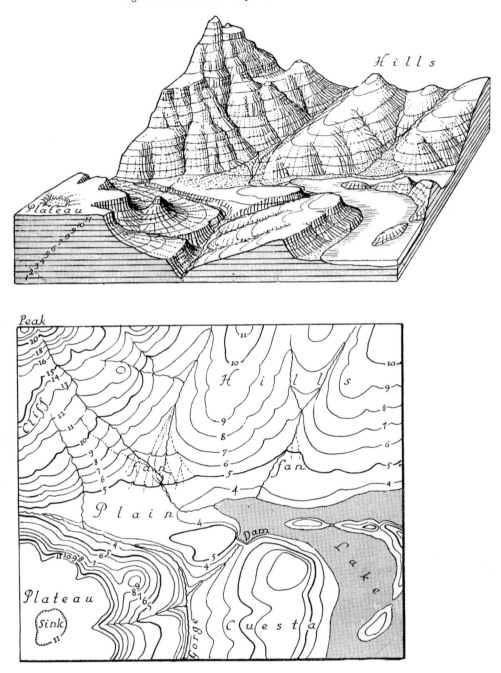

zation in Chap. 3. At still smaller scale (maps of countries and continents), the generalization needs to be so great that the original character of the lines is lost and the contours do not express more than the altitude above sea level. Even so, they are difficult to read; they are almost always layer-tinted for visual impact.

The conventional tints for various altitudes (expressed in feet) are as follows:

Land below sea level	Olive green
0–500	Green
500–1,000	Yellow-green
1,000–2,000	Yellow
2,000–3,000	Orange
3,000–8,000	Light brown
5,000–8,000	Brown
8,000–12,000	Red-brown
12,000–16,000	Red
16,000 and over	White

This scheme is far from universal. We have to scrutinize the key carefully, as the colors may mean different intervals and different elevations. In most foreign maps, the intervals will be in the metric system. For conversion, see Table A.3 in Appendix 1.

The underlying idea of this scheme is twofold. In Austria, where this color scheme was first used, the valleys are green in contrast with the barren, brownish mountains. A better explanation, however, is psychovisual. A map is seen from above; thus the mountains are nearer the observer. Artists will paint the nearby objects in warm, long-wavelength reds, browns, and oranges and the farther objects in cold pale blues, grays, and greens, as the warm colors are filtered out by the atmosphere. An additional advantage of this color scheme is that black printing is more readable over green and yellow than over brown, and most of the small printing for towns and rivers is in the valleys.

This color scheme is now over 100 years old and is used on almost every atlas map, wall map, and aeronautical chart. Yet this method has liabilities. To dissociate the green color from vegetation is difficult, and, for instance, we have the opposite of the conventional colors in arid regions with brown valleys and forested mountains. To find vivid green patches in the Sahara may mislead people. One may wonder whether the time has not arrived yet for a change, and for using colors for something more significant than altitude (see Chap. 8).

Hachuring

In the early nineteenth century, hachuring was the common way to show relief and was used until recently on black-on-white maps because it is less confusing than contours.

The hachures are short parallel lines,

Fig. 7.9. Nineteenth-century hachured map. This map also has 100-m contour lines. The small circles indicate a forest. Note the frequent spot heights. (*From* 1:75,000 *Specialkarte of Austria-Hungary.*)

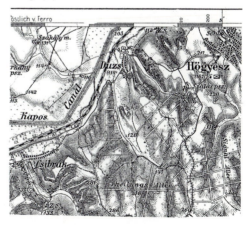

thick on steep places and thin on gentle slopes. As hachures are drawn in the direction of slope (as water would run on the surface), they conform well with natural lines of erosion and are understood by the layman better than contours. From hachuring, however, we cannot tell the exact slope or the elevation; thus they are less useful than contours for many purposes. In addition, on small-scale maps, hachuring often deteriorated into the "hairy-caterpillar-and-pine-branch" method, which fortunately is rarely used at present.

In the exact Lehman system of hachuring, there are exactly the same number of lines for every centimeter, and the thickness of lines is mathematically prescribed for every steepness. This is most difficult and laborious, and highly skilled hachuring as a practice is now dying out. Hachuring has been rarely used since halftone printing was invented and is now usually replaced by plastic shading.

Fig. 7.10. Plastic shading lighted from the northwest. First a paper or plastic base is gray-coated; then shading is drawn with pencil or crayon and highlights are produced with an eraser. (*By Richard E. Harrison.*)

Plastic Shading

A good way to show relief is to photograph a relief model which is lighted from the side. The same effect can be achieved by simple shading with wash or carbon pencil. This needs to be reproduced by a halftone screen (see Chap. 14). If we want to eliminate the undesirable screen effect over the white areas, we may use Ross board (see Fig. 1.6) or we may "opaque" the highlights on the negative (see Chap. 14). For the best results, we use a gray-coated paper or plastic corresponding to the shade of the horizontal areas. The southeast-facing slopes are shaded, while the northwest-facing slopes are highlighted with an eraser. Southwest- and northeast-facing slopes are parallel to the source of light and would look uniformly gray. To show these slopes it is permissible to shift slightly the source of light, usually to the west.[1]

The obvious question arises: Why do we shade the sunny south slopes, and keep light the shady north slope? For some psychovisual reason, we see, any unfamiliar landscape reversed if the shadows are away from us. The hills look like valleys, and vice versa (see Fig. 2.6).

Combinations

Both hachuring and plastic shading are used in combination with contour lines with or without layer tints. A combination of contours, gray-green to red-brown merged layer tints, and purple plastic shading is becoming now increasingly popular in atlases and medium- and small-scale maps. Some maps of the U.S. Geological

[1] Ruth Mean, "Shaded Relief," U.S. Aeronautical Chart and Information Service *Tech. Man.* RM-895, 1958.

Survey have contour lines and plastic shading combined with a gray-green "valley tint" or a red-brown "upland tint." The effect is most striking, as shown in the Valdez, Alaska, sheet or the sheets of the Yosemite Valley. The Swiss topographic sheets combine contours with plastic shading and a generous amount of cliff drawing and symbols for moraines, talus slopes, rock outcrops, bluffs, etc., with excellent effect.

The Tanaka Kitiro Method

This method, proposed by Prof. Tanaka Kitiro in Japan, transforms a contour map in a kind of plastic shading. The contour map is ruled with closely set horizontal lines, as in Fig. 7.11. Beginning with the intersection of a horizontal line with the lowermost contour, he connects this with the intersection of the next higher horizontal line with the next higher contour line. He continues this until the highest contour is reached and then we descend the opposite way. This he did with every horizontal line. The resulting wavy lines are profiles, produced not by a vertical but by an inclined plane. The success of the system depends on the choice of the interval of horizontal lines, the thickness of the profiles, and the nature of the terrain. Trials in the low, irregular country of New England were not very successful.

The Kitiro system was recently revived by Robinson and Thrower,[2] who used the Kitiro method for a mountainous terrain. On the basis of the forms obtained, Professor Thrower drew a highly effective landform map. A somewhat similar effect can be

[2]Arthur N. Robinson and Norman J. W. Thrower, A New Method of Terrain Representation, Geog. Rev., October, 1957.

obtained by the Dufour method (see Chap. 22).

Professor Kitiro also prepared very successful maps with illuminated contour lines (see Fig. 7.12). Upon a gray base the con-

Figs. 7.11 and 7.12. Kitiro method of showing relief. Oblique profiles are drawn by connecting closely set horizontal lines with each next higher contour line. (*Courtesy of Jour. Geog., vol. 79, 1932, pp. 213–219*).

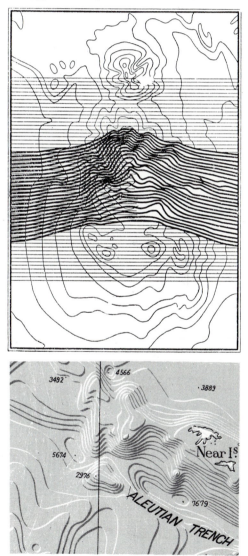

tour lines on the northwest-facing slopes are white, while on the southeast-facing slope they are black. The lines are wide, but thin out where the black and white meet. The method is particularly effective for submarine topography, using white and dark-blue lines upon a medium-blue base.[3]

---

[3] Tanaka Kitiro, The Relief Contour Method of Representing Topography on Maps, *Geog. Rev.*, vol. 40, 1950, pp. 444–456.

Exercise 7.3

Draw a contour-line map of an imaginary region on 1:62,500 scale with contour interval of 100 ft. It should be a coastal region with deep bays, lowlands, plateaus, mountains 1,000 ft high, many rivers, lakes, salt marshes, and sand bars. A harbor city of about 3,000 people, roads, railways, and airport should complete the map. Use pencil and colored crayons or paint. It is advisable first to make a rough sketch of the region as seen obliquely from above.

8

Land Forms
and Land Slopes

During the late nineteenth century, the geological exploration of the United States was at places ahead of its topographic mapping. Some geologists, not having detailed maps, used to express land forms by sketches and by block diagrams (see Chap. 22). On the latter the geological structure was drawn in the front, and the land forms were sketched on top with more emphasis on the type of land forms than on perfection of detail.

It was found that most people could understand a block diagram much more easily than a contour-line map (see Fig. 7.8). Some geomorphologists, like William Morris Davis and A. K. Lobeck, experimented with the idea of applying the obliquely viewed landform drawings to vertically viewed maps. Although this is bad geometry, it turned out to be good psychology, as such drawings show the land more or less as we are accustomed to seeing it.

Symbolization

A landform map does not claim to show every hill and valley exactly in place.

Rather, it shows the general type of topography with symbols. The symbols are derived from the oblique view of the type of land seen from about 45° above, as shown in Figs. 8.1 and 8.2. It should be emphasized that landform symbols are used for small-scale maps only. A map larger than 1:500,000 should be transformed into an actual block diagram.

Symbols can be varied in size; for larger mountains we use larger symbols in a somewhat logarithmic proportion. A mountain that is four times higher will be drawn only twice as high. If the correct proportions were used, the small but important land forms would not be shown at all. There will be some displacement of mountain peaks because of the oblique view, but in small-scale maps this is rarely more than a few millimeters. Symbols also can be used transitionally. There can be a difference in the amount of symbolization. A. K. Lobeck's maps are symbolized so much that he calls them "physiographic diagrams," while Fig. 8.3 shows the country more nearly as it appears in air photos.

The information given on a landform map is much richer than that given on a contour map. A 1:4,000,000 landform map can show all the relief features contained on a 1:1,000,000 contour map and a great deal more. Elevation above sea level, however, has to be given by frequent spot heights. Landform maps can be reproduced by inexpensive offset printing in black alone.

Figs. 8.1 and 8.2. Physiographic symbols as applied to landform maps and block diagrams (see Chap. 22).

Plains blank

Undifferentiated Tundra Boreal forest Wet taiga Bush

Forest Grass Dry land Sand Gravel Hamada Savanna

Palms Jungle Selva Rice Plowed land Corn Grain Tree crops

Dissected — rolling land Cuestas & flatirons Flood plains Fans

Plateaus
low — high — cut-up Canyon land — mesas — badlands

Syncline Anticline

Folded ridges — Dome — Basin ridge — Arched basin

Block Mts. — / reduced   Complex mts. — glaciated

Complex Mts.— reduced — peneplain — rejuvenated —

gneissic — schistose — slaty mts. Glaciated shield   fjord

Volcanic forms

Volcano   Caldera   Volcanic necks Lava plain Lava capped plateau

Limestone

Sinkholes Knobs Lapies Bastions Karst Polje   Mogotes Coral

Glacial deposits

Moraines   Drumlins   Kames   Eskers

Information for landform maps comes from topographic maps, air photos, geologic maps, soil maps, and descriptions. Flying over the area with observant eyes and making notes is most helpful in planning a landform map (see Geostenography, Chap. 23).

Every effort should be made to obtain oblique air photos of the area. The U.S. Map Information Office may help in this. The trimetrogon pictures are best for our purpose. They are usually taken from 20,000 ft with flight lines 15 to 25 miles apart, which provides ample coverage. Instead of looking at each overlapping picture, time is saved by examining about every fourth or fifth picture closely, as these will also overlap a few miles away from the flight line. Even so, thousands of pictures have to be scrutinized. We sketch the land forms on the corresponding aero-

Fig. 8.3 Part of a landform map.(*From G. B. Cressey, "Land of the 500 Million," McGraw-Hill Book Company, Inc.*)

nautical charts. We may use such annotations as are described in Chap. 23 on Geostenography. For instance, *Fl* indicates flat or featureless; $3M^r$ indicates matureland, 300' relief, with rounded slopes; $2GP/$ stands for gneissic peneplain of 200' relief, with a northeast-southwest trend. Markings should be explained on the side of the map. In general, capital letters indicate type of land; a number in front of the letter means height in hundreds of feet; a mark on the upper right indicates the nature of the dissection. A great deal more can be noted, as explained in Chap. 23. Besides markings, we use simple sketches. The distance of a feature can be found by noting its nearness to the horizon in the photo.[1]

The Drawing of Land Forms. The symbols in Fig. 8.4 are drawn by hachuring. The direction of hachure lines is the direction in which water would flow on that surface, but they are not seen vertically from above as on a hachured map but as they would appear seen obliquely. Figure 8.4 shows some of the elementary forms. Some variation of these will produce all the land forms of Fig. 8.2. In this figure the right-facing slopes are drawn heavier, giving a certain amount of plastic shading, which makes livelier maps.

Plains can be left blank or, if rolling, they can be shown with short horizontal or slightly curved lines. Plains give an excellent opportunity to show vegetation and cultivation, as shown in Fig. 8.1.

Plains with slightly incised rivers are shown in Fig. 8.5. We lay out our river sys-

[1] For further detail, see E. Raisz, The Use of Air Photos for Landform Maps, *Annals Assn. Amer. Geog.*, vol. 41, 1951, pp. 324–330; and E. Raisz, Direct Use of Oblique Air Photos for Small-scale Maps, *Surveying and Mapping*, vol. 13, 1953, pp. 496–501.

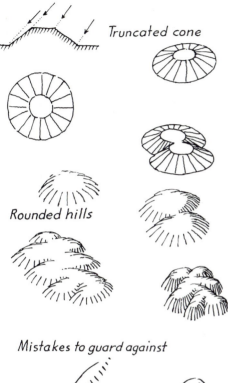

Truncated cone

Rounded hills

Mistakes to guard against

Wrong   Right

Fig. 8.4. Land forms drawn with obliquely seen hachure lines. Note how the height and steepness of hills can be indicated.

Fig. 8.5. Rolling plain. First the actual river system is drawn and the slopes adjusted to them, assuming even, symmetrical slopes of the same height.

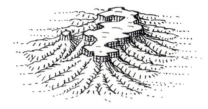

Fig. 8.6. Lava-capped mesa with dissected slopes. The system of hill hachuring is here applied to a surface sloping in all directions.

tem and then draw the same hill symbols on it as those in Fig. 8.4, but adjusted to the rivers. The proper design of the head of the river is critical. We should not forget that the river diminishes with height and there are just as many lines in the back as in the front, but part of them are not visible. The success of landform draw-

ing depends on the "scalloping" of back slopes, as is shown in Fig. 8.4.

Plateaus of horizontal rock structure have even height, even symmetrical slopes, and are evenly dissected by dendritic rivers. They differ in the amount of remaining upland, the sharpness of the edges, and the height and steepness of the slope. The height, slope, and shape of lines can express the various types of plateaus, as in Fig. 8.1. If no upland is left and the whole land is in slope, it is called *matureland*. Maturelands can also develop on any other rock structure.

In dry regions or where there is a very hard cap rock, we have a sharp edge between upland and slope, forming cliffs, canyons, mesas, and buttes. Here we apply the cliff symbol, a jagged line with horizontal axes for each indentation, as in Fig. 8.6.

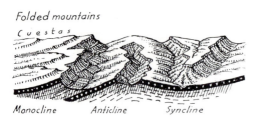

Fig. 8.7. Folded mountains are drawn by the combination of the ridges and slopes, but with varying steepness. Note the dissection of the gentle back slopes.

Cuestas are ridges where inclined strata jut out, making either elongated ridges or short coffin-like forms. A cuesta is rarely alone; there is usually a whole series of them corresponding to the various strata. In the front of the cuesta, we draw a cliff symbol as before, but the axes of indentations are slightly oblique in the direction of the bedding. On the back slope of the cuesta, we draw straight, parallel, "consequent" or "resequent" rivers running in the direction of the dip of the strata, as in Fig. 8.7.

Folded and domed mountains usually consist of a combination of cuestas and elongated hills, as in Fig. 8.1.

Complex mountains are so called because of their highly complex structure.

Most of the world's mountains are in this group. For practice in drawing them, it is easiest to start with a steep oceanic island, as in Fig. 8.8. Proceed as follows:

1. Draw the river system as it appears on a vertical map.
2. Locate every peak as it is on the map. From this, draw a vertical line and mark the top of the peaks elevated according to their height. Some vertical exaggeration is permissible (see Fig. 22.5).
3. Elevate the headwaters of the rivers toward the peaks, as in Fig. 22.5.
4. Draw a system of crests.
5. Draw slope lines in the direction the water would run on that surface.
6. Shade the southeast-facing slopes, the steeper the darker. Heavy lines and even solid black can be used with good effect.

Elongated mountains have a definite trend. This shows up in the river system, and the ridges can be drawn accordingly. Gneiss, sedimentary rocks, and slate—each has its characteristic form, and some examples of these rocks are shown in Fig. 8.2.

Crystalline peneplains compose a large part of the earth's surface, which consists of ancient complex crystalline rocks peneplained and, to various degrees, redissected.

Fig. 8.8. To practice drawing high, complex mountains, it is best to make them into islands. Note the sharp crests and strong shading.

Fig. 8.9. Glaciated mountains have mostly concave forms. Note the cirques, troughs, tarn lakes, arêtes (jagged crests), etc.

The characteristic is a strong joint control which shows up particularly in glaciated areas such as the Canadian Shield and Scandinavia. The rock, weakened along joints, will be more easily eroded, forming valleys. If such peneplain is elevated, glaciers may occupy the valleys and the result is a fiord region, such as those in Norway and British Columbia (see Fig. 8.2). In unglaciated areas, the joint structure is mostly mantled by a heavy cover of residuals, resulting in an undulating terrain with occasional knobs, like much of Africa or the Brazilian Upland (see Fig. 8.2).

Glaciated mountains, a result of local glaciation, may transform the valleys of a mountainous region into glacial troughs and the headwaters of the rivers into cirques, and may sharpen the crests into arêtes. The prevailing forms will be more concave than convex. These forms are shown in Fig. 8.9. Note the heavy shading of the upper slopes.

Deserts may have all the land forms of other lands but of greater angularity and with a prevalence for waste-filled basins and large areas of sand dunes. The various forms of sand dunes may be shown by dots, as in Fig. 8.1.

Limestone is eroded by solution and its forms are highly irregular. As long as the limestone layers are horizontal, the drawing of the various forms of sinkholes is relatively simple, but in karst regions of folded and faulted limestone interbedded with other rocks, the forms can be extremely complex. Some of these are shown in Fig. 8.2.

Volcanoes, for the most part, have a typical shape; by drawing a somewhat exaggerated crater they can be made readily recognizable. Extinct volcanoes, however, may look very different, as the Three Sisters in Oregon, for instance. While young lava is dark in color and shows a flow structure, old lava regions often do not differ greatly from a dissected plateau or mountain region. Lava-capped mesas, however, are characteristic features, as shown in Fig. 8.2.

Exercise 8.1

Copy the geological section in Fig. 8.10 and draw a strip of corresponding surface landscape as seen somewhat from the right and above.

Landform, Landscape, Land-use, and Land-type Maps

These terms represent four basic approaches of cartography. Landform maps have already been discussed. Their purpose is to show relief—the lay of the land. The landscape approach is the artist's way to

Fig. 8.10. *Exercise.* Copy these geologic structures and draw the corresponding land forms behind, as seen obliquely from above. You may assume that the hard sandstones will make cliffs, and cuestas over the shades. The schist forms a strongly elongated island, and granite tends to make rounded cupolas.

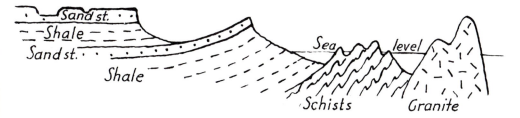

represent the earth in its true colors and patterns. A colored air photo with symbols and lettering added would be the best example of the landscape approach. For a small-scale map, of course, the colors and patterns would have to be standardized but kept as close as possible to the natural colors. The land-use map shows the works of man. It shows crops, pasture, forest—all with contrasting colors. It emphasizes roads, railways, cities, manufacturing and mining centers, ports, shipping routes; and it is likely to have statistical insets. This is a highly important type of map and will be discussed in Chap. 23.

The land-type map is a combination of the three basic approaches. It shows land forms by contours plus plastic shading on large- or medium-scale maps, but on small-scale maps the landform symbols of Fig. 8.1 can be used with advantage. The basic coloring, however, is not of layer tints but

derives from the landscape approach. The natural color and pattern of the land are standardized for easy recognition. Thus we would show green pastures, rusty desert, white ice, mottled dark-green forest, patchy savanna, etc.

As maps are for human use, land use is emphasized. Cultivated areas stand out and products are differentiated by colors and patterns or even index letters. Golden wheat, closely dotted green corn, light-green pastures, red factories, distinct road and railway symbols—all these will make a vivid map. On a small scale we can show the checkerboard pattern of cultivated fields in America, the more irregular patterns of Europe and Asia, etc.

Several notable land-type maps have already been made. The Swedish "Folksko-lans Atlas" carried this idea to the grade-school level. The maps by Hal Shelton of Denver, Colorado, closely approach the

Fig. 8.11. Slope categories were delimited with the help of a slope indicator (see Fig. 7.5). The areas of different slopes were measured with a planimeter and laid out to form a "general land-slope curve."

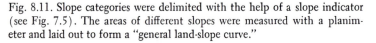

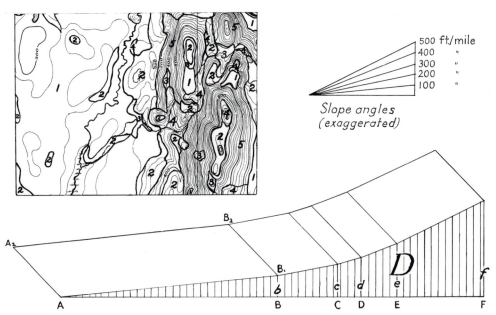

idea, chiefly from the landscape point of view.[2]

The land-type map is the map of the future. The high expense of color printing can be put to better use than showing elevation above sea level. Showing the various types of land works on the imagination and gives a more correct image of a region. We need to close the too large gap between land and map.

Land-slope Analysis

Traveling in the eastern United States, one cannot fail to notice the close correspondence between slope and land utilization. The flat valley bottoms are plowed, and gentle hillsides are pastured until they give way to forest on the steep slopes. At other places the fields are on the flat uplands, and the steep valley slopes are forested. The slope conditions are important for understanding any land, and their representation presents interesting cartographic problems.

Large-scale Land-slope Maps. On a large-scale map, we have not much problem. A contour map with closely spaced intervals gives most of the answers. Relative relief can be read directly, and average slopes found with a slope indicator (see Fig. 7.5).

For medium and small scales, a contour map is of less use. Within a contour interval of 500 ft we may miss many steep knobs several hundred feet high which render the land unsuitable for plowing. We have to look for other methods to express slope conditions.

Slope-category Maps. We divide the topographic sheets, with the help of the slope

indicator, into patches of four to six slope categories, as in Fig. 8.11. We trace the outlines of the patches and place index numbers on the patches. This can be photographed to smaller scale, and if it is too intricate the map can be generalized in the same way as contour lines are generalized (see Chap. 3). We tint or shade the patches increasingly darker where the slopes are steeper and we have a map similar to the one shown in Fig. 8.12.

Fig. 8.12. Slope categories were laid out on topographic sheets and generalized to form a medium-scale map. (*By Raisz and Henry, Geog. Rev., vol. 27, 1937, pp. 467–472.*)

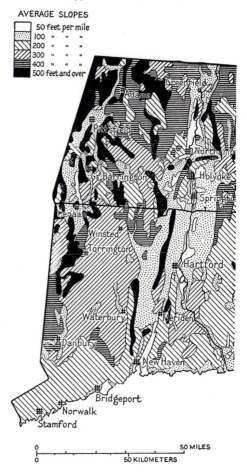

AVERAGE SLOPES

	50 feet per mile
	100 " " "
	200 " " "
	300 " " "
	400 " " "
	500 feet and over

Greenfield

Adams

Pittsfield

Northampton

Gr. Barrington

Holyoke

Springfield

Canaan

Winsted

Torrington

Hartford

Waterbury

Meriden

Danbury

New Haven

Bridgeport

Norwalk

Stamford

0   50 MILES

0   50 KILOMETERS

---

[2] The author's early attempts are the end-paper maps in D. H. Davis, "Earth and Man," Macmillan Company, New York, 1942.

Outlining patches on topographic sheets with the slope indicator depends on individual judgment. For more exact procedure, we may use the Wentworth method. We divide the topographic sheet into mile (or smaller) squares and draw diagonals. We count the number of contour lines crossing the diagonals and the sum of these will give an index number of steepness for each square. From this we can generalize a small-scale slope-category map.

Relative relief maps are used on lowlands or dissected plateaus. On merged mountainous land, the system does not work so well unless it is on a very small scale. The topographic sheets are divided into 5-minute (or smaller) rectangles and in each the difference between the highest and lowest elevation is plotted. From this, in turn, a small-scale color-patch map can be generalized.

Trachographic Maps. This method shows

Fig. 8.13. Simplified trachographic symbols for small-scale maps showing ruggedness of land.

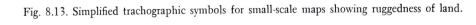
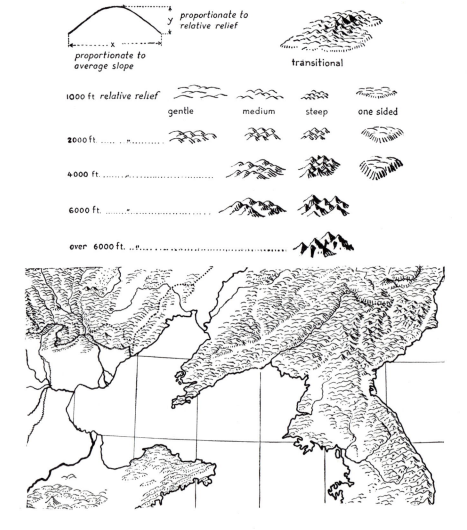

the ruggedness of the land without regard for its genesis; it is used only on small-scale maps. As mentioned before, ruggedness is made up of two major elements: relative relief and average slope. The fundamental symbol of the trachographic method is a hill-shaped curve, as shown in Fig. 8.13. Variations of this curve can express hills and mountains and, to a lesser degree, dissected plateaus. Trachographic symbols are easier to draw than a landform map and, if combined with some symbols of special land forms like sand, lava, sinks, etc., they make quite a good record of relief. Care must be taken that the lines should not be placed so regularly as to look like fish scales. Heavier lines or shading on the right-facing slopes will enhance relief. It is quite possible to draw one shading line for 1,000 ft of relative relief, two shading lines for 2,000 ft, and so forth.

Superimposed Curves. Land slopes can be expressed well by curves. We may take, for instance, a state map and design a curve for each county. These curves are drawn over each county, as in Fig. 8.16, and we obtain an easily comprehensible map. Several types of curves are in use.

Hypsographic curve (see Fig. 8.14) shows how much land lies between two contour lines. We draw horizontal lines corresponding to the contour intervals. We calculate how much land is above 8,000 m and plot this on the 8,000-m line according to a scale. Similarly, we lay out the area of land over 6,000 m, above 4,000 m, and so forth. The hypsographic curve gives a good idea of how much land is at certain elevations. For instance, Fig. 8.14 shows vividly how little land is over 2,000-m elevation. It shows also at what elevations the land is generally steeper. Note the continental shelf and slope in Fig. 8.14.

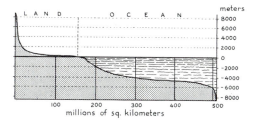

Fig. 8.14. Hypsographic curve showing the distribution of the earth's surface according to altitude.

Hypsographoid curve was proposed by F. Uhorczak of Poland in 1931 and later by J. Hanson-Lowe under the name of *clinographic curve*. The construction is similar to that of the hypsographic curve. We calculate, as in Fig. 8.15, how much land is above 900 ft, take the square root of it and lay this out on the 900-ft line. After all square roots are thus plotted, we obtain a curve. This can be regarded as the half-profile of a solid, looking somewhat like a cone or cupola of a mosque, which is derived from circles, each proportionate to the area of the land above. Hence the name "hypsographoid." The hypsographoid curve shows the average slope of the land at each elevation, and plateaus, peneplains, and other breaks of uniform surface show even better than on the hypsographic curve.

General Land-slope Curve. In Fig. 8.11 we divided the land into slope categories and measured with a planimeter the area

Fig. 8.15. The hypsographic curve shows the amounts of land at each elevation. The hypsographoid curve shows the radii of circles representing the amount of land at each elevation. The two curves are in quadratic relationship.

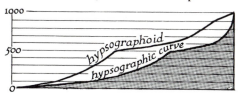

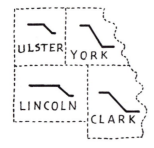

Fig. 8.16. Flatland ratio maps show flat upland, flat lowland, and slopeland by superimposed diagrams.

of each category. With this information we lay out the horizontal component, while the vertical components were obtained by drawing parallels to the slope angles with some vertical exaggeration. This curve informs us how much land is available in each county or other division for plowing, pasture, and other uses controlled by slope.

Flatland Ratio Maps. These maps have only three categories: (1) flat uplands, (2) slope lands, (3) flat lowlands. Figure 8.16 shows the layout. The area of each category is laid out as horizontal component. The slope land in the middle is shown by an oblique line, the vertical component of which is the average difference in height between the flat uplands and the flat lowlands.[3]

Other Methods. Several other methods were proposed to show slope conditions, and some of them are very ingenious. Arthur H. Robinson divided the topographic sheets into $\frac{1}{10}$-mile squares. He estimated the average slope and placed one dot for each degree of slope into the square. This map not only shows the average slope

at every place, but makes an effective shaded relief map.[4]

Bogdan Zaborski shows slope conditions with lines superimposed upon maps. The slope of lines indicates the average slope; the thickness of line, the relative relief; the distance of lines from each other, the placing of rivers; and the proportion between black and white indicates the percentage of flatland. The map is most informative but too laborious to prepare for general use.

Closely related to land-slope maps are the *geomorphological maps* discussed in Chap. 22. The various papers of Prof. Wesley Calef of the University of Chicago throw light on many problems of land-slope mapping. The European concept of relief energy is summarized by Walter Thauer.[5]

Measuring Areas on a Map

This is a frequent task of a cartographer not only for land-slope analysis, but also for statistical maps. The simplest way is to use a *polar planimeter*, which should be standard equipment of the laboratory. According to Robert L. Williams a U-shaped iron, with one sharp point and the other flattened as a hatchet, does almost as well.[6]

If a planimeter is not available, areas may be measured by simpler methods. The easiest is to place over the area to be measured a transparent section paper divided

[3] J. O. Veatch, Graphic and Quantitative Comparisons of Land Types, *Jour. Am. Soc. Agronomy*, vol. 27, 1935, pp. 505–510.

[4] A. H. Robinson, A Method for Producing Shaded Relief from Areal Slope Data, *Surveying and Mapping*, vol. 8, 1948, pp. 157–160.

[5] Neue Methoden der Berechnung und Darstellung der Reliefenergie, *Petermanns Geogr. Mitt.*, vol. 99, 1955, Heft 1, pp. 8–13.

[6] Robert L. Williams, The Hatchet Planimeter, *Prof. Geog.*, no. 2, 1954, pp. 14–15.

into square of $\frac{1}{10}$ in. We count the squares completely within the area and average the ones divided by the outline of the area. If greater accuracy is required, we may use a dotted Zip-a-tone pattern and count dots. We may save a great deal of counting if we divide the Zip-a-tone into squares containing 100, 1,000, and 10,000 dots with colored ink. If we paste the Zip-a-tone over a cellophane sheet, it will not stick to the paper.

Exercise 8.2 *

For practice in land-slope analysis, divide topographic sheets among several students

* This exercise is optional.

giving each about 4 in. square. Give the following directions:

a. Prepare a hypsographic curve.
b. Prepare a clinographic curve.
c. Divide the area into slope categories and make a general slope category curve. Shade the slope categories on an overlay.
d. Divide the area into half-inch squares and give index number by the Wentworth method. Draw choropleths and shade the slope categories on an overlay. Compare this with *c.*
e. Make a relative relief map indexing the same divisions. Draw choropleths, shade, and compare it with *c* and *d* maps.

Individual items may be assigned to individual students.

9

Government Maps

In the United States a cartographer has a better chance of being employed by the government than by a map company, school, or business. Government work is usually highly specialized, strictly standardized, and requires greater technical experience than the average cartography student may hope to acquire. Chances to carry a map from beginning to end are rare. The work usually is broken down into compilation, layout, fine drawing, inking or scribing, checking, plate making, printing, distribution, etc.—each phase carried out by experts. Yet the principles of good cartography are the same as elsewhere, and a man trained in general cartography will have little difficulty fitting in anywhere. Special training for the more technical phases is given in most large government mapping agencies.

Government cartography is wide and varied. Topographic sheets and sea charts are fundamental but, besides these, various kinds of maps are published by scores of agencies in Washington. Add to this the production of states, cities, military establishments and the production of foreign countries, and we have before us a field of such complexity and magnitude that we can deal only with the most important phases here. The Map Information Office of the U.S. Geological Survey can be consulted for any question on our government's maps.

Topographic Sheets

In 1747 César François Cassini accompanied Louis XV on his campaign to Flanders. Here he prepared two large-scale maps based on triangulation and exact surveying. The king was so pleased with them that he ordered the whole of France mapped thus. Due to the state of finances, this monumental work, the "Carte géométrique de la France," 1:84,600, was completed only after more than forty years and then by private subscription. Nevertheless, it served as a pattern for other countries. In the midnineteenth century all of Europe was topographically mapped, and by the end of the century most countries had their topographic sheets. Interestingly enough, the colonies of major powers were much better mapped than were independent countries. The present state of mapping the world is shown in Fig. 9.1.

The average topographic sheet is a trapezoid rarely exceeding 2 ft in either direc-

---

NOTE TO THE INSTRUCTOR: This and the next two chapters are best presented in the map room, if possible.

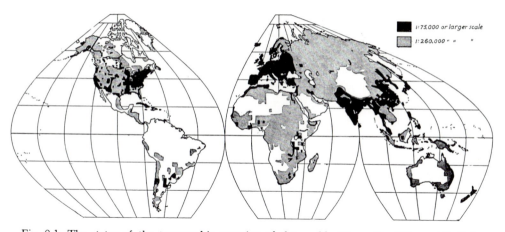

Fig. 9.1. The status of the topographic mapping of the world. (*From the "Times Atlas,"* by *John Bartholomew, ca.* 1955.) A considerable number of maps have been published since this date all over the world.

tion. Its scale varies from 1:10,000 to 1:500,000. It may be on the polyconic, Bonne, Cassini, Transverse Mercator, polyhedric, or trapezoidal, or other projections. It matters little, because the curvature of the earth on such small areas is hardly noticeable. Most topographic sheets are laid out centered on their own central meridian. We cannot mount a number of them together without discrepancies. Small countries, however, have one central meridian for the whole country, and the individual maps are laid out left and right in rectangles, cutting the parallels and meridians obliquely, but the sheets of the whole country can be mounted on a wall.

As the map covers the entire trapezoid, all the legend is outside and is called *marginal data,* consisting of the title of the series, name and number of sheet, scale, date, contour interval with datum level, explanations, authorities, neighboring sheets, relative reliability, grid information, magnetic north, name of projection, etc.

The numbering of sheets is different in every country, but recently the interna-tional system of numbering has been spreading (see Fig. 9.2). The number gives the longitude and the latitude nearest to the intersection of the prime meridian and the equator. To this we add the width and the height of the map in degrees and minutes. Scales, projections, symbols, and grids are explained in their respective chapters.

U.S. Geological Survey (Department of the Interior). The systematic topographic

Fig. 9.2. The International Numbering System for maps. The map in the lower left has its upper right corner at S7°45′ and W14°15′ and it is 20′ high and ½° wide.

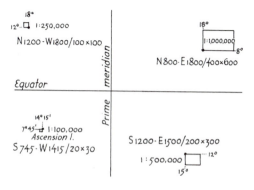

survey of the United States started when the many individual military and other governmental surveys were combined in 1878 into the U.S. Geological Survey. While in most European countries the military offices mapped the land, here the people were interested in the great mineral deposits of the West. The geologists needed a base map on which to lay out the information, and thus the Topographic Branch of the U.S. Geological Survey was established. Basic triangulation and leveling was and is still performed by the U.S. Coast and Geodetic Survey. The main instrument of the U.S. Geological Survey was the plane table. In the early days, there was some trouble with the Indians because it was difficult to explain what the people with weird instruments were doing in their tribal lands. Many surveying parties went out with rifles on their shoulders or with Army protection.

If one looks at the map of the U.S. Geological Survey called "Status of Topographic Mapping," one sees a patchwork of various stages of mapping (Fig. 9.4). About one-third of the country has no large-scale topographic maps at all. About one-quarter has up-to-date modern maps. The rest of the mapping is in all stages in between. The usual scales for our topographic maps are 1:24,000, 1:31,680 (New England), 1:62,500, and 1:125,000. The new 1:250,000 sheets, with or without plastic shading, are most helpful to geographers. About twenty thousand quadrangle maps have been published, including maps for Alaska, the Hawaiian Islands, and Puerto Rico. The yearly distribution reached about four million sheets in 1956.

The other publications of the U.S. Geological Survey are varied. Cities, irrigated regions, national parks and monuments, and mining areas have special large-scale maps. River surveys are often the only information on unsurveyed areas. Base maps for all states are published on 1:500,000 and 1:1,000,000 scales. Many kinds of maps are available for the country as a whole. Geological maps, oil or gas investigation maps, geological folios, water-resources maps, and maps in the various books and pamphlets are listed in the yearly "Publications of the Geological Survey," obtainable from the Map Information Office.

Our basic maps are the quadrangle sheets of the U.S. Geological Survey. The newer 1:24,000 sheets, based on air photography, with great detail and intricate contour lines, reveal in some respects more about the land than actual travel. William B. Upton, Jr., made selections for "A Set of 100 Topographic Maps Illustrating Specified Physiographic Features," which can be ordered at reduced price [1] from the Washington or Denver office of the U.S. Geological Survey.

The Bureau of Land Management, formerly the General Land Office, was and is responsible for the layout of townships and sections of public lands. This includes most of the land west of Pennsylvania, Kentucky, and Tennessee, as they were part of the public domain at the birth of our country. The system of 6-mile-square townships and 1-mile-square sections started in the early days of the Republic and is shown in Fig. 9.3. Of each township, a "plat" was prepared by the surveyor, generally on 2 in./ mile scale. These plats may be consulted in the Interior Building in Washington; for some parts of the country, they furnish the only information.

[1] At present, $15.00.

The Bureau of the Census compiles maps for census purposes, showing townships, wards, census tracts, etc. The Minor Civil Division maps of the states are frequently used by geographers as base maps for statistical layouts. The Bureau is also instrumental in the production of the national atlases of several Middle American republics and participates in preparing the National Atlas of the United States.

Bureau of Reclamation. In cooperation with the U.S. Geological Survey, this section of the Department of the Interior maps Federal irrigation projects, based upon air photographs, in great detail. The maps are published on various scales. The extensive photo coverage of the Bureau, particularly in the Western states, can be of great use for geographers.

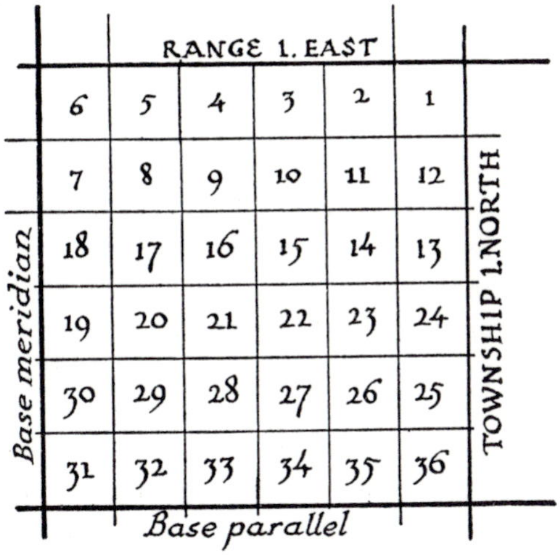

Army Map Service. This office prepares and prints more varied maps than any other agency, yet we find relatively few of them in geography departments. Only some of the maps are for sale; others may be acquired through official channels. Maps issued by the Army Map Service, particularly in wartime, range into hundreds of millions of copies.

Most of the Army Map Service maps are compiled from existing maps, but the Army Engineers have made many surveys, particularly of military establishments here and abroad. Publications on sale include such items as "Glossary of Cartographic Terms," "Geographic Place Names," "Russian Short Glossary." These can be ordered, together with the "Public List," from the Army Map Service.[2] The bulletins contain basic information for map makers. Outstanding research is conducted in scribing,

[2] The address is 6500 Brooks Lane, Washington 16, D.C.

Fig. 9.3. The township-and-section system was introduced at the end of the eighteenth century to survey and divide all public lands. Now it gives the characteristic pattern of our landscape west of the Appalachian Mountains.

engraving and printing, drawing instruments, and the preparation of terrain models. The map collection is the second largest in the country and probably in the world.

Department of Agriculture. A geographer naturally will be interested in the map production of this department. The half-inch and quarter-inch maps of the *Forest Service* cover our national forests and are often the only source of detailed information, particularly in the West and in Alaska. Most maps are without contours but have frequent spot heights.

The *Soil Conservation Service* publishes detailed soil maps and descriptions of about two thousand counties, besides small-scale maps and a "Soil Atlas." The Service is particularly active in air-photo coverage and mosaics. Their detailed maps show land-use categories and capabilities, in addition to soils, slopes, and soil erosion.

The *Bureau of Agricultural Economics*

publishes some of our most important maps, such as the 1:5,000,000 land-use maps by F. J. Marschner and the "Generalized Types of Farming" map. For the various maps of agricultural production and farm statistics, the lists of the Department of Agriculture should be consulted. Air photos of all the farming areas of the country can be ordered. A map of coverage can be obtained from the Map Information Office.

Department of State. The Cartography Branch prepares hundreds of maps yearly for the use of the Department. They deal with economics, ethnography, demography, social data, etc., and are based on considerable research. The maps are generally not available to the public, but copies may be obtained on request for special studies.

Department of Commerce. The Transportation Map Series of the country is published by the *Bureau of Public Roads* in cooperation with the U.S. Geological Survey. The maps are on various scales. The new 1:250,000 maps show highways, railways, airports, canals, navigable rivers, etc. The *Highway Planning Survey* prepared a number of traffic-flow maps. Their large-scale street maps of cities showing the type of pavement, central business districts, and other information for urban geography are very useful.

The International Boundary Commission. This agency was created to define and map in detail a strip along the boundaries of the United States and Alaska, a strip which was usually 10 miles wide. As much of this boundary runs through wilderness, these maps are of great value for considerable areas.

The Tennessee Valley Authority produces detailed planimetric maps of the watershed

and a land-use map of great detail of part of it.

Special Maps by Federal Agencies. The Federal Communications Commission produces maps of the telephone and telegraph lines and broadcasting stations. The Post Office Department produces state maps of postal routes. The Rural Electrification Administration maps show electric facilities, both public and private. The Federal Power Commission [3] publishes state maps of electric facilities. The Bureau of Indian Affairs has maps of the reservations.

This list is most incomplete. To cover all map-making agencies and their work would require volumes. This is just a general idea of the subject; if any problem arises, it is best to write directly to the agencies.

State and City Agencies. Several states publish maps for local use and for planning purposes. The best way to obtain them is to write to the state planning boards. Almost every city has a map prepared by the city engineers. Departments for utilities, water, sewage, and electricity often have their own maps. Several states have atlases showing mostly economic and social features. The atlases of Illinois, Kansas, Missouri, Nebraska, New Mexico, New York, Arkansas, the Northwestern states, the Intermountain states, etc., are outstanding. Chambers of commerce often have maps which note historical spots as tourist attractions.

Charts

The term *chart* is used variously for maps, large layouts, single-subject maps, etc. Here we limit the use of the term to

---

[3] The address is 1800 Pennsylvania Avenue, N.W., Washington, D.C.

maps relating to navigation. We usually think of charts for navigation over the waters, but in recent years charts for air navigation have become equally important.

Charts for Water Navigation. When the first portolan charts were issued in the thirteenth century, people marveled at their accuracy. While on land we may inquire our way, on the open sea we rely on charts, a compass, lights, and other aids for navigation. A modern chart, besides showing accurate shores, coastal features, and soundings, has a great deal of other information, such as tides, currents, shoals, and harbor facilities. Scales vary from large-scale harbor charts to small-scale charts of seas and oceans.

U.S. Coast and Geodetic Survey. The triangulation and leveling work of this agency of the Department of Commerce will be discussed in Chap. 15. Its original function was charting the coasts of the country. Established by law in 1807, the first charts were not published until the 1840s. Since that time, the U.S. Coast and Geodetic Survey has published thousands of charts of the coasts of the United States, Alaska, the Philippine Islands, and some rivers, canals, and estuaries. The catalogue is an impressive volume and contains lists of charts of all scales, aeronautical charts, magnetic, gravitational, and astronomical charts, handbooks, etc. Several of the special publications such as "Elements of Map Projection" and "Cartography," by Charles H. Deetz and many others, are fundamental for a cartographer.

Hydrographic Office of the Navy. Charts of foreign coasts are prepared, primarily for the use of the Navy, in the Suitland, Maryland, office. The charts look similar

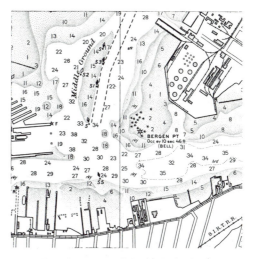

Fig. 9.4. Part of a nautical chart.

to those of the Coast and Geodetic Survey. As other nations chart their own coasts, the Hydrographic Office does much compilation. Revising inadequate charts and producing new ones involves considerable work; and the list of publications includes charts, aeronautical charts, Loran charts (see Chap. 22), Current charts, Timezone charts, Magnetic charts, Sailing Directions, books, etc. Monthly Pilot Charts are published jointly with the Coast and Geodetic Survey and the Weather Bureau. These charts—one for each ocean—are veritable storehouses of information. Winds, currents, fog, ice, storms, sailing directions, etc., are included. On the reverse are listed various articles on maritime matters.

The Office also publishes various atlases such as the "Ice Atlas of the World" and "Atlas of Sea and Swell." The "Hydrographic Manual" is a handy reference book. The Sailing Directions and Pilots are often consulted. At present the Office is engaged in preparing a large atlas for each ocean (see also Chap. 22).

Other Map-making Agencies. The survey of the Great Lakes is carried on by the Corps of Engineers in Detroit, Michigan. The charts are similar to the Coast and Geodetic Survey charts and cover not only the Great Lakes, but also the connecting canals and rivers. The Mississippi River Commission, with its main office in Vicksburg, Mississippi, publishes a complete set of maps on various scales of this turbulent river every few decades. Comparing current maps with older editions, we see the changes of a great river somewhat as in slow motion. The huge working model of the Mississippi River in Vicksburg is most interesting. Charts of the upper Mississippi, the Missouri, the Ohio, the Illinois, and other rivers are also published by the Corps of Engineers.

Aeronautical Charts. The navigator of a plane requires charts too. For him, however, the altitudes of mountains, landmarks, landing fields, beams, and runway beacons are important. Because the plane may fly at hundreds of miles per hour, the scale of the chart is small. Thus an air chart has strongly colored layer tints, and easily visible features, such as airways, shapes of lakes, outlines of cities, etc., are emphasized.

The scale varies with the purpose of the chart, and each scale has a name indicating its special use:

1:5,000,000–World Planning Charts
1:3,000,000–Long-range Navigation Charts
1:2,000,000–World Radio Direction-finding Charts
1:1,000,000–World Aeronautical Charts, referred to as WAC
1:500,000–Pilotage Charts
1:200,000–Approach Charts
Large-scale Landing Charts and Target Charts, on various scales

As most of these charts are based on air photos, they contain much new information and are fundamental for geographers.

All the services, such as the U.S. Geological Survey, the U.S. Coast and Geodetic Survey, and the Hydrographic Office, cooperate in the production of these charts, but the chief producer is the Aeronautical Chart and Information Center (ACIC) in St. Louis, Missouri. The projection of most charts is Lambert Conformal Conic. Neither compass directions nor great-circle directions are quite correct in this projection, but within the limits of a WAC sheet, the projection closely approximates both.

For fast and high-flying jet planes, the ACIC is now publishing the 1:5,000,000 Global Navigation and Planning Series in 26 charts and also some strip charts for the most frequented routes. No contour lines and altitude tints burden the charts, but a light green tint indicates the plains; mountains are shown by plastic shading over a buff base. Lately a number of special charts have been prepared to help our astronauts in orbital flights.

Foreign Maps

The review above gives some idea of the magnitude of the official cartography of a single country. The mapping programs of other countries are in no way behind, and it would be futile to attempt to describe them all here. The issues of *World Cartography* of the United Nations contain reports of official map publications of foreign countries. Changes are frequent, but the United States embassy or consulate of the country can give up-to-date references. It may be noted that in most countries the army is the main map-making agency.

10

Private Maps

This chapter deals with maps made by private map companies or individual cartographers. To sell, these maps must be useful and attractive. Rarely does a map company make expensive surveys; most of the maps are compiled.

Atlases

A collection of maps of uniform design and published in bound or loose-leaf volumes is called an *atlas*. The name is taken from the Titan who held the earth on his shoulders, and whose pictures often decorate the title page of Renaissance atlases. The first atlas known was contained in Ptolemy's "Geography," which was republished repeatedly during the Renaissance. Atlases in the modern sense start with the "Theatrum Orbis Terrarum" of Ortelius in 1570 and the contemporaneous La Freri atlas published in Rome. These were followed by a golden age of the grand atlases in the seventeenth century. Beautifully executed, hand-painted in rich colors, and bound in parchment, these folio-

sized atlases are now collectors' items of great value. In the eighteenth and nineteenth centuries the grand atlases became less decorative but more exact.

World Atlases. The typical *grand atlas* (folio sized) shows all the lands of the world in varying scales. They usually contain a number of world maps, continent maps, hemispheres, polar regions, even star maps and explanations of map projections. These are followed by detailed maps on varying scales with insets of cities and important places.

A gazetteer or index of all the places shown on the maps, located by coordinates, is rarely missing. Relief is usually shown by hachures, altitude tints, or both. Its purpose is chiefly reference; thus, it has a great number of place names, roads, and railways. Some atlases, particularly American atlases, carried this so far that there was no room for anything else, and relief was left out completely.

At present the public expects and demands more than this. The latest grand atlases show relief; but they also include a number of special maps covering vegetation, population, climate, and economic facts. Production of a world atlas is costly in time and money. A new atlas produced anywhere in the world may make the older ones obsolete, as a grand atlas has an international sale.

NOTE TO THE INSTRUCTOR: This chapter, like the previous one, can best be presented in the map room. The laboratory hour can be used for finishing the large-scale map.

Fig. 10.1. Old atlases usually had a very decorative title page often displaying Atlas, the Titan, holding up the earth. (*Photograph from the Library of Congress.*)

Notable among the prewar atlases is the Stieler atlas, produced by Justus Perthes in Gotha, Germany. Since 1825, this has been published in 11 editions; the last international edition was interrupted by the war. Among other German grand atlases, Andree, Debes, Velhagen und Klasing, and Keyser are of very high quality. In Great Britain, George Philip & Son, Ltd., of London and John Bartholomew in Edinburgh have published several editions of good world atlases. The "Touring Club Atlas of Milano" of 1938 was one of the richest grand atlases, and it was last published in 1951. The Agostini atlases of Navarra are also of high quality. Excellent French atlases have been published by Vivien de St.-Martin and by F. Schrader.

Among the postwar atlases, the Dutch and Swedish publications should be mentioned. The "Atlas Mira," published by the U.S.S.R. in 1954 through the collective effort of their best cartographers, is a good source for contour lines in foreign lands. The greatest atlas of recent times is the "Times Atlas," produced by Bartholomew in five volumes. In the United States, Rand McNally & Company in Chicago, C. S. Hammond & Company, Inc., in New York, and others produce large world atlases.[1] The excellent maps of the *National Geographic Society* can be collected into a loose-leaf atlas.

National Atlases. Our storehouse of geographic knowledge is filling up to the brim. A world atlas can present but a small part of it. Much of our information comes from the various nations' atlases of their own land. They are published either by government agencies or by private concerns

[1] World Reference Atlases, *Prof. Geog.*, vol. 3, 1945, pp. 1–8.

subsidized by the government or business firms. Besides their topographical pages, they contain maps dealing with climate, population, economics, social conditions, and other topics.[2] The first ones appeared around the turn of the century and their number is constantly increasing. Almost every country of Europe, in addition to Mexico, the Central American republics, Cuba, Brazil, Argentina, the French colonies, Morocco, Tunis, the Congo, Egypt, India, Japan, China, Australia, Tanganyika, Ghana, etc., published such atlases.

The new "Atlas of Canada," published in 1958, deserves especial praise. Some atlases, such as those from Bengal and Pakistan, are solely for agriculture. The "Grand Soviet Atlas" has all the special maps of the U.S.S.R., as well as remarkably up-

[2] Ena L. Yonge, National Atlases: A Summary, *Geog. Rev.*, vol. 47, 1957, pp. 570–578.

to-date maps of world distributions. Interestingly, the United States does not yet have ready a national atlas, but the project is under way, and a committee worked on it for many years. The project is so large and the statistical data are so rich that an atlas would be out of date before the work was published, so the present plan calls for irregular issues of loose-leaf maps which will go on for years. The various organizations in Washington, scientific societies, universities, even private individuals, are invited to contribute to the national effort. At present, the U.S. Geological Survey is in charge of coordination.

Only the style and the page size are uniform, so that the publication could be either bound together or placed in atlas-shaped boxes. This is a healthy approach, because in our rapidly changing conditions the emphasis must be on up-to-date informa-

Fig. 10.2. The average amount of space devoted to various subjects in ten single-country atlases: Belgium, Brazil, Canada, Egypt, Finland, France, Netherlands East Indies, Russia, Tasmania, and the United States. (*From Norman L. Nicholson, A Survey of Single-country Atlases, Geog. Bull. Ottawa,* 1951.)

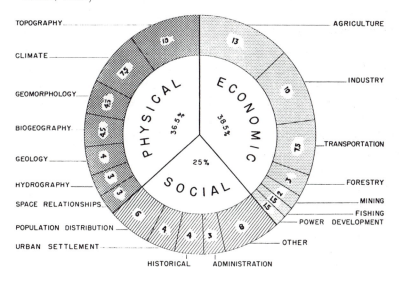

tion, and it is easier to replace a map with a new edition than a whole volume.[3]

England and Australia also preferred large sheets of uniform-scale maps for their atlases; these can be filed in a box or loose-leaf binder.

Educational Maps

We can hardly imagine a schoolroom without multicolored wall maps and globes. We learned our geography from maps in books and from school atlases. Our concept of the world is molded by these maps, and it is the combined task of the cartographer and the educator to make them better. There is plenty of room for improvement; companies turning out maps are reluctant to tread new paths, teachers are not always ready to accept new ideas, and most American products are behind European standards.

Wall Maps. Wall maps have three functions: (1) They have to present an easily visible and easily remembered picture of a country, continent, or the world; (2) they serve as background for the teacher's instruction; and (3) to a certain extent, they should serve as reference. They should not be too complex; colors should be bold, lettering and symbols large. The idea that lettering should be readable across the classroom had to be abandoned; letters 1 in. high at a minimum would be required for legibility from 40 ft (see Chap. 5). Much detail has to be omitted. Relief is usually shown with altitude tints, often combined with bold plastic shading. Only a few producers have as yet broken away from altitude tints, although colors could

be used to better advantage for the various land types (forest, pasture, cultivation, and so on). In order to be visible in the average classroom, the width of wall maps varies from 3 to 7 ft.

Wall maps are used often at close range, and it would be a mistake to omit so much detail that the map would be useless for reference. Thus most wall maps have considerable small lettering and insets. There is a growing tendency to include pictures, boxes with explanatory material, statistics, etc. These not only stimulate interest but also help the teacher to point out pertinent facts.

School Atlases. In America the custom is to include all the maps for instruction in the geography or history textbooks. In Europe, separate atlases are used, and the best map companies compete with each other for this large and ever-changing market. One can buy for a few dollars a beautifully rendered small world atlas with maps of climate, vegetation, cultivation, and economics for the world in general and any one country in particular. Equally excellent are the school history atlases. The best German, Italian, English, Dutch, Swiss, and Russian school atlases may give stimulus and a rich source of reference to our students. Particularly commendable are the Swedish atlases, which broke away from the altitude-tint method and presented the world in more realistic colors. The French and Spanish school atlases have so much text that they could almost be classed as textbooks.

Commercial Maps

City Maps. Good city maps are found often on the reverse side of the auto road maps and as insets in atlases. Detailed maps can be obtained from the city engineer's of-

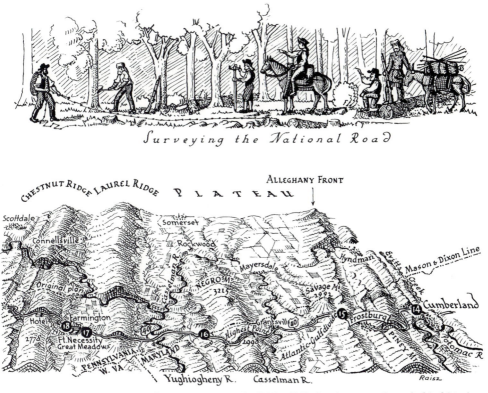

Surveying the National Road

Fig. 10.3. Travel map from George R. Stewart's "U.S. 40," showing a section of this historic transcontinental highway. (*Courtesy of Houghton-Mifflin Company.*)

fices. The Sanborn Map Company in Pelham, New York, is specializing in detailed city maps originally made for fire insurance companies but useful for other purposes, too. Sometimes the chamber of commerce will supply maps. Plans of foreign cities are often found in guidebooks. The chief purpose of these maps is to assist people in finding their way in the maze of streets and they usually have a street index. Functional city maps are discussed in Chap. 23.

Auto Road Maps. More auto road maps are used in the United States at present than all other types combined. Some 150 million are distributed in a single year free of charge. One of the peculiar results of this is that when a map maker, not hav-

ing the backing of oil companies, wants to sell his products for whatever moderate price, people are almost indignant.

The purpose of the auto road map is to help motorists to find their way and to stimulate travel by indicating places of interest. To show the nature of the land, hills, and mountains may make the map too complicated. (Yet some European road maps seem to be able to show them without crowding.) American road maps are produced in a few large establishments, based largely on the maps of the Federal government's Bureau of Public Roads (see Chap. 9). They also include local information from every filling station and company surveyors. New editions are published every year; thus these maps in their limited way give the most up-to-date information available. Some of our companies produce road maps of the Latin-American republics, Canada, and even other continents.

Travel Maps. Airlines are beginning to be aware of the appeal of views from great altitudes. The windows of planes are still far too small, the wings are usually in the way, and the maps given out are too small in scale to allow easy recognition of the features below. Some of the airlines now use Hal Shelton's beautiful "landscape" maps, and this may raise the standards of this neglected industry. The airborne geographer needs to locate himself at any moment from the visible features; for this the 1:1,000,000 WAC charts or even auto road maps are far better than maps found in planes.

Very few railways produce good maps of the land visible from trains, and a great educational opportunity is thus lost. To attract tourists, some European railways,

steamship lines, and travel agencies publish beautifully executed and informative pamphlets with maps, pictures, and notes.

Illustration Maps. Some of the most stimulating jobs of a cartographer are maps for book and magazine illustrations. To a certain extent, maps are also used in advertising. Unfortunately, these maps are often made by commercial artists with no training in geography, let alone in cartography, and they often violate the most fundamental principles of our profession. Unfortunately, too, our public is undiscriminating and readily accepts the gaudily colored, incorrect, and unreal presentations found even in some of our leading periodicals. Fortunately, there are many notable exceptions, such as Richard E. Harrison's maps in *Fortune* (see Fig. 1.6), R. M. Chapin's maps in *Time* (see Fig. 14.3), and others.

Personal Maps. Suburban or country living often presents problems to our friends. How are they to find us? Small maps like the one shown in Fig. 10.4 are most help-

Fig. 10.4. This kind of informal map makes an interesting task for a student of cartography.

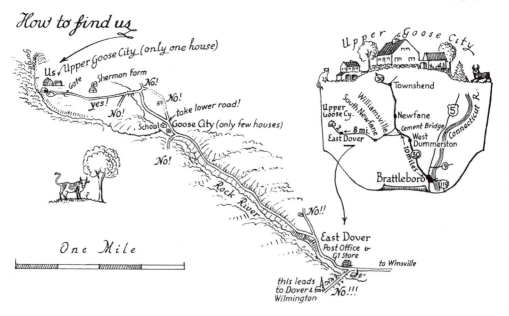

ful. The purpose is clarity, yet a few artistic touches are not out of place.

Geographical Society Maps

A number of organizations publish outstanding maps. Descriptions are given briefly here; organization addresses are listed below.[4]

American Geographical Society publishes the *Geographical Review* with many excellent maps in it. The society also publishes the Latin-American sheets of the Millionth Map of the World, several maps of the Americas, Antarctica, and the world. The "Atlas of Diseases," under the direction of Dr. Jacques M. May, uses the ingenious world projections of William A. Briesemeister (see Chap. 18). The periodical *Focus* excels in concise cartography by Vincent Kotschar.

National Geographic Society maintains a large staff of cartographers under the direction of J. M. Darley, and has published over a hundred large maps of continents and countries and historical sheets. The maps are excellent in giving the maximum information with the minimum of crowding. The more recent maps are of uniform size, to be assembled into a loose-leaf atlas. The lettering of the maps, designed by C. E. Riddiford, is justly famous. Most of our

[4] American Geographical Society, Broadway and 156th St., New York 32.
National Geographic Society, 16th and M St., N.W., Washington 6, D.C.
Association of American Geographers, Library of Congress, Map Division, 1785 Massachusetts Ave., N.W., Washington 6, D.C.
American Congress on Surveying and Mapping, 733 15th St., N.W., Washington 5, D.C.
Geographical and Map Division, Special Libraries Association, 31 East 10th St., New York 3.
Royal Geographical Society, Kensington Gore, London, W.2.
Petermanns Geographische Mitteilungen, Justus Perthes str., Gotha, East Germany.

newspapers and periodicals accept the society's spelling of geographic names as standard. The *National Geographic Magazine* is a rich source of first-class photography, historical paintings, and text maps.

Association of American Geographers publishes maps and articles on cartography both in the *Annals* and in the *Professional Geographer*. The Cartography Committee presents a panel, papers, and exhibits at the yearly meetings.

American Congress on Surveying and Mapping publishes *Surveying and Mapping* and many articles on cartography, largely from the surveyor's point of view. Its yearly meetings are highly informative.

Geographical and Map Division, Special Libraries Association, publishes a list of new books and maps in its quarterly bulletin.

The Royal Geographical Society has excellent maps and photographs in its *Geographical Journal*. Emphasis is on exploration, and more original traverses have been published by this organization than by any other private society. The maps have a traditional flavor; they are lettered by hand with quill pens, yet excel in clarity (see Fig. 5.1).

Deutsche Gesellschaft für Kartographie arranges professional meetings and publishes the "Kartographische Nachrichten" (C. Bertelsmann Verlag, Gutersloh, West Germany).

Petermanns Geographische Mitteilungen has had a high reputation for text, pictures, and maps since its founding in 1856. Though it operates behind the Iron Curtain, it has regained much of its former excellence and originality.

There are hundreds of map-publishing houses and geographical societies in the world, all publishing maps and pictures. Many are of high standard and deserve attention. Lists of them can be obtained from the Map Division of the Library of Congress. For periodicals with articles on cartography, see the Appendix.

11

Map Collections and Compilation

Maps have been made for thousands of years and there is hardly any part of the earth for which no map is available. Some areas are well mapped on a large scale; others are just barely sketched. Before making any kind of map, we have to find out what has been done before. We rarely have a chance to make the first map of a region. Sometimes we lay out our own surveys or work from aerial photos, but most of the time we take existing maps and redesign them with new data or with a new point of view for our purpose.

Map Collection

To find all the existing maps of a region is not easy. No map library has all the maps, even of its own region. Some very large map collections, such as those of the Library of Congress, the American Geographical Society, the New York Public Library, and most of the large universities,

are made available by the intercollegiate loan system, but only extra copies or photostats are loaned. The two greatest collections, those of the Army Map Service and the National Geographic Society, are not easily accessible.

One map at hand is worth ten elsewhere, and the expense involved in building up a good college collection is repaid in efficient work. The college library, of course, should have all the good local city and state maps as well as all of the U.S. Geological Survey topographic sheets of the state or region. The new 1:250,000 set for the whole United States and Alaska should be available. Geologic, climatologic, economic, and all other special maps of the state and Federal government are relatively inexpensive and could be even more important than the topographic sheets. Many students will consult marine charts of nearby waters.

The Map Information Office of the U.S. Geological Survey gives full information on what is available in government maps. For Canada, the 1:250,000 and the 1:500,000 maps of the National Topographic Series and the excellent general maps of the country showing vegetation, minerals, hydroelectric power, etc., can be ordered from the Department of Mines and Technical Surveys in Ottawa.

For foreign countries and also for the

NOTE TO THE INSTRUCTOR: This chapter is primarily for the map librarian and can be assigned to students for reading only.

United States, a complete set of the *National Geographic Magazine* maps are good for reference. For more detail the 1:1,000,000 World Aeronautical Charts, the 1:500,000 Pilotage Charts published by the Aeronautical Chart and Information Service, and the sheets of the "International Millionth Map of the World" are necessary. About 180 colleges are depositories of many thousands of war maps, including captured enemy maps. Some colleges were overwhelmed by this windfall, but most were stimulated into building up a good collection. For special studies, however, it is best to order up-to-date maps. Addresses from which the most important topographic sets of foreign countries may be ordered can be obtained from the United Nations.

Atlases give fast and well-organized reference. They usually have an index by which places can easily be located. They are described in the previous chapter. The map librarian may find that ordering school atlases of many countries and one large detailed atlas, such as the "Times Atlas," is the most economical procedure. The "Atlas Mira," for which a translation is published, is excellent for Asia. Most informative, however, are the various national atlases, and as many of them should be acquired as the budget allows. Obviously one should subscribe to the forthcoming sheets of our national atlas. The "National Atlas of Great Britain" has a series of excellent maps on 10 miles/in. scale and they can be obtained from Edward Stanford, Ltd.[1] The various state atlases of the United States have to be bought individually.

There are many other desirable maps. Road maps, which can be obtained from oil companies, often have the most up-to-

[1] 12 Long Acre, London, W.C.2.

date information. Wall maps and globes are necessary for instruction, and the catalogues of our map publishers list an ample selection. Newspapers and magazines often have maps of recent events, which should be cut out and preserved. Particularly useful are the maps and graphs in the News of the Week section of the Sunday *New York Times*. Weather maps, pilot charts, geological maps, landform maps, etc., will be discussed later. City maps can often be obtained from the chamber of commerce or the office of the city engineer.

Many universities have accumulated a great number of old maps which are of value for historians. For geographers, however, they may be a liability because of their bulk. The map librarian might do well to ship many of them to the college library.

Just as important as maps is a good collection of aerial photographs. The Map Information Office sends on request the yearly edition of "Status of Topographic Mapping" and "Status of Aerial Photography." The Photographic Records and Services Division of the U.S. Air Force has a selected list of air photos for sale called "Index to Aerial and Ground Illustration of Geologic and Topographic Features throughout the World." State highway departments and state and city planning boards may also be consulted concerning air photos.

The map librarian does well to subscribe to the periodical *Surveying and Mapping* and to the bulletins of the Geography and Map Division of the Special Libraries Association, where new maps are listed and described. Also, one should be on the mailing lists of all important domestic and foreign map-publishing houses for their annual catalogues. A list of the addresses of

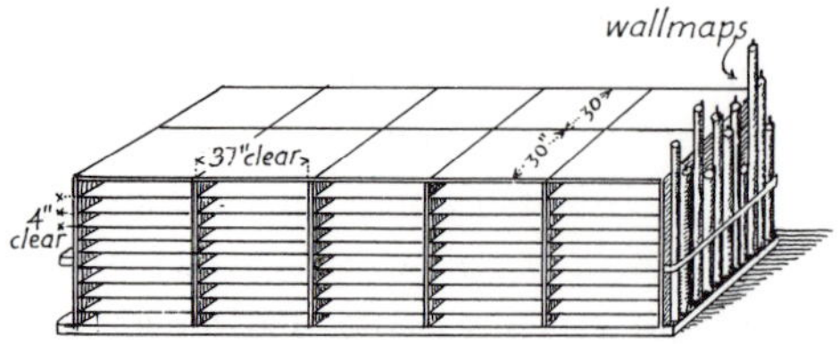

Fig. 11.1. These five duplex cases can easily store 30,000 loose maps without using wall space. Top of case can be used for spreading out maps.

important map publishers can be obtained from the Map Division of the Library of Congress.

Storing of Maps. Most colleges use ready-made steel cases with drawers for their maps. These are usually 48 in. wide, 32 in. deep, and 2½ in. high. They usually come in units of five drawers and each one can store 500 to 1,000 maps. Wooden cases with 36- by 24- by 4-in. compartments are very much cheaper (see Fig. 11.1). In either case, the maps should be kept in smooth and strong manila folders. Each folder may contain 30 to 50 maps. The folders may buckle in handling if they are too large and too heavy. It is better to fold the oversized maps than to handle folders more than a yard wide. One of the dangers of map drawers is that people sometimes simply throw the maps in, instead of putting them in the proper folders. In open map cases, this crime is immediately apparent. About five folders are placed in a compartment and each folder is numbered. If numbers are not consecutive, but run 780, 782, 784, 786, and 788, this allows for future growth. In replacing the folders within a compartment, it is better to leave them on the top of the pile than to squeeze them under. The number and name of the folder should be visible in large letters at the edge of the

folder. Here again, the open cases have the advantage. If there is danger of dust, sliding doors may be provided.

Maps may be divided into two groups: map sets (sheets of a larger series) and individual maps. Set maps are numbered with their folder numbers only, since they already have individual numbers. Individual maps should each have a number in addition to the folder number. Serial maps should always be put in their proper place within the folder. Individual maps are better replaced on the top inside the folder; thus the most frequently used maps will be on the top.

Folders should be in regional order. The sequence of any American school atlas is a good guide in this, starting with world maps and progressing from the United States and Canada through the various countries and continents. Atlases, books, and pamphlets should be placed as near their regional folder as possible. For instance, the "Atlas do Brasil" should be placed near the map folders of Brazil, or possibly on top of the case. Charts and weather maps, however, seldom fit into regional order and have their own cases. War maps can be placed in umbrella stands or hung by hooks from the molding.

If there are several copies of the same map, it may be best to place them in three separate collections: a complete A collection for consultation, a B collection of second copies for loaning, and a C collection of extra copies for use in class instruction. The C collection is important, particularly when 40 students clamor for the same map for a certain assignment.

Map Cataloguing. Maps are generally catalogued for the convenience of other departments. It is much easier to find a good

U. S. *Geological Survey.*
    Philadelphia and vicinity, east, 1955 [and west, 1956]
(Pennsylvania-New Jersey) Mapped by the Geological
Survey. Washington, 1958.
    2 col. maps 167 x 101 cm.
    Scale 1 : 24,000.
    "Polyconic projection. 1927 North American datum."
    "Contour intervals 10 and 20 feet with supplementary contours
at 10 foot intervals."
    "Red tint indicates areas in which only landmark buildings are
shown."

    1. Philadelphia — Suburbs and environs — Maps, Topographic.
    2. Philadelphia—Maps.   I. U. S. Coast and Geodetic Survey.

    G3824.P5A1 1956.U6   Map 60–269

    Library of Congress   [3⁄5]

AREA SUMATRA	SUBJECT (MAJOR CLASSIFICATION) RESOURCES	SCALE 1: 750,000	CLASS. NO. J520-27
EXACT TITLE ONDERNEMINGS - EN		DATE 1928	22360
OVERZICHTSKAART VAN		EDITION X	LANGUAGE (TEXT) DUTCH
HET EILAND SUMATRA			S C R F Ⓤ NO. FILE COPIES ✓
PUB. AUTH. AND NO. COMMISSIE VOOR DE WELVAARTSPOLITIEK			
COORDINATES	SOURCE OUTPOST		

PROJECTION X		DIAGRAM	ONE PIECE	COLOR ✓	PHOTOGRAPH NEG.
STRAIGHT MERIDIANS	STRAIGHT PARALLELS	PICTURE	MULTIPLE ✓ 6	MONOCHROME	PHOTOSTAT NEG.
CURVED MERIDIANS	CURVED PARALLELS	PROFILE	SET	MANUSCRIPT	POSITIVE
STUB COORDINATES ✓	NO COORDINATES	AERIAL PHOTOGRAPH	SERIES	PRESS RUN ✓	BIBLIOGRAPHY
AGRICULTURE ✓	MILITARY	POWER	CLIMATE	SOILS	VEG'T'N. COVER
Animal Industry	Collation	Amounts	FISHING	SURFACE ✓	Complete
Crops ✓	Admin. Areas	Lines	GEOLOGY	Contours ☐	Incomplete
Regions ✓	Military Grid	Plants	HISTORY	Form Lines Geod	WATER SUPPLY ✓
AIR NAVIGAT'N ✓	OIL ✓	Types	ROADS ✓	Geomorphic	WATERWAYS ✓
Distances	Fields ✓	RAILROADS ✓	Bridges	Hachures ✓	O. I.
Landing Areas	Pipe Lines ✓	Bridges	Distances	Pictorial Layer	Depths
Routes	Refineries	Distances	Traffic	Shading Color	Distances
BOUNDARIES ✓	Storage	Traffic	Tunnels	Spot Heights ✓	Port Facilities
Internal ✓	PEOPLES	Tunnels	Surfaces ✓	TELECOMMUN.	Routes
INDUSTRY	Distribution	Electrified ✓	Other Categories	Cable	Traffic
Areas	Ethnology	Gauges	MINERALS ✓	Radio	Navigability ✓
Plants	Linguistics	Multiple Tracks ✓	Deposits	Telegraph	Canals
Types	RELIGIONS	Traffic Facilities	Mine Locations ✓	Telephone	G C I P S

                                              DATE CAT. DEC. 16 '46

                MAP CATALOG CARD   RED   GREEN   Ⓑ BLACK

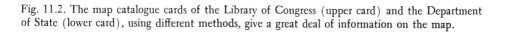

Fig. 11.2. The map catalogue cards of the Library of Congress (upper card) and the Department
of State (lower card), using different methods, give a great deal of information on the map.

map of Guatemala by looking through the
folder on Central America than by trying to
visualize it from a catalogue and then going
to the folder. A sample of catalogue cards is
shown in Fig. 11.2. One may subscribe to
all the catalogue cards of the Library of
Congress and thus keep up to date with
what has been published. Even where maps

are not catalogued, an acquisition book should be kept.

Map Preservation. Unless a map is expensive or irreplaceable, it is hardly worth mounting. For the price of mounting, one can buy from two to ten extra copies of the map. Mounting distorts a map, makes it bulky and useless for the light table.

Mounting. Two methods of mounting are in use. In the wet method, a sheet of muslin is stretched over a curtain stretcher and spread with flour paste, or Dextrol, Arabol, etc. Then the map is moistened, laid on, and the extra paste rolled out, starting from the center. The proper amount of stretching and wetting can be learned only by experience. If misjudged, the map may pop off the muslin or it may have wrinkles after it is dry.

The dry method is safer but more expensive. It is commonly used for mounting photographs and maps on cardboard also. A specially coated sheet of Parawax, Chartex, etc., is placed between the map and the cardboard and pressed with a hot iron. Photo-supply stores can furnish the materials. Coated muslin sheets are on the market to which the maps can be dry-mounted directly with a hot iron.

Embedding. Precious maps are best preserved by embedding in plastic. The map is sandwiched between two transparent plastic sheets and strongly compressed under heat. Very fragile old maps can thus be preserved. The work has to be done in special workshops, such as those in the Library of Congress or the National Archives in Washington. There are specialists in this art in most large cities. This method is obviously expensive, but safer and cheaper than framing maps under glass. Embedding frequently used maps is some-

what dangerous on account of possible breakage.

Microfilms. Microfilming maps is less successful than filming books. Most maps are colored; they are too large; and many have too fine lettering. They can be reproduced better by 35 color film, but to be legible, a single film rarely can show more than a square foot of map. To use a map with a projector is sometimes awkward. Yet often this is the best way to obtain a copy of a map from a distant library.

Map Compilation

Given a map problem, sometimes we are so eager to start drawing that we may skimp on our preliminary studies. Yet work and time spent on proper compilation amply repay us by producing a better map. Too often we start with an imperfect base map, find later that much better sources are available, and have to start the map all over.

Much of the latest information may be in books and periodicals. For instance, for a map of the Syrian Desert, there were useful references in a book of Musil's travels, in the various maps of the *Geographical Journal* of the Royal Geographic Society, and in a British air pilot's handbook. For base maps, the World Aeronautical Charts were helpful. These references, combined with air photos and field notes,[2] contained enough material to prepare a 1:750,000 map. If less data had been available, a map on a smaller scale would have been chosen.

For landform map of Mexico, the author's approach was different. First, all avail-

[2] Henry Field, The North Arabian Desert, Archaeological Expeditions of Peabody Museum, 1956, Cambridge, Mass.

able governmental and private maps of Mexico were collected here, as well as abroad. A surprising amount of information came from school textbooks of the individual state sof Mexico. The "Atlas de México," by Jorge Tamayo, was acquired. The volumes and maps of the Sociedad Mexicana de Geografía y Estadistica were consulted and some of the old maps had excellent topographic detail of the Valle de México. The Instituto Geológico and the Instituto Geográfico of the University of Mexico had many articles, a new geological map, and some relief models. But the chief source of information was a complete set of trimetrogon air photographs. They were used as described in Chap. 23, by transferring the information with geostenographic symbols to WAC sheets and from there to the 1:2,500,000 manuscript map.

Every map problem is different and one cannot be thorough enough in advance research. This takes less time than changing the completed map when new material is found.

At present the mapping of the world is going through a thorough change brought about by aerial photography. Every effort to obtain maps containing this new detail must be made. If no such maps are available, we may try to obtain the air photos and use them directly.[3]

[3] E. Raisz, Direct Use of Oblique Air Photos for Small-scale Maps, *Surveying and Mapping*, vol. 13, 1953, pp. 496–499.

Collecting Maps

Collecting maps is a fascinating hobby. For the value contained in most of our government maps, particularly in the topographic sheets, they are almost a giveaway Our home region acquires a new meaning if we buy all the topographic sheets and geological maps pertaining to it. A geography student should never travel without maps covering the area of his travels.

Useful maps can be obtained free of charge from chambers of commerce. Maps from newspapers and periodicals are useful. Older editions of the finest atlases can be bought cheaply at second-hand bookstores. Many libraries give away duplicates. Auto road maps can be obtained from the AAA or from any filling station. Many large industrial companies have maps for free distribution which are listed from time to time in the *Journal of Geography*. The tourist bureaus of the various nations have beautiful maps for the asking.

Many of us may be interested in collecting old maps. Some of them are of great artistic and historical value. One may pick up gems at second-hand book stores cheaply. Certain caution is necessary: Europe is flooded with cheap reproductions of old decorative maps, mostly works of the sixteenth- and seventeenth-century Dutch masters, and many of these have been brought over here by tourists. They can be easily detected, however, by their harsh coloring and coarse lines.

12

Map Design and Layout

Most people are fascinated by maps. It stirs the imagination to see thousands of square miles of land pictured on a small piece of paper. People like to hang maps on walls for their scientific and artistic quality. We may count on the interest in maps, but we also have to present the information truly and harmoniously arranged, with clarity, simplicity, and a touch of beauty. All this depends on the map maker's ability in design and layout. The two are closely interrelated. *Design* means the over-all plan and style of the map, which we have to decide before we draw a line; *layout* means arranging on paper what has been already planned. The two cannot be separated. For some, the imagination does not work unless a few lines have been sketched on a piece of paper—even the most imaginative artist gets new ideas while laying out the map.

Design

The Mental Map.[1] When we explain to a stranger how to get to the railway station,

[1] John E. Dornbach presented an excellent paper on this subject at the Pittsburgh meeting of the Association of American Geographers in 1959.

we translate into words the map which is in our minds. We are likely to accompany our words with gestures similar to drawing a map. When a blind person moves around in his room without hitting any furniture, he uses a mental map formed by many previous bumps. When the Eskimo made his relief model, he made a mental map of hundreds of visual images (see Fig. Int. 2). In each of these, the map was formed by personal experience synthesizing hundreds of impressions of distances, directions, turns, and landmarks into a mental image. There are big differences between people in their ability to form mental maps. Very few of us would do as well as the afore-mentioned Eskimo.

When we have to form a mental map of large regions mostly outside our personal experience, the process is more complex. Only people who travel by land or air a great deal think of large areas as actual expanses of land. Most people visualize a region as they have seen it on existing maps. When they made a map, explorers and map makers always thought in terms of actual distances and directions. They usually oriented their maps in the direction away from themselves, as John Foster pictured New England with the West on top (see Fig. 14.1). When this author was asked to draw a panoramic section of Mexico, he did it with Monterey on the left and Lower California on the right because that is the way he looks at Mexico from Cambridge, Massachusetts. He was asked, however, to redraw the section because this is not the way people see Mexico on maps. When maps are on such small scale that they are beyond personal experience, people do not regard them as pictures of land but as diagrams of locations. That map and reality became so far apart is partly due to the fact that our maps do not suggest the forms and colors of actual country very much. When Hal Shelton or Franz Hölzel painted their maps, they actually flew over the land and studied its appearance. One cannot fail to get the impression of real country from their maps.

The Mental Design. We were asked, for instance, to draw a map of Egypt for a college textbook in geography for one 4- by 6-in. page. The first question is: What should it show? What are the most essential geographic factors of Egypt? We know that it is an oasis in the desert along the Nile Valley and Delta. No less important is the location of the Suez Canal. These two features have to stand out. Immediately, the rather square shape of Egypt as a political unit comes to our mind; if drawn on the elongated page, this would leave much unused space, and the Nile Valley would appear very small. Could we regard Egypt as a historical and geographic entity, omit a part of the western desert, and draw the Nile Valley on the full height of the page? For this we must consult the author. With his agreement, we can start thinking about how to plan the map. First the fertile area has to stand out; we may try a symbol suggesting plowed fields. This should be in contrast to the desert, which should be light-colored and show clearly that it has, fortunately for Egypt, relatively little sand. The sharp edge of the plateaus and the Nubian Mountains can well be shown by simplified landform symbols.

What technique should we use? We un- derstand that the map will be reproduced by letter press, on which halftone reproduction often appears somewhat dull; thus, pen and ink, with not too fine lines, is preferable. How should we show land and sea? Here we decide upon the greatest possible contrast between desert and water and make the sea black. This would also emphasize the Suez Canal. There are only a few names on the map, and hand lettering seems to be the most economical. We will need a cartouche for title, scale, and legend. This has to be placed where it interferes least with the important features. Actual size and shape have to be left for the layout.

What style should be used? A text for college students should have clarity without much artistry, but we may make the cartouche slightly resemble an Egyptian column. All these decisions were made mentally before a line was drawn. Next, we collect source material (various topographic and geologic maps), make a list of names, and are ready for the actual layout.

Thus, our design or over-all plan for the map is formed by facing the following questions and problems:

1. Considerations
   a. The task. Is it definitely prescribed or can we suggest improvements?
   b. The interests of the users of the map. What are their ages and education?
2. General outline
   a. What area should we include?
   b. How should we fit the area into the available space, and what will be our scale? What projection should we use?
   c. What are the essential features to emphasize?
3. Technique
   a. What technique should we use?
   b. What style should we follow? What kind of border, legend, insets, etc., should we use? How should we show relief, land, and sea?

Fig. 12.1. In the design of a map, size, shape, content, style, and many other factors must be considered before the actual layout. (*From Preston E. James, "Outline of Geography," Ginn & Company.*)

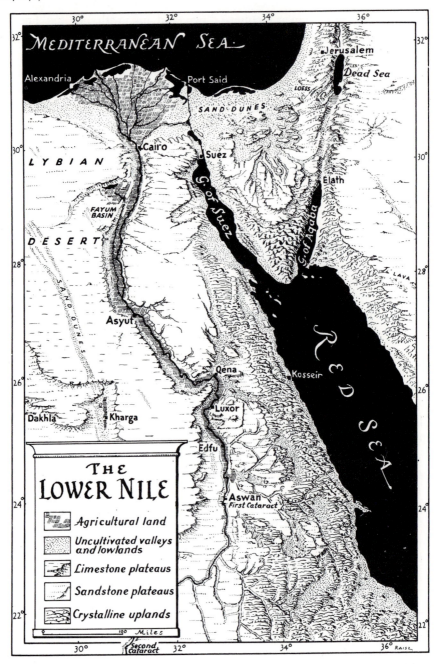

c. How much and what kind of lettering should we put on the map?

d. How will the map be reproduced? Are shades and colors permissible?

e. How much larger than publication size should we draw the manuscript?

4. Where can we find material for compilation?

5. How much time is available?

These considerations are not necessarily made in the above order. For instance, to use color or not may be one of our first considerations. Very often our choice is not free. The map has to fit detailed specifications or conform to a preestablished style.

The most common shortcoming of students is that they sit down right away to draw a map without thinking over all the factors required to form a mental map. For an experienced cartographer, all the considerations are more or less instantaneous and, given a task, he seems almost immediately to take a piece of tracing paper and start to block out a map. Yet, while blocking it out, he goes through all the above questions in his mind, supported by the visual image on the paper. Nothing is wrong with this, as long as he is willing to discard his first design and try a different arrangement if it is warranted. This sketching process helps us to achieve *balance*, a harmonious arrangement of light and shade. We will further develop this in the choice of our symbols, lettering, legend, and border as the layout progresses.

Layout

After all decisions are made, we compile material as described in Chap. 11 and are ready for the layout of the map. First, with pencil or charcoal a rough layout is made on tracing paper. The following pages are intended to give some directions, but it is more important for the student to think

over his special requirements than to follow strict directives. The first considerations have to be size and shape.

Size and Shape. A map may be limited by the page size of a book or periodical; it may have to fit the typewritten sheet of a thesis. Even if our choice of size is free, we have to consider such things as the practical number of folds if the map goes in a book, the size of the printing press, or the size of the usual library folder. Large maps which are awkward to handle may better be divided into separate sheets. Maps to be held open in the hands should not be wider than 2 by 3 ft, but charts and plans can be larger. Wall maps are limited by the available wall space in schools, which is rarely more than 5 by 7 ft.

The shape of a map depends on the shape of the area. Here we may run into problems. For instance, we have to fit a map of Chile onto a book page. The maker of small-scale maps is always eager to "gain scale." Shall we cut Chile in two and show it as in Fig. 12.2b, or shall we preserve the

Fig. 12.2. Two different rough layouts for a map of Chile. The choice depends upon the purpose of the map and the amount of detail to be shown.

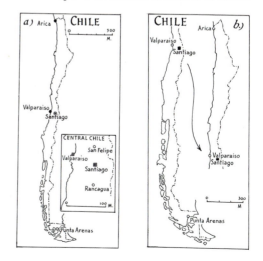

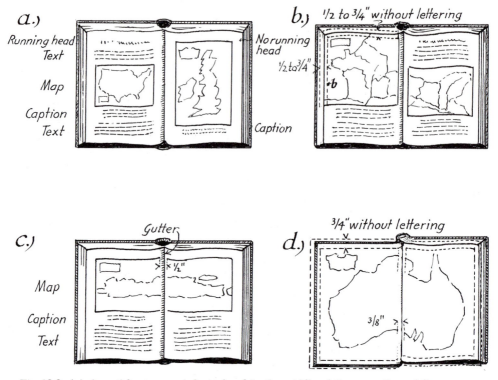

Fig. 12.3. (*a*) A partial-page map is best placed in the middle of the page. Some full-page maps will go up and down the page, others will go sideways.

    (*b*) The map on the left is "bled," carried out to the edge of the paper. The map is drawn at least ¼ in. larger. No lettering or important feature should appear in the outer ½ to ¾ in. The map on the right is "built out" beyond the type. This should be done only in emergency.

    (*c*) Very wide maps can be put across two pages, allowing for a ½-in. pulling apart in the gutter.

    (*d*) End-paper maps are often bled all around. A ⅜-in. interruption for gutter is sufficient.

unity of the country and add some insets as in Fig. 12.2*a*? Our decision will depend upon the detail and type of information required.

    Page sizes of books and periodicals allow for some latitude. Sometimes only the type page can be used, as this is easier for the printer and does not present problems in recutting and rebinding school books. Sometimes, however, it is possible to "bleed" the maps, to bring them out to the very end of the paper. In this case the map is extended ¼ in. (printing size) beyond the page size for possible cutting. However, no lettering should appear in this extension or in another ¼ in. inside the bleed line; this allows for some latitude in cutting. All the essential parts of the map have to be inside *b* of Fig. 12.3*b*.

    In the case of double-page maps, we extend the map in the middle as far as ¼ in. from the "gutter," as in Figs. 12.3*c* and 12.3*d*. The eye readily carries over, and the interruption is not a serious liability.

No small lettering should be divided by the guttter, but large spread lettering reads well over to the other side.

It is quite customary to place maps on the *cover linings (end papers)* of books, as in Fig. 12.9. There they are easy to find and have a larger spread. They can be printed separately by offset, and time is gained because the maps are not required until the binding of the book. Librarians do not like them because library cards cover up part of the map, and in rebinding the map is lost. Thus it is good to have the same map at both ends and to supply some extra copies for rebinding. These maps are usually bled, allowing a ¾-in. safety margin. The color of end-paper maps should harmonize with the binding of the book. The gutter can be narrow—⅜ in. is enough.

Folded maps can go in a pocket or be bound into the book. Pocket maps are easier to handle but often lost. Bound-in maps are awkward and tear easily. For folding, the general rule is to fold the map accordionwise, first vertically, then horizontally.[2]

Guide Maps. If no serviceable existing map can be found of an area, maps are laid out by parallels and meridians in a suitable projection. The method for doing this is discussed in Chaps. 17 and 18. In the author's experience, the time spent in drawing one's own projection is usually repaid in a

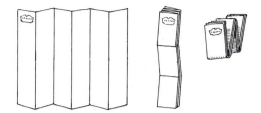

Fig. 12.4. Maps should be folded like an accordion, first vertically, then horizontally.

better map. Often, however, we find an older map of the area which is good enough for use as a guide. We first look for government maps which are not copyrighted. If we use private maps, we have to write for permission from the copyright owners. If the map has been used for an unpublished thesis, this is not necessary, but courtesy demands at least an acknowledgment.

We look for guide maps on a larger scale than our own. Imperfections of shorelines, rivers, and locations will show less if reduced. Rarely is the best larger map of the area on the chosen scale. More often we will have to use photostats, pantograph, or the square method for changing the scale. As a rule, we lay out a map with the central meridian vertical. Our central meridian may appear on our guide map as a curved line. In this case, we have to redraw the desired area, changing the parallels and meridians. It helps to draw diagonals connecting the intersections of parallels and meridians on both the guide map and our own. These diagonals may also be curved lines conforming to curved meridians and parallels.

Rough Pencil Layout. Whether we construct a projection or copy its outlines from an existing map, it is best to make a rough layout on tracing paper. Rarely will the

[2] In "General Cartography," the author recommends folding the map horizontally first and then vertically. It has been found, however, that it is easier to hold a vertically elongated map than a horizontally elongated one. It fits a pocket better, and one may open the map at any place and page it like a book. For more elaborate foldings, consult Helmut Muhle, "Folding of Maps," Second International Cartography Conference, Rand McNally & Company, Chicago.

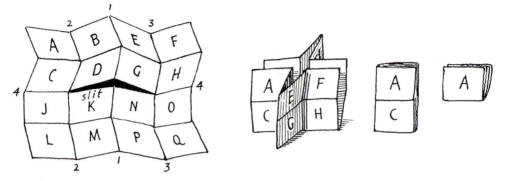

Fig. 12.5. British Army road maps can be opened at any page or at any two or four adjacent pages. The reverse side is printed too.

desired area fit into the available space, but the empty places often allow for good composition. The examples in Figs. 12.6 and 12.9 show how to use insets, pictures, cartouches, legends, compasses, etc. The tracing-paper layout is not wasted effort. However roughly laid out, it helps to visualize the map as a whole and to make a work plan.

Enlargement. We draw the map 1½ or 2 times larger than publication size so that the map will look more finished in reduc-

Fig. 12.6. The proportions of the enlarged drawing can be figured either by a diagonal or by calculation with the help of a slide rule.

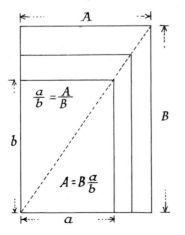

tion. The amount of enlargement depends on the draftsman. The work of some cartographers is so fine that hardly any enlargement is needed; others like to work 2 to 3 times larger than scale. Such enlargement is not only uneconomical in time; it also may result in a map that looks weak in publication size. Maps drawn by experts according to a well-established system are often not enlarged at all, as in negative scribing (see Chap. 27).

When a more or less square map has to be fitted into an elongated page, the width will be the "controlling side," as there is plenty of space for the other vertical side. If we enlarge the width of the map (*b* to *B* in Fig. 12.6) and want to know how large to draw the height, we proceed as in the figure. This can be done with a slide rule.

Title. The title should give the name of the area, the type of the map, the name of the author, the year of publication, the scale, and remarks. It should have letters large enough to be the first feature to strike the eye. The title is best placed at the upper left corner, but it can go wherever there is space.

The lettering is usually centered around a vertical axis, as in Fig. 12.7. We count the letters and lay them out forward and back from the center line. To make the title stand out more, we often encase it in a cartouche. On old maps we often see most elaborate cartouches, decorated with coats of arms and scenes and products of the country. At present, tastes are simpler, but some indication of the style of the region enlivens the map.

Legend. Every symbol and abbreviation which is not obvious should be explained in the legend. It is not necessary, for instance, to explain the usual railroad symbol unless normal- and narrow-gauge railways are differentiated. A glossary of foreign terms and a pronunciation index may be helpful.

Scale. Bar scales are preferable to numerical scales, as the map may change its size by various photographic processes. In maps showing foreign countries, a kilometer scale should be added:

$$1 \text{ mile} = 1.61 \text{ km}$$
$$1 \text{ km} = 0.62 \text{ mile}$$

Scale extensions or very small divisions are only for large-scale maps. On small-scale maps they may give a false impression that one can measure distances accurately on every part of the map.

Border. Except for some informal sketch or bled maps, most maps have a border. The most common type consists of two lines about ¼ or ½ in. apart; between these lines are placed the degree numbers of parallels and meridians. The space between inner and outer border lines is sometimes used for completing spread names of areas, part of which are outside the map, as in Fig. 5.1. The inner border is called the *neatline*, and should contain the map. It is permissible

Fig. 12.7. Pencil layout of a title. Letters are spaced left and right from the center line.

to cut into the inner border with some protruding parts of the map, as in Fig. 12.1, and thus gain scale; the outer lines should remain intact, as a rule. In Victorian times, most elaborate decorative borders appeared on maps, but this is no longer in fashion.

The margin outside the border is at least 1 in. on large maps and ½ in. on smaller ones. On serial maps, where marginal information is given outside the borders, at least 1 in. of clear margin is necessary to preserve the map from damage. A very wide margin is just a waste of paper and may make storing more difficult.

Insets. Inset maps are of different kinds. On most maps there are certain important regions which are often shown in insets of larger scale. If the main map is of an area not generally known or if we want to show general relationships, we often have an inset showing the location of the area within the country. In this case the inset is of smaller scale, as in Fig. 12.9. Another type is the "panhandle" inset. If the area has a long extension, as in the cases of Oklahoma, Florida, or Alaska, we may gain scale in this way. For instance, in a map of Alaska, the outer Aleutian Islands may be shown in an inset. In this case, the inset is usually on the same scale as the main map.

Fillers. A map may be enlivened if airplane views, landscape sketches, or pictures are added. Statistical graphs and tables are common in economic maps. Indeed, on some maps the garnishing may occupy more space than the main map. Decorative border panels with pictures of cities or people were common on old maps, and are still used sometimes if the main map does not fit the page (see Fig. Int. 11).

Compass Roses. On maps oriented with North on top, compass roses are not neces- sary as the parallels and meridians will show the cardinal directions. They are often included as decorations, however. On charts, compass roses help navigation and show the magnetic declination. On most topographic sheets, true, magnetic, and grid north are indicated (see Chap. 17).

The foregoing may be illustrated by analyzing the design and layout of Fig. 12.9. This map illustrated Bruce Lancaster's *Bright to the Wanderer,* a novel based on the story of the Upper Canadian Rebel-

Fig. 12.8. Decorative compass roses have been common on maps since the time of the portolan charts. The compass rose in the upper center contains the initials of the names of winds in medieval Italian: Borea, Grecco, Scirocco, Ostro, Lebedo, Ponente, Maestro. The cross stands for East. On the lower right is a marginal scale indicator. (*From Morison, "Admiral of the Ocean Sea," Little, Brown & Company,* 1942.)

lion of 1837.[3] As this map was to serve as constant reference for the reader, we decided to place it on the cover lining; this also allowed for larger scale.

Most of the story centered on Toronto but the map also had to show Montgomery Tavern and Nipigon Lodge many miles away. Thus we decided to use a perspective map, in which the foreground can be shown on larger scale. For showing the flight of the rebels and the relation of Upper and Lower Canada, we planned an inset. The style of the map should be that of the mid-nineteenth century; the technique—pen and ink, with cellotint on the lake. The perspective view calls for semipictorial handling of symbols and relief. Lancaster obtained an old map and pictures of contemporary buildings from various sources in Toronto. The cartographer read the galley proofs of the book and listed all geographic names. For the actual layout, we took an old map of Toronto. We drew a square grid on the map and laid out a similar grid, but in oblique perspective (see Fig. 22.3); upon this we laid out the map. There was not much choice as to where to place the inset and the cartouche. Not only is the coat of arms of Upper Canada of historical interest; it also helped to balance the map, as the center of interest is concentrated at the bottom of the map. A compass rose in perspective helps orientation. Note that all the lettering is laid down in true perspective. The map was printed in purple, as this is the color of the binding. The paper is cream-colored.

Fine Drawing. After the rough layout on tracing paper is completed, we are ready to make our fine drawing with pencil on drawing paper.

Preliminary Work. We cut the paper to size, allowing for large margins. We first draw the parallels and meridians and the neatline, followed by shorelines, rivers, roads, and spot locations. Then we draw symbols, mountains, and special features. Only after this comes the lettering, as we want to utilize uncluttered places. Letters are laid out lightly, but the individual letters need not be perfect in form; that will come in inking. Next we draw guide lines for lettering. Border, title, scale, legend, and insets will conclude the pencil drafting. Border, title, scale, and perhaps horizontal and vertical parallels and meridians are laid out with a T square and triangle on fixed paper; otherwise the map need not be attached to the drawing table. All lines are drawn very lightly with a pencil that is not too hard, but sharp. Lines wedged into the paper cause trouble in inking. *No inking should be attempted before the map is completely ready in pencil.*

Inking. We clean the pencil drawing with eraser powder, dust it off, remove any thumbtacks so that the paper moves freely, and we are ready for inking.

Inking is done in the opposite order from the penciling. First we ink the lettering, as this has the right of way over everything else. Then come borders, title, legend, and scale, followed by symbols, shorelines, rivers, boundaries, and relief; and finally, parallels and meridians. The sequence is important; a meridian is interrupted to give way to a river, and a river is interrupted to give way for a name. On colored maps, however, the lines are not interrupted, but the darkness of colors is

[3] Little, Brown & Company, Boston, 1942.

Fig. 12.9. Cover-lining map with inset. The original drawing was ½ in. larger all around to allow for cutting. (*From Bruce Lancaster, "Bright to the Wanderer," Little, Brown & Company, 1942.*)

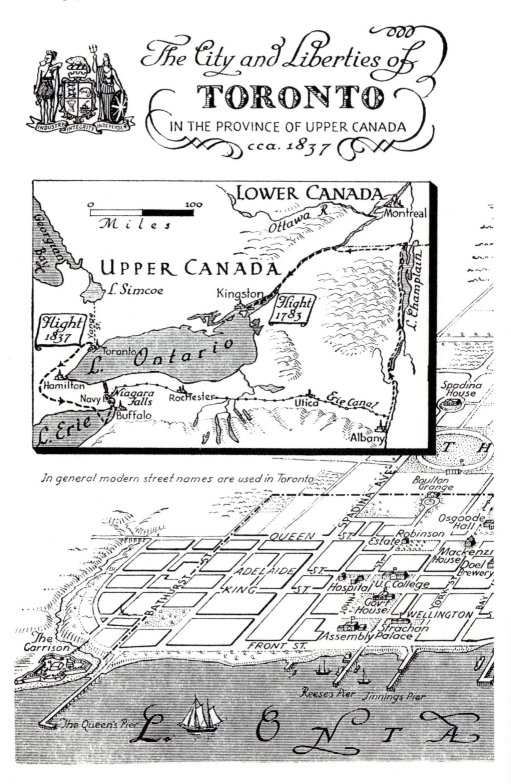

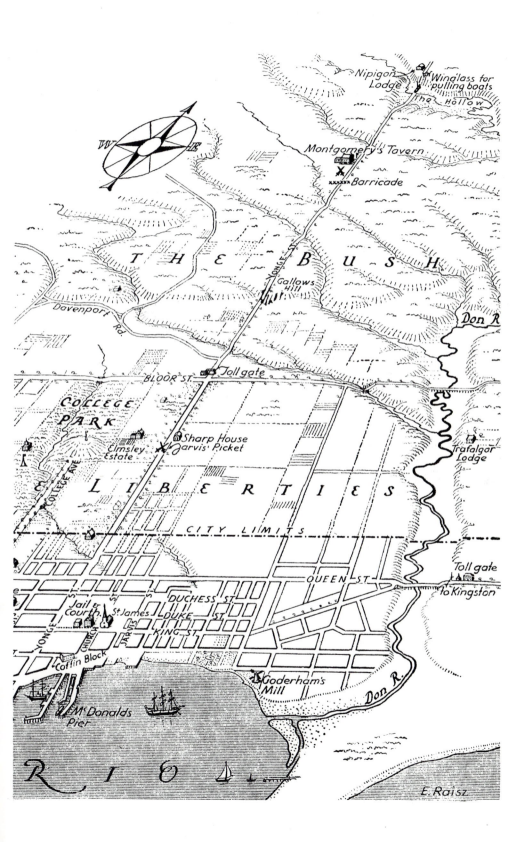

Nipigon Lodge

Windlass for pulling boats

The Hollow

Montgomery's Tavern

Barricade

W E

THE BUSH

Gallows Hill

Don R.

Davenport Rd.

BLOOR ST. Toll gate

COLLEGE PARK

Sharp House
Elmsley Estate
Jarvis' Picket

Trafalgar Lodge

LIBERTIES

CITY LIMITS

COLLEGE AVE

Toll gate
to Kingston

QUEEN ST.

DUCHESS ST.
DUKE ST.
KING ST.

Jail & Courth. St. James

YONGE
CHURCH
JARVIS

Coffin Block

Goderham's Mill

Don R.

McDonalds Pier

RIO

E. Raisz.

chosen in the above sequence. Black letters will show over red boundaries or blue rivers. Inking is done with the tools described in Chap. 1. As dozens of tools are used, the importance of wiping them clean immediately after use cannot be overemphasized. A copper wire mesh, used for scrubbing dishes, immersed in a small glass of water is good for wiping pens.

Completing the Map. After inking, the pencil lines are erased. This is done gently, with a soft eraser; otherwise the ink lines may pale too. It is not good to use very smooth paper because ink lines rub off easily. Next, we touch up imperfections and paint out unnecessary ink lines with white. Only after this can we apply cellotints; these make further corrections on the covered areas almost impossible.

The map is trimmed to size, mounted lightly on cardboard, and covered with cellophane. We mark the reduction to publication size on one side, and note "to 7½ in.," for instance. We mark only one side, as the reduction is done photographically. For the engraver, we indicate the number of copies required, the grade of paper, the size of margins, the address for delivery, and where to send the manuscript and the negative. If the map is to be folded, a folding scheme is included.

Exercise 12.1 *

Make a general medium-scale map of a selected region 50 to 100 miles wide on a 1:250,000 to 1:1,000,000 scale. Make your own compilation, design, and layout. Relief can be drawn either by landform symbols or by contours; in this case, add plastic shading. Add insets of interesting places. Make a well-balanced composition of the main map, insets, title cartouche, and border. Parallels and meridians are indicated on the borders only. The sequence of preparation is as follows:

a. Lay out the composition of the map roughly in pencil on tracing paper.
b. Transfer the composition to drawing paper and, with pencil, draw every detail of the map. Use very light lines.
c. Clean the map with eraser powder.
d. Ink in lettering and borders with black ink.
e. Ink in rivers, lakes, and shorelines with blue waterproof ink.
f. Ink in relief with brown.
g. Erase all traces of pencil lines.

---

* NOTE TO THE INSTRUCTOR: It is better if each student does a diffeernt region, so long as the tasks are comparable in size.

13

Lines, Shades, and Colors

A map is composed of dots, lines, shades, and often colors. They express land and water, mountains and valleys, roads and cities, and sometimes such items as changes of population. The mental connection between geographical features and their expression on maps is a complex one. Not only must the cartographer make this connection to his own satisfaction; he must make it so that others will react to his drawing in the right way.

Lines

One can make a map without shades and colors, but not without lines. The earliest maps had little but lines. Lines are used for rivers, shorelines, roads, borders, scales, parallels and meridians, and also for letters. Lines can be thick or thin, straight or curvy, full or dashed, single or double; thus the map maker has a wide choice. Yet maps often consist of monotonous, confused sets of lines, very often without variation in thickness or in shape.

Some choices are obvious, and many of

them we discussed in connection with symbols in Chap. 3. Parallels and meridians are hairlines, but not so thin that they should fail in printing. Rivers on small-scale maps are irregular, curvy lines, growing thicker downstream; on large-scale maps, however, they have their true shorelines. Roads are single or double lines, straight or of even curvature. Boundaries are usually dash-dot lines. Neatlines are thin; outer-border lines are often heavy or even double. Contour lines, however, are even in thickness, except every fourth or fifth is made heavier for easier reading.

Shorelines must be thin; one could not show small islands or small lakes and off-shore bars if the lines are wider than the feature. Artists, unfamiliar with geography, like a "raised and shaded" line for shores; the lines on the South and East shores are heavier, so that the land areas look as if they were cut out of cardboard and pasted over the sea. The result is that places like Miami, New Orleans, or Venice appear to be on top of a 1,000-ft cliff. As much of the land is bordered by flat coastal plains, a geographer frowns on this device.

The thickness of a line may express the importance of the feature. A state boundary will be heavier than a county boundary, and a large river will have a heavier line than will a smaller river. The thinnest line we may use on maps is about $\frac{1}{250}$ (0.004) in. Offset printing (see Chap. 14) can produce even thinner visible lines than this, but we cannot rely upon it. In a zinc cut (see

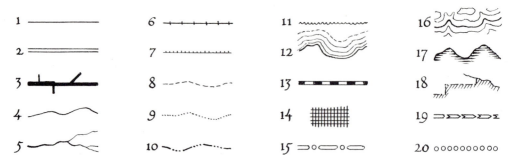

Fig. 13.1. Here are 20 lines of different type, shape, and thickness. For what can each of these be used on a map? Make a list.

Chap. 14), the very fine lines may fail to reproduce at all, and we should not go under 0.006 in. Too fine lines tend to gain thickness in printing, as the engraver has to overexpose the whole map. Colored lines need to be heavier than black lines in proportion to their darkness. Dashed lines, and dotted lines particularly, must be heavier; otherwise they may fail to reproduce when cut into zinc. White lines on black need to be twice as thick as black lines on white. As we make our maps 1½ to 2 times larger, we have to draw proportionately heavier lines. The engraver can make all the lines heavier or lighter by photographic methods if desired.

Our tools are all grades and kinds of pencils and pens, ruling pens, pivot pens,

Fig. 13.2. The thinnest line which is safe to use on a map, drawn 1½ times publication size, is 0.008 in., or 0.2 mm, thick. The numbers below the lines indicate millimeters; the ones above are hundredths of inches. (*Courtesy of Pelican Graphos Pens.*)

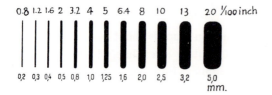

road pens, Payzant or Leroy pens. The finest and most even line can be made by negative scribing on coated plastics with special tools (see Chap. 27). They are, however, rarely available in departments of geography. Pivot pens make good contour lines but are hard to handle for other kinds of lines. Straight lines are easily made in any thickness with a ruling pen and straightedge. Leroy or Wrico pens are graded by numbers, but the thinnest line is only about 0.01 in. Thus, for most line work, the cartographer still relies on his set of pens, graded by thickness and each in its own holder. A little ink-retainer wire (see Fig. 6.2) will help keep the lines fine and even. Here are some rules for achieving fine and even lines:

1. Keep the pen clean. Wash the pen thoroughly before starting, and use india-ink solvent if available. Before each filling or dip, clean the pen with a rag or wet copper wire mesh.
2. Dip or fill often in order to have an even amount of ink in your pen. Some keep the ink bottle closed and fill with a dropper; some refill a tiny bottle of ink every week to only ¼ in. and dip the pen, thus taking a small amount of ink each time.
3. Pens wear down and their points get stubbed, making heavier lines on the downstroke. A honing stone will help this.

4. *Do not press on the pen.* Clean and dip instead.
5. Try not to go over the same line twice. Use distinct, definite strokes and fill in gaps later, rather than painting lines.

The little sketch in Fig. 13.3 shows how many features can be expressed by lines. This is so much more remarkable, as nature does not contain many lines. For instance, the shoreline and the edge of a cliff are not real *lines* in nature, but the intersections of two surfaces. A landscape painter may record the same picture as the sketch shows without any lines, but his painting would be much larger. It is a psychovisual fact that we can recognize features from a few characteristic lines which remind us of the shapes and textures of familiar objects. Cartographic symbolism is possible because of this ability.

Shades

However judiciously the lines on a map are chosen, it still looks flat and unattractive unless the areas are differentiated by shades or colors. Colors are obviously preferable, and we use them whenever we can afford it. As geographic publications rarely can afford colors, we try to express as much as we can with shades. By "shades," printers mean the varying degrees of darkness. A photographer would call them "tones." Shades in printing are reproduced by dots and patterns; thus they always have an element of texture.

On an air photo, tones are usually more apparent than lines. Here the tones mean the light-reflecting quality of the surface. The textures of the tones, such as forest, grass, etc. (see Chap. 2), differentiate between various types of lands, and should be studied when making shading patterns.

Fig. 13.3. Lines on a landform drawing may mean many things: (1) river, (2) profile of hills, (3) edge of cliff, (4) structures, (5) hachure lines, (6) shoreline and reflection, (7) cultivation, (8) vegetation. Note the difference in handling these lines.

Shades on maps may show variation in certain elements of the landscape—in land use, in vegetation, in political or social divisions, or gradations of one factor, such as altitude, rainfall, temperature, population.

Gradations. Gradations are best expressed by steps of even flat shades of increasing darkness, separated by isopleths (lines of equal values; see Chap. 14) or contour lines. Actual gradation from light to dark without steps may be nearer the truth, but it is difficult to produce and reproduce, as it requires the skill of an artist with a brush or airbrush, and has to be reproduced by the halftone process. As neither the application nor the reproduction ensures complete results, it is rarely used by cartographers. The use of even dots or parallel lines is much more common. We can easily make about ten gradations between white and black. Figure 13.4 shows such gradations using cellotones.

One would think that if we carefully measured the areas of black in relation to white, we could arrive at an even gradation of shades. This is not true, however, according to tests made by Robert L. Williams.[1] The eye can differentiate the lighter

[1] R. L. Williams, Map Symbols, Equal-appearing Intervals for Printed Screens, *Annals Assn. Am. Geog.*, vol. 48, 1958, pp. 132–139.

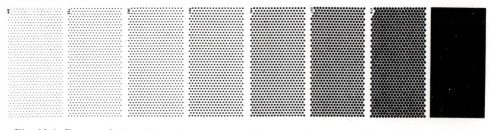

Fig. 13.4. Even gradations. Since there are imperfections in the commercially available tints and distortions in the reproduction processes, it is most difficult to achieve perfect gradations. (*By Robert L. Williams.*)

more easily than the darker tones; thus in the lighter portions a smaller percentage change is necessary than in the darker shades. For example, for 10 equal-appearing shade differences between white and black, the percentage value of black areas ranks about 2.5, 8, 17, 28.5, 43, 58, 73, 85, 94, 100. These values are expressed in his "curve of the gray spectrum" in Fig. 13.5.

Further tests by George F. Jenks disclosed that map users prefer fine textures, and that dot shadings are more pleasing than line shading. Irregular shadings were liked the least. One should be particularly cautious about close crosshatching. In re-

Fig. 13.5. Curve of the gray spectrum shows how to proportion black and white space in dot or line shading to achieve even-looking gradations. (*By Robert L. Williams.*)

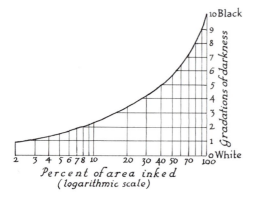

production it easily darkens or clogs, making black patches.

On nonstatistical maps, gradations are less important than the contrast in character. Here we may combine shading with symbolization. Figure 1.5 shows some mechanical patterns for sea, forest, bush, swamp, granite, etc. See also Fig. 8.1 for hand-drawn patterns for vegetation and cultivation. Cellotone companies offer a rich assortment of map patterns, yet the poorest hand-drawn pattern which expresses the nature of the land is worth more than a perfect mechanical pattern of indifferent nature. At one meeting, a geographer presented a map on which orange groves were represented by one kind of oblique lining and tobacco by the same line pattern slanted the other way. He did not have enough imagination to draw circles for oranges and leaves for the tobacco.

Application. Stippling and patterns are applied by hand or mechanically. Dot patterns are easily drawn by hand. The dots are best placed irregularly but with even density. One way of doing this is to draw a wavy line of evenly placed dots around and crosswise through an area, then to fill in the remaining empty places evenly by eye. Parallel lining is more difficult to draw freehand, but quite possible. It is safer, however, to rely on a *section liner*, a mechanical device by which a rule can be set back parallel to itself by even steps.

Most patterns are applied by using various cellotone sheets. The patterns are printed on the underside of cellophane which is coated with adhesive, such as Zip-a-tone, Craftone, Contak, etc. Application is described in Chap. 2. Each company produces a great variety of patterns. When the tints are applied to an enlarged manuscript map, we may have surprises in the printed reduction. Pattern books usually show how the design will look at various reductions. Patterns that are too closely dotted may show up not only darker, but patchy. A hundred dots per inch in publication size is about the limit.

Fig. 13.6. Relief of Baluchistan shown with Craftint Doubletone paper. Two developers bring out two hidden patterns. Some hand dotting is added. (*Courtesy of Henry Field.*)

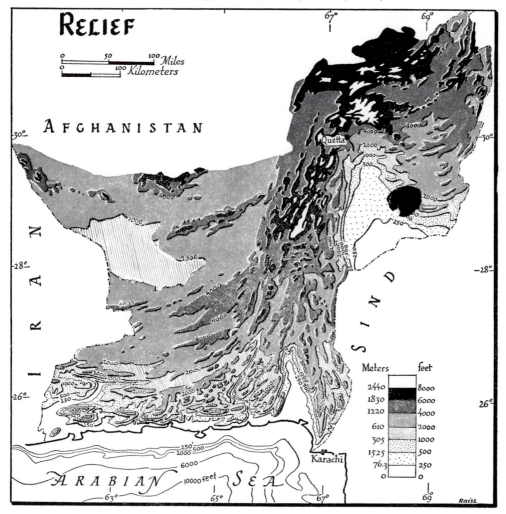

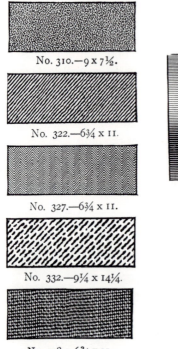

No. 310.—9 x 7½.

No. 322.—6¾ x 11.

No. 327.—6¾ x 11.

No. 332.—9¼ x 14¼.

No. 338.—6¾ x 11.

Fig. 13.7. Ben-day shading is applied by the engraver to the metal printing plate itself. The cartographer indicates the patterns on an overlay.

Craftint Singletone and Doubletone drawing papers have an invisible pattern printed on them which can be made black by special developers. The advantage of these papers is that one can easily make intricate designs of shading by applying the developer with pen or ruling pen. The line work and lettering are inked on the same paper. Singletone has one pattern while Doubletone has two patterns brought out by different developers. Ross board (see Chap. 1) is better for shading than for even tones.

Ben-day shading is applied on the metal printing plate by the engraver by rolling ink over various templates. All the cartographer has to do is outline on a trans-

parent overlay the areas to be shaded and mark the index number of the pattern desired. Ben-day pattern books can be obtained from an engraver. This method is somewhat more expensive, but the patterns can be most delicate. The operator is able to produce transitional gradations in the shading by a slight twist of the template.

The simplest method of shading is to paint or airbrush the original drawing to any shade. This, however, has to be reproduced by halftone engraving, which has no clear white or black (see Chap. 14).

Vignetting (fading out colors) can be done many ways. The simplest way is to use hand dotting or lining and gradually thin out the dots or lines. We can also use pencil, charcoal, or airbrush and reproduce it in halftone. The Ben-day operator also can do it. Air charts utilize photomechanical vignetting, a complicated process.[2]

Colors

Color attracts the eye, and a colored map is much better understood and remembered than a black-and-white one. A map publisher had a set of colored wall maps which sold for $15 each. He published an uncolored edition of these for only $1.50 per map, yet this did not sell. Most educational and commercial maps are colored, in spite of the fact that each additional color increases the cost about 1¼ times. The average scientific map in books or periodicals seldom has color, but an unpublished thesis map may be colored by hand. Serial maps are colored according to a fixed scheme and most atlases have their own color systems. Rarely has a cartographer a chance to use color freely. Yet

[2] Described in Aeronautical Chart and Information Service *Tech. Man. R. 61,* 1955.

he should be familiar with the fundamentals of coloring, as he may be called to
design a scheme.

Physics of Colors. This book will not go
deeply into the physics of colors, particularly on account of new discoveries which
challenge the classical theory.[3] It is enough
to remember that, from the enormous range
of radiant energy which makes our uni-
verse, our eye can perceive an extremely
narrow band of wavelengths varying from
780 to 380 millimicrons (millionths of
millimeters).

According to the classical theory, the
white light of the sun can be divided into
the full colors of the rainbow by passing
through a prism, which refracts the long-
wavelength colors less than the short-wave-
length colors. Thus we have a sequence of
colors: infrared 780, red 630, orange 600,
yellow 560, green 490, blue 440, violet 380,
and ultraviolet, where the numbers mean
the limiting wavelengths in millimicrons.
Infrared and ultraviolet radiation cannot
be seen but can be photographed. Each
wavelength of light produces the same
color. All colored lights can be mixed from
the three primary colors, red, yellow, and
blue, and the mixture of all three produces
white light. Why should these three colors
be primaries? It is usually explained that
the eye has specially tuned receptors for
these colors but no such differences have
been discovered. The new theory relates
them to certain areas of the brain. According to the new theory, we need at least a
pair of wavelengths a certain distance apart
to produce stimuli for colors, and color
vision is more complex than previously
thought.

[3] E. H. Land, Experiments in Color Vision,
Scientific Am., no. 5, 1959, pp. 84–99.

Fig. 13.8. Vignetting (fading out of colors) can
be done by hand dotting, lining, the use of a
halftone screen over pencil, or airbrush work.
Sometimes a complex process is used by lighting
a dot screen sideways.

We are, however, less interested in colored lights than in reflected colors produced
by pigments, which act somewhat differently. The color of each object depends on
the color of the light that illuminates it
and the light it reflects. In white light, yellow pigment completely absorbs or sub-
tracts from white light the blue color,
magenta pigment absorbs the green, and
blue-green pigment absorbs the red. These
colors are called "subtractive-primary" or
"complementary" colors. Magenta and yellow printed over each other produce red as

Fig. 13.9. The color triangle. Each circle is
printed in the subtractive primary colors—yellow,
magenta, and blue-green. In printing two of these
over each other, we obtain the three primary
colors of the spectrum—red, green, and blue. The
combination of all three produces black.

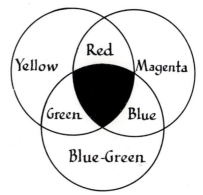

the other two primary colors are subtracted; yellow and blue-green produce green; blue-green and magenta theoretically should produce blue. (Our usual pigments produce more of a violet.) The mixture of yellow, magenta, and blue-green absorbs all primary colors and appears black. Brown can be produced by mixing orange with some black, and khaki color by adding some green to the orange.

Defining Colors. This is usually an individual matter. Different people will define "buff" or "mauve" differently. Besides, 1 man in 25 (but very few women) is color-blind to a certain extent. Most of them cannot tell red from green.

A map maker has to define his colors precisely for the printer. This he can do by painted samples or by choosing from the printer's color chart. According to the Munsell Color Chart, every color can be defined by three factors: hue, darkness (value), and chroma (intensity). There are 10 hues: yellow-red (orange), yellow, yellow-green, green, blue-green, blue, purple-blue, violet, purple, and red-purple (magenta). Sometimes, for greater refinement, the shades between the 10 hues are further subdivided into 10 steps, making 100 hues in all. Each hue is divided vertically, from white on top to black on the bottom, into 10 shades of darkness, and horizontally into from 1 to 10 shades of chroma (meaning brilliance of color), ranging from gray to the pure color. Thus about 5,000 useful colors can be defined, more than enough to satisfy a map maker. For instance, a certain turquoise blue can be specified as B.G., 5/6, meaning a blue-green halfway in darkness between black and white and of the sixth gradation of brilliance from gray to full blue-green.

The Effect of Colors. We have learned to attach certain significance to various colors. Thus, yellow is "bright"; we usually symbolize the sun as a yellow disk, in spite of the fact that sunlight is composed of red, yellow, and blue. Red stands out boldly, while blue has a certain quiet flatness to it. Brown stands out over green, and green over violet. This is not all imagination. We painted one copy of the same relief model blue and the other orange. No one hesitated to say that the orange showed the relief more boldly than the blue. One of the interesting facts is that black letters are legible over yellow more than over white.

Choice of Colors. We choose colors for truth, emphasis, contrast, and beauty, but we have also to consider the conventional symbolization. We almost always paint water blue, although the water of the Rio Grande is more like chocolate. Nobody would hesitate to paint ice white, forests green, or deserts in more reddish colors. Only recently cartographers turned to the true colors of nature. Most maps are printed in the conventional altitude tints which may place a brilliant green in parts of the Sahara Desert. This problem is discussed in Chaps. 7 and 23. In the conventional altitude tints, there is usually too great a contrast between the greens and the browns. This makes for vivid maps, but gives too much prominence to certain contour lines.

With colors we may emphasize features that are important for our maps. Thus cities, roads, and arrows for movement are usually shown in red or black, which stand out vividly and cover the other colors. Generally lines are apt to be black or red, while large areas are painted more frequently in pastel colors of lighter shades and of lesser chroma. A map in which all colors are brilliant will resemble some early geological maps, where colors run riot but are hard to read. Color emphasis can be misused. German maps showing the languages of Europe almost invariably show German-speaking areas in red and the others in less brilliant colors, while Italian maps show their own language in red.

For easier understanding, colors should have a certain amount of contrast. Seen from great height, our earth is somewhat drab, with gray-greens prevailing; even deserts do not show much brilliance. To give more distinction to fields, forest, cultivation, etc., we use more vivid colors. We even have to change the hues slightly. For instance, fields of wheat, corn, and hay are not too different in color, but in mapping them we probably will choose yellow for wheat, green for corn, and lighter green for hay.

In political maps we color the various countries in all available hues, chosen so that adjacent countries do not have the same color. We may not use a separate color for all 100 nations, but with varying hue, chroma, and darkness, we need to repeat very little. Conventional political coloring has its dangers; some children have actually been surprised when sailing to Great Britain to find that it is not pink. The gaily colored political maps appeal to the public, but they are not always an educational asset. A better method is vignetting, coloring just a narrow strip along the boundary lines, as is done on the *National Geographic Atlas* maps.

In choosing contrasting colors, we should not go too far. That which belongs together should remain together. For instance, on a general geological map, all Tertiary formations should range from an orange (Eocene) to a light yellow (Pleistocene). If we want to emphasize the difference between two Oligocene formations we may use a slight difference in shade or overprinted patterns.

All colors have to *blend in harmonious beauty*. Our maps often have harsh and clashing colors with little attention paid to the above principles. Let us study the early Dutch maps which used colors sparingly but with striking effect. The cartographer has to choose between using all colors and shades transitionally as in a painting, and using a patchwork of distinct flat colors. Fully painted maps require so-called four-color-process printing (see Chap. 14). This is more expensive, somewhat risky as to results, less distinct in appearance; it also requires a greater artistic skill from the cartographer. Most of our maps are a patchwork of distinct flat colors. Their application is discussed in the next chapter. Yet with the perfection of process printing, some very excellent maps were painted in full colors, and we hope a better portrait of the earth will emerge from these efforts.

Color Regions of the World. The Quartermaster's Research and Development Center, in its "Color Atlas of the World," divided all land into color areas differing every month (see Table 13.10) Cartographers will find this a great help in true coloration of their maps and globes. In the Munsell numbering, the Y., R., and G. designate yellow, red, and green. The numbers before these are the intermediate hues ranging from 1 to 10. The first letters mean the hue; the last two figures mean darkness and chroma. For instance, 7.5 Y.R. 6-6 reads as follows: Hue is 7.5 tenths nearer to

TABLE 13.10. COLOR REGIONS OF THE WORLD *

Color	Munsell	Per cent of world land area	Type of land
White		2–35 (Mar. to Jan.)	Snow and ice (Antarctica excluded)
Partly white		2–6 (Oct.)	Melting snow
Tan	7.5 Y.R. 6–6	29–34 (July)	Desert and semidesert (one-half in Africa), laterite, mountain soils
Earth red	2.5 Y.R. 4–6	1½–7 (July)	Cold desert (four-fifths in Central Asia), barrens
Earth brown	10 Y.R. 3–2	1–13 (May)	Grassland in winter; tundra in fall (seven-tenths in Asia)
Green	5 G.Y. 4–5	24–32 (July)	Broadleaf, trees, crops, grass in summer
Olive green	10 Y. 4–4	8–16 (Aug.)	Crops, pastures, bush, forests in fall, conifers (temperate zone)
Olive drab	3 Y. 4–4	2–4 (Aug.)	Tundra in summer, Mediterranean lands
Forest green	7.5 G.Y. 3–2	2–3 (July)	Mixed forest
Color varies		Less than 0.1	Urban

* According to the Quartermaster's Research and Development Center.

yellow-red than to red; darkness is $\frac{6}{10}$ nearer to white than to black; and chroma is $\frac{6}{10}$ nearer to full brilliance than to gray.

How to Paint an Area Evenly. This is a common task of a cartographer and not always easy. The following hints may help:

Fig. 13.11. Painting maps with transparent water colors. Numbers and arrows indicate sequence and down-slope direction of painting.

1. Divide the area to be painted in sections which can be painted rapidly, separating embayments and peninsulas.
2. Mix enough paint and try out its color. Clean your paper.
3. Place map on an inclined board and turn the map so that you paint downhill from a wide part to a narrow one.
4. Carry a "wet front" with a well-filled brush progressing first on the sides, then in the middle. Always paint downhill. Pick up extra paint at the bottom with a blotter.
5. Proceed to the next section, but do not turn the paper before paint is dry.
6. Do not overlap sections; rather leave a white space which can be filled in later with a half-dry brush.
7. Imperfections can be doctored best with an eraser.

Exercise 13.1

Add shades and colors to the map exercise of the previous chapter. Make four contact prints according to Chap. 26. To one of them apply various cellotones to show cities, forests, fields, water bodies. The second photostat should be politically colored with flat tints.

The third photostat should be hand-colored, showing land use. Contact prints sometimes repel water color and need to be rubbed with pounce before painting. The fourth map should be colored with colored sheets of cellotints. These can be combined with black patterns by having two sheets of cellotones over each other.

Fig. 13.12. Use of screen to show negative values. (A) Drawing for black lines; (B) drawing for white lines; (C) some screened; (D) combination of A and C. B can be screened by white cellotone or by making a film negative and from this a film positive over which a halftone screen is used (see Chap.14).

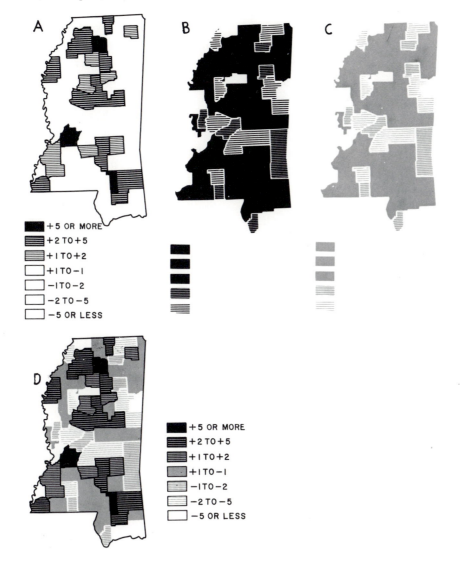

14

Map Reproduction

How often a cartographer looks with distress upon proof prints of his map! He feels that his beautiful creation has been mutilated by heavy lines, patchy surfaces, and incorrect colors in bad register with each other. His distress, however, is more than matched by that of the engraver looking at the shortcomings of the cartographer, such as too fine lettering or too dense screens, faint lines and too much reduction of the original drawing. Much of this distress could have been avoided if the two had each known more about the other's work.

History

Woodcuts. Since the late fifteenth century maps have been cut in wood and engraved in copper. Cutting into wood is laborious, but this method was used for centuries for maps which were printed with the text of old geography books. Printing was done from a line or surface which stood out from the interline areas, which were cut lower. Lettering was often stamped in. Woodcut maps are crude (see Fig. 14.1), but they have a certain charm and quality.

Copper Engraving. Atlas maps and loose maps were cut into copper with much less effort (see Fig. 10.1). The map was drawn reversed (in mirrorlike fashion), and the lines and letters were cut into a polished copper plate with a graver. Ink was rubbed into the grooves and the rest of the plate was wiped clean. The plate was pressed against damp rag paper on a flat press. Beautiful maps with exquisitely fine lines were produced, and they are well preserved on the fine paper. But only a few thousand prints could be made without recutting, so the maps were expensive. The number of copies was increased when electrocopies of the plates were made, but the process is not used for maps in America at present, although there are still a number of copper cutters in Europe.

Lithography. As the story goes, in the 1790s the young Aloys Sennefelder wrote a bill for his mother's laundry work with soot on a slab of dense limestone from Kelheim, Bavaria. Noticing a peculiarity of this limestone, he discovered the principle of lithography. Grease and water do not mix. If we draw or paint anything on this kind of limestone with a greasy ink and wet the stone, we can print on paper. The impression, however, will be reversed, so we

---

NOTE TO THE INSTRUCTOR: If photographic equipment is available Chap. 26 can be studied after this one.

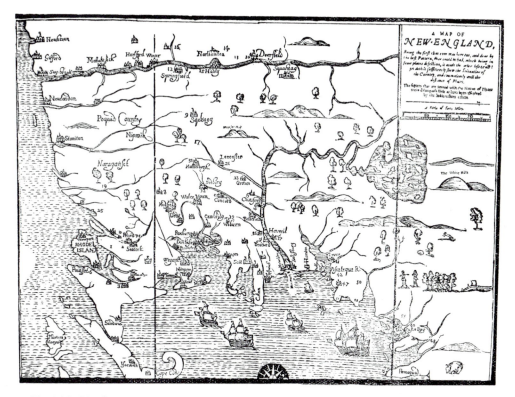

Fig. 14.1. The first map drawn, engraved, printed, and published in the American Colonies. (*John Foster's Woodcut, Boston, 1677.*)

have to draw on the stone in a mirrorlike fashion. If the stone is rolled over again with greasy ink, the ink will adhere only to the lines; thus, by alternate inking and wetting, copy after copy can be made. Lithography was soon used to reproduce maps. In America it became popular in the 1830s, and grew into one of the great national industries.

Wax Engraving. Invented by Sidney E. Morse in 1842, wax engraving was the favorite American method until offset printing replaced it. The map is drawn or photographed on a thin layer of wax upon a copper plate. Lines are cut into the wax with a stylus, not scratching the copper. The plate is dusted with graphite to make

it conductive and is placed in an electrolytic bath, where copper will be deposited upon it. After some strengthening this copper is lifted from the wax. It will make a good printing plate, printing from the raised lines, and it can be printed together with text. The main advantage of the process is that lettering can be set up in type and pressed into the wax, and lined patterns can easily be produced by a grooving machine. Only a few good wax engravers are active at present, yet the method has advantages.

At the present time, none of these methods is frequently used. Offset printing and photoengraving are the most commonly used methods of reproduction for publica-

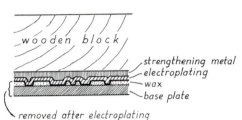

Fig. 14.2. Wax engraving. The base plate is sep-arated from the printing plate by melting the wax.

tion. For small nonprofessional printings, various duplication processes may be used.

Offset Printing

The present offset printing is the direct descendant of lithography. We do not use the heavy Kelheim stones because a finely kerneled aluminum or zinc sheet has the same property. These metal sheets can be curved around a cylinder and put on fast rotary presses which produce thousands of copies in a few hours. The printing is indi-rect. The zinc sheet prints on a rubber cylinder, and the rubber prints on paper. This is why the process is called "offset" (see Fig. 14.3). Coarser and more ab-

Fig. 14.3. Offset printing presses can print the finest lines fast and economically. The printing is carried over to the impression cylinder by a rubber blanket. (*Courtesy of the U.S. Army.*)

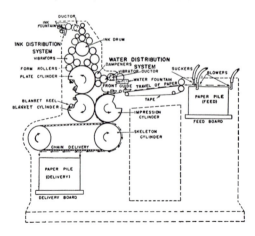

sorbent paper can be used which gives fine-ness to the lines.

The greatest advance, however, came with the use of photography. There is no more need to work on lithographic stone in printing size; instead we can draw the map conveniently on paper on a larger scale. The importance of this is greater than it seems at first. The actual drawing of the map is done by a geographer or cartogra-pher in an office or university instead of by a highly specialized craftsman at the print-ing establishment. Thus the making of maps gradually shifted from a skilled carto-technician to a geographer-cartographer with a less formal but more scientific at-titude.

Procedure. The cameraman takes the manuscript and photographs it to publica-tion size, producing a negative. In the dark-room he has a zinc plate ready, sensitized with an albumen compound which has the property of hardening when light strikes it. He places the negative on the zinc plate in a vacuum frame to make firm contact; then he throws strong light on it. The al-bumen becomes insoluble where the light strikes it—that is, where the black lines of the original drawing were. The plate is cov-ered with developing ink and the design appears. The unexposed albumen is washed off. The lines are treated with resin to add strength. The zinc plate is wetted with a weak solution of acid and gum arabic which slightly etches the interline areas, and the plate is ready for the rotary press. Offset printing of maps is fast and inexpensive, yet the lines are of fine quality. Some presses can print maps 5 ft wide, but the usual size is 2 by 3 ft.

For happier cooperation between printer and cartographer there are some points to remember and many of them apply to work which will be done by photoengraving as well as by offset.

1. The printer needs solid black lines. Faded, half-erased, or gray lines will not print. On the other hand, any fairly dark smudge or pencil mark will show up in the negative.
2. Don't use yellowish drawing paper, as yellow photographs easily.
3. Mark clearly on one side of the map to how many inches this side should be reduced. Do not mark the other side; the camera takes care of that. Indicate also the width of the margin you want. Give the number of copies desired and specify the kind of paper.
4. If you use cellotype or Zip-a-tones, add a note to avoid overheating them. They may fall off.
5. If colors or overlays are used, have at least four register marks on each drawing. These consist of small crosses outside the margin. Their exact center is pricked through with a needle.
6. Art work supplied by the engraver is very expensive, so if corrections are to be made, ask for an estimate or do it yourself.
7. Ask for the date of delivery. Printers are very apt to put aside smaller jobs if a large job is urgent.
8. Don't expect delivery within a week. The printer has to put several small jobs on the cylinder to make the large rotary press pay. Special paper or special color will cost more.
9. Ask for the return not only of the original drawing, but also of the negative; otherwise the printer will keep the negative. You may want both to make corrections for a later edition.
10. You may ask for a proof print in offset but this will be expensive, as the press has to be set up especially for it. A photostat or contact print is much cheaper and will give some idea of the appearance of the map when reduced.

Photoengraving

In this process, maps are printed from raised lines and can be printed together with type on a flat letter press. Maps in books are likely to be reproduced this way. The manuscript map is photographed, and the negative is lighted through upon a zinc plate exactly as in offset printing. However, the lines are strengthened with acid-resisting resin, and the interline areas are deeply etched by nitric acid. The plate is mounted type-high (0.9186 in.) upon a block of wood and delivered to the printer. The photoengraver usually provides some proof prints but does not do the printing.

The manuscript map for photoengraving should be a little coarser than for offset, as the lines cannot be quite as fine. Dense, cross-line patterns should be avoided.

Halftones are often produced by photoengraving; copper instead of zinc plates are used and delicate inking is necessary. Copper engraving with delicate inking may be three to four times as expensive as zinc. Colors can be handled by photoengraving just as well as with offset.

Types of Map Manuscripts

Map drawings can be prepared for reproduction in one of the following ways:

1. *Line.* This may have lines, letters, or solid surfaces.
2. *Line and shades.* This includes cellotones and Ben Day.
3. *Halftone.* This has merging shades like a photograph.
4. *Flat colors.* This consists of lines and patches of even colors.
5. *Merged colors.* This is like a painting.

Each type has its own problems and is discussed here. Line work has relatively few problems, and was discussed in Chap. 13.

Shades and Patterns. In the previous chapter we discussed how to use Zip-a-tone, Craftint, Ben Day, etc., for even shades. Difficulties may arise if the map has such fine patterns that the printer has to use special papers and inks to avoid patchy spots. A surprising number of dust spots may show up which were hidden under the Zip-a-tone. It is good to consult the offset printer before applying fine tints. Hand-dotted or lined maps may look quite uneven in printing if the ink was not black enough or the lines faded under erasing.

Halftones. These are maps painted with varying shades. They are not very common, as the cartographer needs to be an accomplished painter or airbrush expert. Yet maps painted in grays are effective and are often

Fig. 14.4. Halftone "combination" plate. This map was prepared on two sheets, one with airbrush, the other containing lines and lettering. The airbrushed sheet was photographed through a halftone screen, and the screen effect over the seas was opaqued out on the negative. Another negative was made of the lines and lettering, and the two negatives were lighted over the sensitized zinc plate. (*Courtesy of Richard Chapin, Time.*)

used in magazines. The maps of R. M. Chapin in *Time* are famous. If we look at Fig. 14.4 with a magnifying glass or under magnification, we will notice that it is composed of tiny dots, larger ones in the dark areas and smaller ones in the light places. There are exactly the same number of dots per inch. No place is without dots, even those which are supposed to be white or black.

This effect is produced by a halftone screen. The screen consists of two glass plates into which fine parallel lines are grooved; then the plates are pasted together crosswise. The screen is placed in front of the photographic plate. This will dissolve the light reflected from the drawing into tiny rays; more light will pass through where the map is lighter, and vice versa. The usual screens range from 75 lines per inch for rough newspaper pictures to 120 or 133 for an average textbook or magazine and 200 for shiny coated paper. Airplane photographs can be reproduced by a special method with a 350-line screen.

The disadvantage of the halftone process is that there are no clear whites or blacks, and the tones in general are duller than in the original. Clear whites can be obtained by opaquing the negative. If, for example, the relief on a map is painted or penciled in plastic shading and we want all water areas clear white, we may opaque the negative over the water areas ourselves. Or, if we cannot obtain the negative, we can make an overlay with all the water black. From this the printer will make a diapositive or autopositive film (see Chap. 27) in publication size and place this over the halftone negative to mask the halftone dots from the water area and light the two together over the zinc plate. If we want parallels and meridians and the names of seas in the water, we reverse the process. We make an overlay on which we blacken all land surfaces and ink the lines or letters over the water. The printer makes a negative on which the land is transparent and the sea has transparent letters on black; then he lights the two negatives together over the zinc plate.

Combination or Surprint Plates are often used for maps. Relief is painted or penciled and reproduced by halftone. Lettering, symbols, and everything that should be clear black is drawn on an overlay. The printer makes a separate negative of both in reduced size. Both negatives are lighted in succession over the zinc plate; thus the screen effect in the lettering is eliminated as the full amount of light will go through the negative of letters. The purpose of this is to get sharp black letters and symbols without the fuzziness inherent in the half-tone process. White letters on black or gray can be obtained by lettering in black on the negative. Cellotype companies also furnish white letters with both transparent and black backgrounds. A more detailed discussion of the reversal processes is found in Chap. 27.

Flat Colors. Most topographic sheets, atlas maps, wall maps, and road maps are prepared in flat colors. The engraver has to prepare a separate printing plate for each color, and one color is printed over the other. Color separation can be done in many ways.

1. *Separate drawing for each color.* The map maker prepares a master drawing in color and makes a separate overlay for each color. It is best to prepare the master drawing and the overlays on dimensionally stable plastic for good register. Each overlay—whatever its color—is drawn with black ink. For lighter blues, greens, reds, etc., the lines have to be thicker than for dark brown or black. The color-separation plates can also be produced by negative scribing in publication size (see Chap. 27).

2. *Light-blue pulls.* We submit a pencil drawing which has all lines on it, whatever the color. The printer prepares nonphotographic light-blue prints of the same size on blue-sensitized drawing paper, one for each color. The cartographer inks in only those lines which appear in each color. Patterns can be applied on their respective color plates. The printer will reduce each color plate to publication size.

3. *Silver prints.* The master drawing is prepared as before, and as many faint positive photographs are made as there are colors. The cartographer inks in one photo after the other. The photos are bleached so that only the inked lines remain. Zip-a-tone patterns can be applied after bleaching. As this is a wet process, register may not be good. This is a cheaper process than blue pulls, and is used for smaller, simpler maps.

4. *Filtering.* We prepare a master map in flat colors but replace blue with olive-green and black with purple. All lines should be of clear, high-chroma inks and paints. The map is photographed; one negative for each color is taken with a filter which will cut out the other colors. Usually considerable retouching is necessary on the negative. As this method requires considerable photographic skill, the engraver should be consulted for the choice of colored inks.

5. *Opaquing negatives.* The master map containing all lines of all colors is inked in final form. This is photographed to publication size, and as many negatives are prepared as colors are needed. On each negative only those lines are kept which appear in a single color; all other lines are opaqued with special opaquing paint or india ink.

The choice of color-separation method depends on the nature of the map. For multicolored maps, a fully colored original and separately drawn color overlays are recommended. If the colors are simple and well established, only a pencil original is drawn and light-blue pulls or silver prints are used. Blue pulls are more expensive but give better register. If only two or three colors are used, filtering is the most economical. Negative opaquing is used where the map consists of lines only—an auto road map, for instance. It would be difficult to opaque a black line crossing a red area.

Four-color Process. A map can be painted in rich full coloring and can be reproduced the same way as oil or water color paintings are printed. As every color can be mixed from red, yellow, blue, white, and black, we can separate each component color by photographing the painting through color filters in addition to a halftone screen. A magenta filter is usually used for the yellow negative, an orange filter for the blue, and a blue-green filter for the red. The black is obtained through a light yellow filter. The negatives are placed over the respective sensitized printing plates, first the yellow, followed by the red, blue, and black. Black lettering and symbols are usually drawn on a plastic overlay of which a separate negative is made to get rid of the fuzz of the halftone screen. This is lighted through to the same zinc plate as the black halftone negative, as in a combination plate.

Registration. The bane of color printing is the problem of registration, to have each color printed in its proper place. Even $\frac{1}{30}$-in. displacement produces unsightly overlapping or uncolored strips. Disregistration can be caused by errors of the color separator, by differential expansion of the film on paper of the color copies, or by a shift of the paper in the printing process. Registration marks on four sides should be pricked through with a needle and a small cross should be drawn centered very exactly on the pinhole. During the work of color separation, registration should be checked frequently. All color copies have to be on the same material and only highly shape-proof and dimensionally stable plastics, such as Vinylite, Mylar, or glass laminates, should be used. Small maps are in less danger; here vellums, even thin drawing paper, can be used.

Process printing requires considerable skill and much manipulation by the printer and the cost is correspondingly high. It is difficult to obtain clear white and the appearance of lines is slightly fuzzy. However, such perfection is obtained by some skilled printers that full-colored and painted maps are becoming more and more common.

Duplicating Processes

If we want the printed map in the same size as the manuscript and not too many copies are required, we can use some of the duplicating processes advantageously. The processes which involve photographic contact printing, such as blueprint, Van Dyke, autopositive, and the various ammonium papers, are described in Chap. 26. These require photo equipment. Here only duplicating machines which are commonly found in departmental offices are described. They are most useful in producing maps for the classroom, for geographic theses, and for the bulletin boards. A great number of these machines are on the market, and a few are described here. (Their selection reflects primarily the author's familiarity with the product.)

The usual office duplicating machines fall into four classes:

1. Gelatin type—Hectograph, etc.
2. Spirit duplicating—Ditto, Standard, and many others.
3. Stencil duplicating—Mimeograph, Auto-copy, etc.
4. Multilith or offset duplicating—Davidson, etc.

Gelatin Process. This is the oldest duplicating process and the hectograph is still found in some offices, although its use is declining. First we prepare a master copy on a paper with a hard surface. On this we can draw with hectograph pencils or inks. Typing can be done through a hectograph ribbon. The master copy is placed face down upon a sheet of gelatin; this readily absorbs the hectograph ink, and the image appears on it in reverse. If we place a sheet of paper over the gelatin and apply gentle pressure, a right-reading impression is made. About 100 to 200 copies can be made on smooth-finish paper. Some machines curve a gelatin pad around a cylinder and print like a rotary press.

Spirit Duplicating Processes. In these processes we copy our maps upon a smooth master sheet with the help of special heavy carbon paper, using a sharp hard pencil or stylus. With the aid of a pressure roller, we bring this master sheet in contact with the slick-finish duplicating papers. Before contact, the duplicating paper is moistened with quick-drying duplicating fluid. With every turn of the handle, a new duplicating paper comes in contact with the master sheet and takes away some of its ink. The potency of this ink, which we apply through the carbon paper, is amazing. Three to seven hundred clear copies can be pulled, depending upon the color. We can draw

on the same master sheet through several different-colored carbons; with a single turn of the handle we can obtain colored maps. The colors, however, are harsh and not very pleasant. The spirit duplicating processes are particularly good for type-sheet-size maps used in class, in typed reports, or on bulletin boards.

Stencil Duplicating Processes. The mimeograph and similar machines commonly found in offices belong to this group. After some practice excellent maps can be stenciled and reproduced. The stencil is a sheet of soft, hairy, fibered paper with a wax coating and a backing sheet. We draw our map with a stylus or special rolling-tip pen which cuts through the coating. The stencil can be fed into the typewriter, and the stroke of the keys also cuts through the

Fig. 14.5. Sketches of some office duplicators, used also for same-size reproduction of simple maps: (1) gelatin process; (2) spirit duplicator; (3) stencil duplicator (mimeograph); and (4) offset process. (*Courtesy of Eastern Corporation.*)

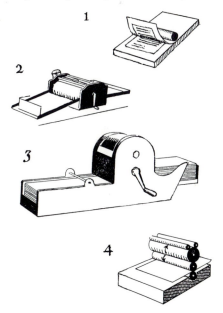

coating. The stencil is reversed and clamped upon a cylinder; the backing sheet is removed. Ink is fed from inside the cylinder. For copies we use absorbent paper. The impression is made by passing the paper between the cylinder and a rubber pressure roller. Several hundred impressions can be made from the same stencil.

Multilith or Offset Duplicating. The principle is the same as that of offset printing —the use of three cylinders, a plate, a rubber blanket, and an impression cylinder. Small offset duplicators for a limited run use a special paper or plastic printing plate. The map can be drawn or traced directly on the paper plate with greasy lithocrayon or with lithoink, using pen or brush. Typing is done through lithocarbon paper. The paper is clamped on the plate cylinder without any further processing. This cylinder revolves under water rollers; the interline areas are wetted, and the lines will receive the ink. The rubber cylinder receives the image from the plate and carries it over the copy paper. Over a thousand impressions can be made from a paper plate. If we want more impressions, we can use a thin metal (zinc or aluminum) plate the same way. Small-size photographic equipment to prepare the printing plate photographically is available from the manufacturers.

15

The Earth

This is the first time in history that man can see his planet somewhat from the outside by looking at photographs taken from rockets at high altitudes. We marvel at the immensity of our sphere, the variety of its land forms, the richness of its colors, and its patterns of sea and land, field and forest, ice and desert. We may be impressed that man was able to put on the map all the shorelines, mountains, and rivers exactly as they appear in the photo without seeing them as a whole.

In these photos the roundness of the earth is obvious. But it is not so obvious to people on the ground. It is of credit to mankind that the spherical form of the earth was recognized more than two thousand years ago, even though to all our senses the earth appears flat. The ancient Greeks not only measured the earth ball with fair accuracy but defined the tropics, the polar circles, and the five zones. Our system of parallels and meridians is credited to the Greeks, too, although it may be of Babylonian origin.

The Size and Shape of the Earth. The shape of the earth is nearly spherical, closely approaching an ellipsoid, slightly flattened at the poles. The equatorial diameter is only $\frac{1}{298}$ larger than the polar diameter, but this makes a difference of 27 miles. On a huge 30-in. globe, this would be only $\frac{1}{10}$ in., not at all perceptible. The shape of the earth, assuming that all the land is reduced to sea level, is called the *geoid*. This has minor irregularities, but an ellipsoid of an equatorial radius of 6,378,-206 m and a polar radius of 6,356,584 m is sufficiently accurate for most calculations. This is the *Clarke spheroid* of 1866 used by the U.S. Geological Survey. Many other spheroids (near-spheres) are in use, but the differences are very small. The latest figures of the earth's dimensions are given in Table 15.1.

The Earth's Motions. The earth makes complete rotation around its axis in 23 hr 56 min. It takes the earth an extra 4 min of rotation to get to noon position again, as the earth moves forward around the sun; thus, our day is 24 hr long. The earth revolves around the sun in an elliptical path in 365.2564 days.[1] The axis of rotation locates the two poles. The equator is cut by a plane which bisects the axis at right angles. We must, however, reduce the South Pole to sea level, as in reality it is on top of a 9,000-ft ice plateau.

The Seasons. The earth's equatorial plane is inclined 23°27′ to the plane of its

[1] Our calendar year is only 365.2422 days long due to the *precession of the equinoxes*, the conical motion of the earth's axis swinging around in 26,000 years.

TABLE 15.1. DIMENSIONS OF THE EARTH ACCORDING TO THE CORRECTED HAYFORD SPHEROID

Equatorial radius: 6,378.3 km, or 3,963.3 miles
Polar radius: 6,356.6 km, or 3,949.8 miles
Ellipticity: $\dfrac{a-b}{a} = \dfrac{1}{298}$
Equatorial circumference: 40,075.9 km, or 24,902.1 miles
Meridional circumference: 40,008.4 km, or 24,860.0 miles
Length of 1° longitude near the equator = 69.17 miles (Clarke spheroid)
Length of 1° latitude at equator = 68.70 miles
Length of 1° latitude at poles = 69.41 miles
Total area of earth (approx.): 196,943,000 sq miles
Radius of the sphere of equal volume: 3,958.8 miles, 1° of latitude on which would be 69.09 miles
Radius of the sphere of equal area: 6,371 km, or 3,956 miles
For conversion into or from the metric system, the following figures may serve for measurements in kilometers:
1 mile = 1.60934 km; 1 km = 0.621 mile; 1 m = 39.37 in. = 3.2808 ft; 1 ft = 0.3048 m

revolution, which is called the *ecliptic plane*. As the axis is always parallel to itself, the earth will be inclined differently to the sun in summer than in winter. This inclination defines the tropics, the place where the noonday sun will be overhead in the North on June 21 and in the South on December 22 (see Fig. 15.2). On the Arctic Circle, the sun will rise only halfway above the horizon on December 22; on the Antarctic Circle, the same condition will occur on June 21. At the poles, the sun will ride a complete circle halfway above the horizon at the time of the equinoxes, on March 21 and September 22 or 23.

Parallels and Meridians. The earth's coordinate system developed quite naturally from the cardinal directions. East, West, North, and South are among the earliest concepts of mankind. The equator is a natural zero line for North and South distances, and the poles are obvious points for convergence of the meridians. It was an early Babylonian idea to divide the equator and the full meridian circle into 360 degrees, and every degree into 60 minutes, and each minute into 60 seconds.

To keep our terminology straight, *parallels are East-West lines and meridians are North-South lines. Latitude is an arc distance measured in degrees both ways from the equator, while longitude is an arc distance measured in degrees from a selected prime meridian.* While a degree of latitude

Fig. 15.2. The Tropics and the Arctic and Antarctic Circles divide the earth in zones and are usually marked on maps.

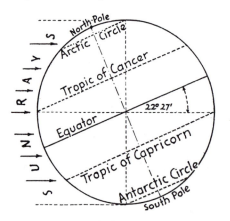

is almost the same everywhere, a degree of longitude decreases from the equator to the pole.

The Length of a Degree of Latitude. This also varies a little—from 68.7 miles near the equator to 69.4 miles near the poles, while at lat 40° it is almost exactly 69 miles. The reason for this is the ellipticity of the meridians. (Fig. 15.3). Latitude is measured by the altitude of the stars or the sun. Near the equator where the radius of the earth's curvature is shorter, it takes less distance to make 1° change than near the poles where the radius of curvature is larger.

The Length of a Degree of Longitude. Longitude varies from 69.17 miles near the equator to zero at the poles. The rule is

$$1° \text{ long} = 1° \text{ lat} \times \cos \lambda$$

where λ is the latitude.

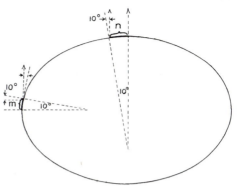

Fig. 15.3. Near the equator a shorter arc effects a 10° change in the altitude of the polar star than is the case near the poles. This makes 1° of latitude shorter nearer the equator.

LONGITUDE VARIES WITH THE COSINE OF THE LATITUDE.

See Figure 15.4. This is the most basic formula of cartography.

As cos 60° is 0.5, the length of a degree of longitude along that parallel is one-half as much as it is at the equator. At lat 40° (near Denver, Columbus, Philadelphia, Madrid, Ankara, Peking) a degree of longitude is about 53 miles; but at Havana, located near the tropic, a degree of longitude measures 63.7 miles. Table 15.7 gives the exact lengths, taking into account the flattening at the poles, while a table of cosines is in the Appendix.

Prime Meridian. Any meridian could be chosen as the prime meridian, and quite a number have been used during the past, including Philadelphia, New York, and

even Hartford, Connecticut. Ptolemy used the Fortunate (Canary) Islands, and on many old maps we find longitudes from Ferro. This is the westernmost island of the Canaries, which is almost exactly 20° west of Paris. At present the Greenwich meridian (based upon the former Naval Observatory in London) is used almost universally, although it is not as desirable as Ferro, since it cuts Europe and Africa in two. However, it makes a good Date Line along 180°. We usually reckon longitude east and west from Greenwich 0° to 180°. Thus the location of Harvard Observatory is expressed as

Fig. 15.4. Longitude varies with the cosine of latitude. This is the fundamental rule of map construction.

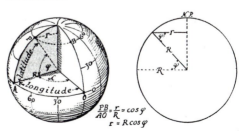

$$\frac{PB}{AO} = \frac{r}{R} = \cos \varphi$$
$$r = R \cos \varphi$$

Lat 42° 22′ 48″ N
Long 71° 07′ 45″ W of Greenwich

which is far from simple. Many attempts
are being made to transform minutes and
seconds into decimal values and reckon
longitudes 0° to 360° starting east from
Greenwich. On French and Turkish maps
we often find *grads*, instead of degrees, dividing the circle into 400 parts.

Measuring Latitude and Longitude.
Measuring latitude is relatively simple,
since our latitude is the same as the altitude
of the celestial pole over the horizon. We

TABLE 15.5. LENGTH OF ONE DEGREE OF
LATITUDE FROM EQUATOR TO POLE

Latitude	Statute miles	Latitude	Statute miles
0°	68.703	50°	69.115
10	68.725	60	69.230
20	68.786	70	69.324
30	68.879	80	69.386
40	68.993	90	69.407

Fig. 15.6. The sextant is used mostly on ships for
measuring the altitude of the sun or stars.

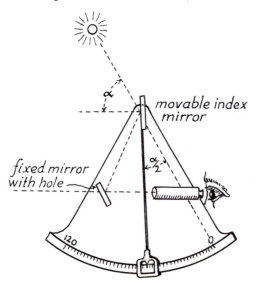

may measure the altitude of the polar star,
as this star is only about 1° away from the
celestial pole. One may observe the altitude
of the sun or of certain stars and calculate
the latitude from an ephemeris. The
"American Ephemeris and Nautical Almanac" is published yearly by the Naval
Observatory.

For longitude we have to determine local
time and compare it with Greenwich time.
The difference gives the longitude, as the
earth rotates 360° ÷ 24 = 15° in an hour.
Greenwich time is carried along with a
chronometer which can be corrected by
radio time signals. The process is not very
simple because we have to correct our
local time by adding or subtracting the
equation of time, which is the difference
between local time, astronomically determined, and Local Mean Time.[2]

In order to find local noon on land, we
lay out the exact North-South line at night
by the polar star and clock the meridian
passage of the sun. On shipboard usually
an observation of the altitude of the sun is
taken at 10 A.M. and at 2 P.M. with a sextant.[3]

For exact determination of latitude and

[2] Because of the inclination of the earth's axis
and the variation in the earth's speed around the
sun, one day is not as long as another. Although
the difference may be just a few seconds, it ac-
cumulates until on February 10, the sun's altitude
is highest at 12ʰ 13ᵐ Local Mean Time; and then
the day begins to decrease until November 2, when
the sun is highest at 11ʰ 43ᵐ. We cannot have
clocks of varying speed, so we use mean time with
days of equal length.

[3] The sextant is an instrument for measuring
angles. Its principle can be understood from Fig.
15.6. The observer looks at both the horizon and
the mirrored image of the sun at the same time. If
the horizon is obscured, as it often is in an airplane,
a bubble sextant is used.

longitude, the observation of the altitude of five to fourteen stars, or the 60° passage of certain "geodetic" stars, will give latitude,

TABLE 15.7. LENGTH OF ONE DEGREE OF LONGITUDE AT DIFFERENT LATITUDES

Latitude	Statute miles	Latitude	Statute miles
0°	69.171	45°	48.995
1	69.162	46	48.135
2	69.130	47	47.261
3	69.078	48	46.372
4	69.005	49	45.469
5	68.911	50	44.552
6	68.796	51	43.621
7	68.660	52	42.676
8	68.503	53	41.719
9	68.326	54	40.749
10	68.128	55	39.766
11	67.909	56	38.771
12	67.670	57	37.764
13	67.411	58	36.745
14	67.131	59	35.715
15	66.830	60	34.674
16	66.510	61	33.622
17	66.169	62	32.560
18	65.808	63	31.488
19	65.427	64	30.406
20	65.026	65	29.315
21	64.606	66	28.215
22	64.166	67	27.106
23	63.706	68	25.988
24	63.227	69	24.862
25	62.729	70	23.729
26	62.212	71	22.589
27	61.676	72	21.441
28	61.121	73	20.287
29	60.548	74	19.126
30	59.956	75	17.960
31	59.345	76	16.788
32	58.717	77	15.611
33	58.071	78	14.428
34	57.407	79	13.242
35	56.726	80	12.051
36	56.027	81	10.857
37	55.311	82	9.659
38	54.578	83	8.458
39	53.829	84	7.255
40	53.063	85	6.049
41	52.281	86	4.841
42	51.483	87	3.632
43	50.669	88	2.422
44	49.840	89	1.211
45	48.995	90	0.000

Fig. 15.8. Gnomonic chart of the North Pacific. On this chart every great circle appears as a straight line. It is used to lay out long sailing and flying routes. (*From E. Raisz, "Mapping the World," Abelard-Schuman, Inc., Publishers.*)

local time, and longitude. The necessary tables and forms are contained in the "American Ephemeris and Nautical Almanac." The determination of latitude and longitude is a complex science and will not be discussed here. Any standard text on astronomy or surveying can be consulted.[4]

Great Circles. We can halve a globe by any plane which goes through its center. This plane will cut out a circle which is a *great circle*, the largest circle possible to draw on a globe. Every great circle bisects any other great circle at two antipodal points. The above rules are only approximately true, as the earth is slightly ellipsoidal in form. All meridians are nearly great circles, and bisect each other at the poles. All parallels, except the equator, are "small" circles.

Globe Distances. The great circle is the shortest distance between two points on the surface. It can be scaled off from a globe with a strip of paper but cannot be

[4] See the "Manual of Geodetic Astronomy," U.S. Coast and Geodetic Survey Spec. Pub. 237, 1952. The small yearbook "Solar Ephemeris" is given on request by the Keuffel and Esser Company, Hoboken, N.J.

TABLE 15.9. VALUES OF VARIOUS PRIME MERIDIANS RECKONED FROM GREENWICH

Cities	Longitude from Greenwich	Map of
Batavia	106° 48′ 38″ E	Java
Padang	100° 22′ E	Sumatra
Sofia	23° 19′ 39″ E	Bulgaria
København	12° 34′ 40.35″ E	Denmark
Helsingfors	24° 57′ 16.5″ E	Finland
Paris	2° 20′ E	France
	2° 20′ 14″	Yugoslavia
	2° 20′ 13″	Poland
Ferro	17° 40′ W	Spain
Athens	23° 42′ 58.5″ E	Greece
Roma (Monte Mario)	12° 27′ 07.06″ E	Italy
Oslo	10° 43′ 22.5″ E	Norway
Lisboã (Castle)	9° 7′ 54.86″ W	Portugal
Bucuresti	26° 06′ E	Rumania
Pulkovo	30° 19′ 38.49″ E	Russia
Leningrad	30° 17′ 15″ E	Russia
Madrid	3° 41′ 14.55″ W	Spain

scaled off from a map (unless the distance is within a few hundred miles) because of the distortion inherent in all projections. If the latitude and the longitude of the two points are known, we have a rectangular spherical triangle, the hypotenuse of which is the distance

$$\text{hav } d = \text{hav } (\varphi_2 - \varphi_1) \\ + \cos \varphi_1 \cos \varphi_2 \text{ hav } (\lambda_2 - \lambda_1)$$

Fig. 15.10. Mercator chart of the North Pacific. This chart will be used for compass navigation after the great-circle route has been transferred here from a gnomonic chart. (*From E. Raisz, "Mapping the World," Abelard-Schuman, Inc., Publishers.*)

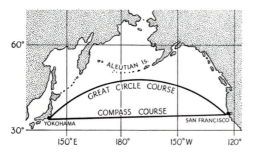

where d is the distance, φ the latitude, and λ the longitude. hav stands for haversine; hav $\alpha = \frac{1}{2} (1 - \cos \alpha)$ is a useful function of geodesy.

Orthodromes and Loxodromes. To go from one point to another, we use either a great-circle direction or a compass direction. If the two points are within a few hundred miles, there is little difference between the two, but at great distances they differ widely. Navigators call a great-circle line on the earth's surface an *orthodrome*.

From Fig. 15.10, it is obvious that the orthodrome cuts every meridian at a different angle; thus its azimuth varies from point to point. As orthodromes are the shortest distances between points on the globe, they are important for air and sea navigation and radio broadcasting. The easiest, but far from exact, way to find the orthodrome is to stretch a string on a globe between two points. The gnomonic projection (see Chap. 17) shows all orthodromes as straight lines. It may surprise many that the orthodrome from Panama to Tokyo will pass near New Orleans, Seattle, and the Aleutian Islands.

To find the compass direction between two points, we have to draw a line which crosses every meridian at the same angle. Such line is called a *loxodrome* or *rhumb line*. We saw these rhumb lines on the portolan charts of the fourteenth century. If we try to draw the compass direction northeast on a globe (see Fig. 15.11), we see that the loxodrome is a curve spiraling around the pole. As most navigation is done by compass, loxodromes are important. The Mercator projection (see Chap. 17) shows all loxodromes as straight lines; therefore almost all charts for navigation are in this projection. Navigators on longer routes take the orthodrome from a gnomonic chart or globe, transfer this line to a Mercator chart, and, by changing the compass direction periodically, arrive at their destination economically by using a compass with due allowance for winds and currents.

Hemispheres. Of the number of ways the earth can be halved, the most obvious is to distinguish between the Northern and Southern Hemispheres, which are surprisingly unlike in the distribution of land and sea. Most atlases usually show the Eastern and Western hemispheres divided at the 20°W and 160°E meridians. This does not coincide with a Chinaman's view; to him we live on the Eastern Hemisphere. It does not agree entirely with the "this hemisphere" of the Monroe Doctrine either, as the Western Hemisphere includes New Zealand, half of Iceland, and a part of the U.S.S.R. The *land hemisphere*, the half with most of the land, is centered in Nan-

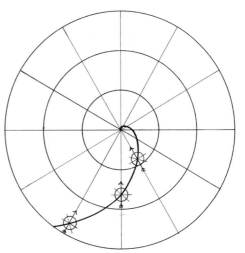

Fig. 15.11. Loxodromes, or rhumb lines, are lines that have the same compass direction all along their lengths.

tes, France. It contains 82 per cent of all lands; if we discount ice caps, the percentage rises to 90. The *people's hemisphere*, centered in the French Alps, has 95 per cent of the population.[5]

Horizon Measurements. From Mount Washington, N.H., can you see the light of the new television tower on Great Blue Hills, near Boston? This and similar questions are often asked a cartographer, and he should be prepared to give the answers. First, let us calculate the radius of the visible horizon from a given height. A trigonometric equation gives the answer, from Fig. 15.12.

$$r = \sqrt{[(R+h)+R]\,[(R+h)-R]}$$
$$r = \sqrt{(2R+h) \times h}$$

As h is relatively small, approximately

$$r = \sqrt{2Rh}$$

[5] Erwin Raisz, Our Lopsided Earth, *Jour. Geog.* vol. 63, March, 1944 pp. 81–91.

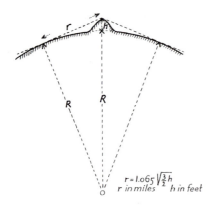

Fig. 15.12. The radius of the unobstructed horizon from an elevated point.

When expressed in miles,

$$r = \sqrt{7,916 \times \frac{\text{height in feet}}{5,280}}$$
$$= \sqrt{\frac{3}{2} \text{ height in feet}}$$

as 7,916/5,280 is almost exactly ¾. Our equation can be improved if we correct for refraction. Refraction is caused by the decreasing density of air with altitude which enables us to see beyond the geometrical horizon. Its value varies with the weather, averaging 1.065r. Thus corrected for refraction, when expressed in miles

$$r = 1.065\sqrt{\frac{3}{2} \text{ height in feet}}$$

The table below indicates the distance we can see from a lookout:

Height, feet	Distance, miles on the horizontal
10	4.1
100	13.0
500	29.2
1,000	41.3
2,000	58.3
5,000	92.4
10,000	130.0

The horizon is slightly larger near the poles than near the equator, as the radius of the earth's curvature is larger. We may now answer our first question about the visibility of the light on the Great Blue Hills television tower from Mount Washington.

$$r_w = 1.065 \sqrt{\frac{3}{2} \times 6,300} = 103.7 \text{ miles}$$

$$r_b = 1.065 \sqrt{\frac{3}{2} \times 959} = 40.5 \text{ miles}$$

This gives us a total of 144.2 miles. As the distance between the two points is only 142 miles, the Blue Hills light would be just visible from the Tip Top House on Mount Washington. It happens, however, that some 250-ft hills in Haverhill, Massachusetts, are in the line of sight just where the two horizons meet. There should, however, be reception of short-wave radio signals, as they bend more than light waves.

The Metric System. On many of the eighteenth-century maps we see a variety of scales of various kinds of units, such as miles, leagues, and toise. To end this confusion, the French Academy designed the metric system. The fundamental unit is the meter, a ten-millionth part of the arc distance from the equator to the pole. Finer measurements later found this distance to be 10,002,286 m. Thus it is more correct to define a meter as the length of the platinum-iridium rod preserved in Paris, copies of which are in the signatory countries (there is also a copy in the office of the U.S. Bureau of Standards in Washington.)

1,000 meters	= 1 kilometer
100 meters	= 1 hectometer
10 meters	= 1 decameter
1/10 meter	= 1 decimeter
1/100 meter	= 1 centimeter
1/1,000 meter	= 1 millimeter
1/1,000,000 meter	= 1 micron

The unit of area is the square meter, and the unit of volume is the cubic meter. The same prefixes are applied to both. Thus a cubic kilometer is a cube with a kilometer side. The unit of volume is the liter, equaling a cubic decimeter; the unit of weight is the gram, which is the weight of 1 cu cm of water at 4° centigrade. A cubic meter of water weighs a ton.

At present all nations use the metric system, except the United States and most countries of the British Commonwealth, but it has been a system recognized by law in the United States since 1860. The use of the metric system is spreading in the United States and even more in Great Britain, but the recalculation of most of our standards for tools, instruments, furnishings, etc., is an enormous and costly task and would lead to some confusion during the transition period.

The Nautical Mile. The origin of the nautical mile is the length of one minute of latitude. As the length of a degree of latitude varies slightly, the various nations adopted different lengths as standards. The U.S. Navy used:

1 nautical mile = 6,080.27 ft = 1,853.25 m
= 1.1516 statute miles

This was changed in 1954 to conform with the International Nautical Mile:

1 nautical mile = 6,076.1 ft 1,852 m
= 1.15 statute miles

Most seafaring nations, except Great Britain, accepted this value. The term *geographical mile* is sometimes used for one minute of longitude on the equator, or 1.1528 statute miles, but the ancient mariner used the term indiscriminately for one degree of longitude too.

16

Surveying

Surveying is taught in every school of engineering, and scores of excellent textbooks are available. Every professional cartographer should have a course in surveying and aerosurveying, but for the geography student a short outline is given here.

Surveying on Land

The simplest methods of surveying, compass traversing, and plane tabling have been described in Chap. 4. More exact measurements are made with a *transit*, an instrument used to read horizontal and vertical angles. Every engineer and surveyor uses it to lay out buildings and roads and to map estates. As the surveyor and cartographer are about the same in the public eye, we would be of low esteem if we did not know how to handle it. The transit has many parts and is described here only in the most general terms.

We look through a telescope which magnifies 10 to 30 times. The telescope can be turned around completely on the horizontal axis. The vertical circle rotates with the telescope. The horizontal axis is supported by legs or *trunnions*, which are attached to

the upper horizontal circle. Thus the telescope can be pointed in any direction and, as the vertical circle and the upper horizontal circle move with it, both angles can be read. The telescope is also equipped with cross hairs; thus we can also read approximate distances by stadia measurement, as described in Chap. 4.

We set up the tripod, so that the legs are about 60° to the ground and the top level is nearly horizontal. We push the legs firmly into the ground. The transit is resting on the *leveling head*. This is placed on the tripod where it can be shoved around slightly; with the plumb bob, we can center the transit exactly over a marked spot on the ground. With three or four screws we level the transit, looking at the two small level bubbles attached to the horizontal circles.

The leveling head has a central socket in which the vertical spindle of the lower horizontal circle turns. This circle is graduated and can be clamped to lock. The upper horizontal circle also has a vertical inner spindle which turns in the hollow outer spindle. The upper circle has no graduation, but two opposite verniers which can be read against the graduations on the lower circle. The vertical circle reads against a vernier attached to the trunnions. After the upper circle is clamped, it can still be moved a small amount with a tangent screw. The vertical circle also has a tangent screw for fine movement. Between the trunnions there is usually a magnetic compass. A sensitive level vial is attached to the telescope.

To read a horizontal angle, we set the vernier at zero and sight on the first point. We clamp the lower motion and read the angle; with the vernier we sight the second point and read the angle. The two readings are subtracted from each other. In case the vernier passes 360°, we add 360° to the second reading; only then do we subtract the first. In reading vertical angles, we set the telescope horizontal with the level bubble. If the transit is perfectly horizontal, the vernier should read zero. If there is a tilt we may read this and, when we make an angle reading, we may add or subtract the tilt.

The vernier is a device to read fractions of the smallest divisions on the main scale. Its operation can best be understood from Fig. 16.2a. On the straight scale the vernier divisions are one-tenth the size of those on the main scale. We read to the last division of the main scale, which is 52 in this case. Then we observe which line on the vernier coincides with a line on the main scale. This reading gives the decimals, so the complete reading in this case will be 52.6.

Similarly, when reading angles, we read to the last line on the main scale, 156° in Fig. 16.2b. A division on the vernier is one-thirtieth of a division on the main scale. As the vernier is divided into half degrees it enables us to read minutes. In the figure the best coincidence of lines is at 17; thus, the complete reading is 156°17′.

Modern *theodolites* are more compact than the usual engineer's transit. The leveling head is usually the three-screw type. The lower horizontal circle has no inde-

Fig. 16.1. A transit can measure vertical and horizontal angles with great accuracy. (*Keuffel & Esser Company.*)

Fig. 16.2. The vernier is an auxiliary scale which enables us to read fractions of the main scale by observing which lines coincide on the two scales. (*After Finch.*)

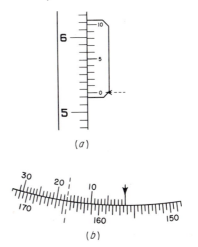

(a)

(b)

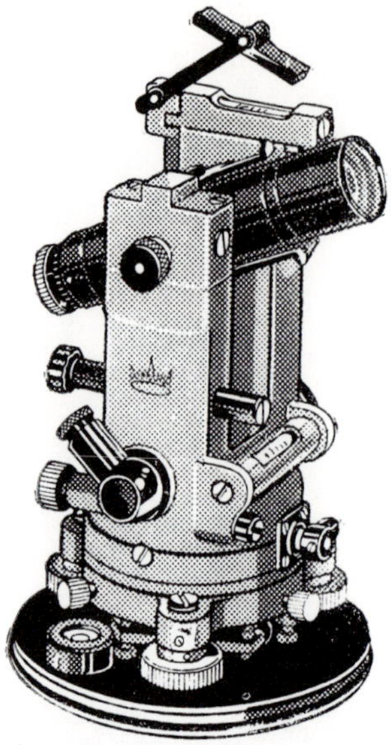

Fig. 16.3. Modern theodolites are more compact than the usual engineer's transit. Some of the finer instruments can be read to a fraction of a second. (*Courtesy of Askania Werke, Berlin.*)

pendent motion, but it has a special clamp which helps to start from 0°. There are no verniers, but with illuminated micrometers we can read on some instruments to

1′ and estimate even smaller divisions. Horizontal and vertical angles appear simultaneously before the eye. See Fig. 16.3.

Geodetic Surveys

The work of the local surveyor is usually on such a small area that the curvature of the earth can be neglected. This work is called *plane surveying* in contrast with *geodetic surveying*. Geodetic surveys are usually organized by the government. In our country, the U.S. Coast and Geodetic Survey does triangulation and leveling, but the detailed surveys for the topographic maps are made by the U.S. Geological Survey. In England, however, the Ordnance Survey does both.

Triangulation. The principles of triangulation were known to the ancients, and the first practical use dates back to the early 1600s. The triangulation of France under the Cassinis in the 1740s was the first great national survey. At present all civilized countries have some triangulation, yet less than one-tenth of the world is actually tri-

Fig. 16.4. In triangulation only a few baselines like *AB* and *QP* are measured. All the other points are fixed by reading angles only. The *QP* type of base produces a greater "strength of figure" than the *AB* base.

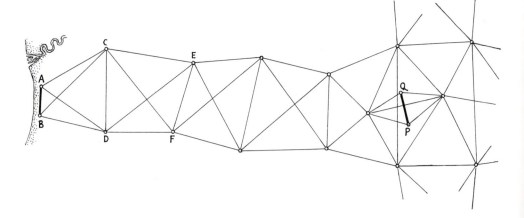

angulated. Parts of our Eastern Seaboard were triangulated by the Coast Survey in the 1830s and the first transcontinental arc of triangles was completed in 1878.

It takes only minutes to measure angles accurately with a very precise transit called a theodolite. Measuring exact distances, however, takes a long time, and in rough terrain, forest, or swamps is very difficult. Figure 16.4 shows the principle of triangulation. A base of several miles is measured very exactly, preferably on a level beach or along a straight road or railway. From the end points of the base, we read the angles to at least two distant points, which fixes their position on the map. Then we set up at these points and measure further angles of points selected to form well-shaped triangles and quadrilaterals. After a dozen or so quadrilaterals we measure a control base. At each base an astronomical fix of latitude and longitude will help not only in placing the triangles in proper place on the map, but also in eliminating the result of the "deflection of the vertical" caused by mountains and by lighter and heavier masses in the earth's interior.

The base is measured with a 50-m *invar tape,* a nickel-steel alloy which expands very little with temperature changes. Stakes are driven every 25 m and a 50-m tape stretched with a pull of 15 kg. Marks are made on top of each stake. Thus measuring proceeds to the end of the base, and then the whole stretch is measured backward. If the difference is over 1 in. in 5 miles, the measuring has to be repeated.[1]

[1] All geodetic operations are classified into four orders according to accuracy. For instance, in triangle closure the permissible error in first order is 3", in second order 5", and in third order 10". Second-order triangulation will find in the spaces between the large triangles of the first order.

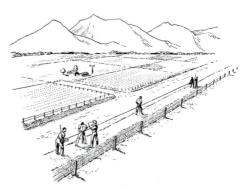

Fig. 16.5. Baselines for triangulation are measured with an invar tape at specified tension. (*From E. Raisz, "Mapping the World," Abelard-Schuman, Inc., Publishers.*)

Stations. First-order triangulation stations are selected about 10 to 25 miles apart, usually on hilltops for good visibility. "Bilby towers," possibly reaching to 120 ft, are often erected in forested areas. These consist of an inner tower which holds the theodolite, and an outer tower which holds the observers, the recorders, and the lightkeeper. An expert crew can erect a tower in about five hours. Angles to the main stations are read at night, but a num-

Fig. 16.6. A Bilby tower consists of two separate towers, the inside one for the theodolite, the outside one for the observers. (*From E. Raisz, "Mapping the World," Abelard-Schuman, Inc., Publishers.*)

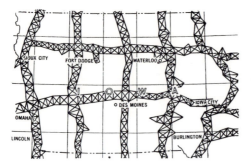

Fig. 16.7. First-order triangulation of Iowa. (*From a map of the U.S. Coast and Geodetic Survey.*)

ber of auxiliary points are read by daylight. The main angles have to be read to an accuracy of about 1 ft in 25 miles.

Office Computation. In the office the triangles are adjusted. The sum of angles of a triangle has to be 180° plus "spherical excess." The length of the sides has to be calculated, and the whole arc of triangles has to be fitted to existing arcs and astronomical (Laplace) stations. The latitude and longitude of each station have to be calculated and fitted to the Clarke spheroid.

The network of triangulation in the United States is so dense that tables are available for almost every region and can be ordered from the U.S. Coast and Geodetic Survey. These tables for each station contain latitude, longitude, sides of all triangles in meters and in feet, and their forward and backward azimuths. These azimuths are not quite exactly 180° apart

Fig. 16.8. The progress of leveling. (*From E. Raisz, "Mapping the World," Abelard-Schuman, Inc., Publishers.*)

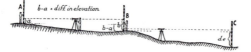

because of the convergence of the meridians.

The U.S. Coast and Geodetic Survey marks at least two triangulated points in every 5″ quadrangle before the parties of the U.S. Geological Survey map the area in detail. These parties make the network even closer by third- and fourth-order triangulation, until all key points, such as hilltops and road intersections, are marked so that they are identifiable from air photographs.

Leveling. Geodetists call triangulation "horizontal control" although the vertical angles are also measured and the height of peaks is calculated. Accurate "vertical control" is done by leveling parties. Triangulation and leveling parties are rarely combined. The former work on hilltops, the latter in valleys along roads, railways, or rivers, with different instruments and at different speed.

Precise leveling is done with a telescope attached to a very sensitive vial of level bubble (see Fig. 16.1). No angles are read. The instrument can be set horizontal with a tilting screw. Some modern instruments have a compensator which levels the line of sight automatically.

The principle of leveling is shown in Fig. 16.8. Rodman A drives a peg into the shoulder of the road or railway, whatever the case may be, and sets his rod on it. Rodman B sets up 100 to 400 ft away, closer when the slope is steep. The instrument man sets up in the middle and reads the height of the horizontal cross hair both forward and backward. Some rods are equipped with a vernier and can be read to a thousandth of a foot within 300 ft. Rodman A now sets up in front of B and the process is repeated. In first-order leveling after a mile or so, they rerun the survey backward, and if the difference is more than ¼ in./mile, they repeat it.

Results. The precision of leveling is amazing. It is possible to level from ocean to ocean and back with a difference of only a few feet. This difference is smaller than that between the Clarke spheroid and the actual mean sea level. In reference to this spheroid, both the Atlantic and the Pacific shores slope up northward and the Gulf Coast rises eastward. The Pacific Ocean stands higher than the Atlantic. Of course, if a different spheroid were selected this might not be so, but there still would remain the variation in mean sea level. Winds and tides can pile up the water near the coast and the gravitational attraction of mountains is considerable. The actual mean sea level is a rather complex surface.

The first transcontinental leveling was made in 1907. At present the network of levels is as dense as the net of triangulation. Leveling reports are sold by the U.S. Coast and Geodetic Survey.

Walking over hilltops or near intersecting roads, we often find bronze *markers* embedded in cement and marking triangulation and leveling stations. These are most important for local surveyors to attach their own measurements. They certainly should not be regarded as collector's items.

Aerosurveying

Almost every large government or private survey uses air photos for detailed maps. Aerosurveying is at present a highly developed science using most advanced equipment. Many excellent textbooks are available, and several periodicals keep us informed about the latest development on

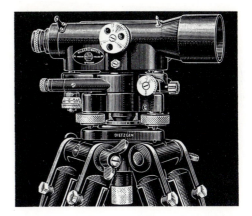

Fig. 16.9. Leveling instrument consists of a very sensitive level bubble vial attached to a telescope. Modern instruments have an automatic compensator to level the line of sight. (*Courtesy of Eugene Dietzgen Company.*)

Fig. 16.10. A leveling party progresses by short stretches. The parasol is not a luxury. Sunshine may offset the delicate instrument. (*From E. Raisz, "Mapping the World," Abelard-Schuman, Inc., Publishers.*)

Fig. 16.11. Markers of the U.S. Coast and Geodetic Survey for triangulation and leveling stations.

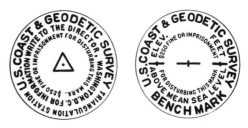

OXYGEN

Fig. 16.12. Vertical air-photo camera mounted in gimbals. View finder is at cameraman's knees. (*From E. Raisz, "Mapping the World," Abelard-Schuman, Inc., Publishers.*)

this rapidly progressing subject. The main principles will be reviewed here.

The Photographic Flight. Air photos for mapping are made from steady big planes with a large fuel capacity; often they are reconverted bombers. They have holes in the bottom for the camera and the view finder. Cameras are large and usually make 9- by 9-in. pictures; their focal length varies from 6 in. to several feet. The longer the focus length, the larger the scale of the picture and the smaller the area covered. The cameras use roll film, some with as many as 500 exposures and automatic control. The photographer sets the clockwork which regulates the shutter, setting the speed with which one picture is taken after the other. He can turn the camera during flight to correct for "crab" caused by a side wind, which tends to deflect the plane from its planned flight line. Then he pushes a button, and the pictures are taken one after the other. With most modern cameras, each photo records automatically a picture of a small clock, an altimeter, and a round level bubble. It also records the number of the flight and exposure. Pictures are nor-

mally taken from great heights, 10,000 to 40,000 ft, unless much detail is needed. Pictures are usually taken with 60 per cent overlap forward and 30 per cent sideways, so that each point appears on several pictures.

The speed of air-photo coverage is amazing. A plane flying at 20,000 ft at 300 miles/hr can cover more than a thousand square miles of territory in an hour. It takes months of laboratory work, however, to produce the maps.

The Laboratory Work. To develop, fix, wash, and dry a roll of several hundred films requires special equipment. The rolls of negatives are preserved in metal cylinders, and positive contact prints are made on demand. A most important and often difficult job is *indexing*, plotting the exact position of the photograph on a reference map. This is done usually on a 1:1,000,000 World Aeronautical Chart. Indexing would be easy if the base map were correct in every detail, but this is rarely the case, as the purpose of photography is to improve a map. Thus, the indexer has to make a compromise between the existing map and the pictures taken at even intervals along a more or less known flight line. Index sheets are usually prepared as overlays on aeronautical charts, and they contain a number of flight lines classified for the Project, Mission, Roll, Flight, and Exposure.

Photo Mosaic Index. To be useful, an air photo cannot be at a much smaller scale than 2 in./mile. It is possible, however, to take all the photographs of an area, such as a county, and make a rough mosaic by overlapping the pictures and mounting them on a large board. The mounting is done so that the number of photographs is visible. A photographic reduction of the mosaic four or five times still shows rivers, lakes, cultivation, and cities and gives a good over-all picture. A photo mosaic index of any study area is most helpful.

Photogrammetry. This is the science of the transformation of photographs into maps. Several textbooks are available on the subject.

No precise map can be drawn without some ground control. In the ideal case, there should be at least three known points for each picture. They are geodetically fixed and marked on the ground with white so that they can easily be recognized on the photograph.

A base map is drawn at proper scale and in proper projection and all measured points are exactly plotted. Identical points on the photographs are pinpricked or marked on transparent positives called *diapositives*. These are now positioned in a plotting machine and adjusted for scale, tilt, and other distortions until the points

on the map and on the photographs coincide. This is usually done on stereoplotting machines. The detail of the photographs can now be drawn on the map either directly or with a pantograph.

Stereoplotting. The houses near the margin of a vertical air photo all seem to be leaning away from the center. This is because only the center of the picture is photographed vertically. The marginal parts are seen somewhat obliquely. Points below and above the datum level are displaced. In Chap. 2 we found that *parallactic displacement* is radial to the center and is increasing in straight ratio with the distance from the center and the height or depth of the object. Parallactic displacement has one compensating virtue. Displacement is different in two adjacent pictures, and, if they are viewed through a stereoscope, we obtain the mental image of a three-dimensional model. This fact is used by the stereoplotting machine, a number of which are on the market. A pair of

Fig. 16.13. A stereoplotting machine. The diapositives are properly positioned on top and projected on platform below. The operator guides a floating needle with the motion of the two wheels. Contours are drawn on the desk at right by a pantograph. (*Courtesy of OMI Corporation of America, Nistri Photocartograph.*)

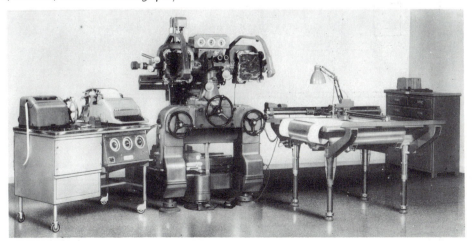

positive prints is used in the relatively simple stereocomparagraph; diapositives are used with the more complicated aero-cartograph, stereoplanigraph, multiplex, etc. In each of them, photographs are positioned for stereographic viewing and mirrors, lenses, and prisms can correct variation in scale and tilt. The operator sees the relief rising before his eyes vertically exaggerated. He focuses at a certain elevation and carries a floating needle point around the mountainside so that it moves along the same elevation. If he loses contact with the mountainside, the needle point shows double. Often a pantograph carries over his motion to a drawing table and a contour line will appear. By lowering and raising the needle point, successive contours can be made.

Most of us have seen three-dimensional slides or movies viewed through red and blue eyeglasses. The pictures are taken from slightly different points; one is projected onto the screen through a blue filter, the other through a red filter. Sometimes the pictures are printed over each other, one in red and the other in blue (*anaglyphs*). With the eyeglasses, each eye sees the picture which was taken from its own side, and the effect is indeed startling.

The Multiplex machine shown in Fig. 16.14 is built on the same principle. The air photos are made into small diapositives which are mounted on a rack in exactly the

Fig. 16.14. The Multiplex aeroprojector uses diapositives with red and green filters and corresponding lenses to produce fused images. The cameras on top have the same orientation as the actual cameras in the plane. (*From Abrams.*)

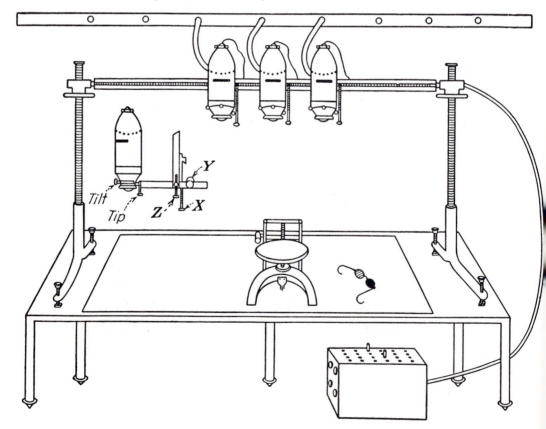

same relative position as that in which the pictures were taken. This is a *stereoscopic model*. Then they are projected, one through a blue-green and the other through a red filter, upon a small platform called a *platen*, which can be raised or lowered. In the center of the platen is a lighted pinhole which serves as a floating mark. Exactly underneath the pinhole is a vertical pencil. The operator looks through red and green eyeglasses at the stereoscopic image on the platen. Then he moves the whole tracing table around to draw the contour lines on the paper underneath. By lowering or raising the platform, successive contours can be made with as little as a 5-ft contour interval.

Radial-line Plotting (Semicontrolled Mosaic). If geodetically fixed positions are scarce, we can make a kind of air-photo triangulation with fair accuracy. Many of the less-developed countries were mapped this way, and it was a common sight in offices to see hundreds of square feet covered by slotted templates.

Even if there were a slight tilt or variation in scale of the photographs, the direction of any marginal point from the principal point is the same. We center the template on the principal point and select eight marginal points at which we direct the radial slotted arms. We do the same on the next photograph, directing the arms to identical points. If there are three points, there is only one position where the two templates can be held together by a peg sliding in the slots. The operation is repeated until the entire area is covered by templates. The assemblage is placed over a base map drawn on the average scale of the photographs on which the geodetically known points are located. If the radial-line

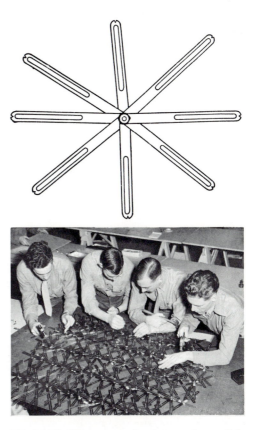

Fig. 16.15. Where control points are few, radial-line assemblies serve as a kind of second-order triangulation. (*From Abrams.*)

plot is good, the assemblage can be slightly stretched or squeezed to conform to these points and the error will be automatically distributed. All points are then transferred to the map, which can now be drawn with fair accuracy.

Mosaics. If all the pictures are rectified—rephotographed for exact scale and corrected for tilt—we can paste together hundreds of pictures and produce a photo map of a city, camp, golf course, or any limited area. The overlapping pictures are cut along roads and rivers where the separation is least conspicuous. A clever operator cuts only slightly into the photograph and "feather edges" the print by tearing the paper obliquely. The photographs are pasted to a board with gum arabic. The features can be named with white or black cellotype on transparent film. Then the mosaic can be rephotographed and printed by halftone. Some topographic sheets here and abroad have air-photo mosaics on the reverse side. In Sweden the mosaics are overprinted with transparent colors and form the base of a large-scale map series.

Trimetrogon Photography. During the war, we found to our dismay that most of the world was indeed very poorly mapped. A method had to be found to make maps quickly. The answer was the "trimet" camera. This is really three cameras mounted in an airplane so that one points directly down and the others point to the sides. This gives a coverage from horizon to horizon, as shown in Fig. 16.16.

The vertical pictures can be used directly, as before; the side pictures are transformed in an "oblique sketchmaster." Of

course, the areas near the horizon are seen so obliquely that they are not of much use in direct mapping, but they help in the orientation of the pictures. The same airplane, however, which covered a 4-mile-wide strip with verticals alone can record a useful strip 20 to 40 miles wide with a set of trimetrogon cameras.

New Developments

Aerophotography is so recent that new methods and new equipment are always presented at the annual meeting of the American Society of Photogrammetry. Many of the new instruments come from Europe. There are cameras which take a continuous vertical picture as the country rolls back below the camera. The Twinplex camera takes two pictures with converging angles. This increases the overlap with the successive pictures, and thus reduces the number of pictures. Color photography is becoming increasingly popular, although colored pictures do not work as well in stereography as black-and-whites. The film coating is thick and the various colors form at different wavelengths. Infrared photography is often used for forests and crops, as green appears light-colored. Similarly various filters may better define the geological formations. Experiments are now in progress toward the automation of aerosurveying using photoelectric scanning devices for plotting contour lines.

Shoran. What was the exact location of the plane when the picture was taken? This is a constant problem in aerosurveying. The answer is short-wave radio. Two intermeasured "transponder" stations on the ground are established along the flight line. The plane sends out short-wave impulses and activates the transponder sta-

Fig. 16.16. The trimetrogon system uses one vertical and two oblique wide-angle cameras so that the three pictures reach from horizon to horizon. (*Courtesy of the U.S. Army.*)

tions, which give a return signal. The time lag between the pulse and the reception of the return signal gives the distance. Thus we have three sides of a triangle, which locates the plane. The plane may even keep itself on course by an automatic device, a *straight-line flight computer*, controlled by shoran. Short-wave signals do not follow the earth's curvature and can be used only for a few hundred miles, depending upon the height of the plane. Shoran is used also for naval surveys of the coast to give the exact location of the surveying ship.

Short-wave radio signals are now used for direct triangulation (Tellurometer and others). Accuracy of measurement of distances by shoran is comparable to first-order triangulation, and may better it in the future. The advantage of shoran is that it can be used for longer distances and in places where visibility is bad.

Geodimeter. This is an instrument developed by Dr. Eric Bergstrand for measuring distances by the velocity of light. It measures the time lag between a light signal and the return from a reflector. It is so accurate that it may even replace the tape in base measurement. The Tellurometer is a similar instrument using short-wave impulses (see Fig. 16.18).

Odograph. This amazing instrument is mounted on a jeep and draws a continuous map of the route taken. The distances are obtained by a speedometer, directions from a magnetic compass, and the course is plotted on paper by a complex unit using photocells and electronic control.

Similar is the "elevation meter" which is mounted on a truck carrying a sensitive pendulum. The angle of slope and distance traveled are integrated electronically. Elevation of points along the route is recorded

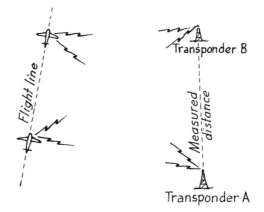

Fig. 16.17. Using shoran the plane gives out short-wave impulses which are returned from two or more intermeasured transponder stations. The time lag locates the plane. Exact location of each picture can be fixed by an airborne digital computer.

with an accuracy of about two feet in 50 miles.[2]

The advance in surveying in the last decades has been great and will be greater. In the future we can expect great progress in measuring and mapping the earth.

[2] J. L. Speert, A New Look at the Elevation Meter. Paper presented at American Congress on Surveying and Mapping meeting in March, 1958.

Fig. 16.18. The geodimeter and similar instruments measure distance by the velocity of light or short-wave impulses. It is used increasingly for city, highway, and construction surveys. (*Courtesy of the Geodimeter Company.*)

17

Map Projections

A globe is nearly spherical; a map is flat. It is impossible to flatten a global surface without stretching or tearing. The science of map projections gives various solutions to the way in which the impossible can be approximated. If a problem has no true solution, it usually has a great number of near solution; thus there are hundreds of different map projections.

As long as the area is small, the problem is not great. Even in a distance of 100 miles the scale error of many projections is less than 1:10,000, which is negligible for practical purposes. The larger the area to be mapped, the more serious is the problem, and it is greatest on world maps.

The term *projection*, though commonly used, is not a fortunate one. Very few of the projections actually used are "projected" in a geometrical sense. *Any regular system of parallels and meridians upon which a map can be drawn can be called a map projection.*

There is no one map projection which is better than all the others; a projection which is good for Asia is not necessarily the best for Africa, and a different projection is used for a navigation chart than for a stati.tical map. Each projection has certain virtues and certain limitations.

On *equal-area* or equivalent maps, every part, as well as the whole, has the same area as a globe at the same scale. To accomplish this, the shapes and angles have to be badly distorted. Note, for instance, at what angles the parallels and meridians cut each other in Fig. 17.1.

A *conformal* map is one on which any small area has the same shape as on a globe, and one point is in the true direction from any other as long as the points are close together. At any point the scale is the same in every direction, although it may change from point to point. In conformal projections the parallels are at right angles to the meridians, and in any small area the length of a degree of longitude is in true relationship to a degree of latitude, namely, $1°$ long $= 1°$ lat cos φ. To make a conformal map of the world, we have to distort the scale. On a Mercator, a conformal projection (see Fig. 17.9), Greenland is bigger than South America, although in reality it is only one-tenth as large. Conformal maps are suitable for navigation, as directions are

Fig. 17.1. Equality of area is often achieved by bad distortion of shape.

Fig. 17.2. On conformal projection any very small area has the correct shape. This can be achieved by varying the scale.

generally not greatly distorted. They are used for the various grid systems, as they are easy to calculate mathematically (see Chap. 18). No map can be equal-area and conformal at the same time; only a globe has true shapes and true areas.

Equality of area and conformality are valuable mathematical concepts; often, however, we prefer projections midway between the two, particularly if they have small scale distortion. The amount of deformation at any point of a map projection can be measured and expressed by Tissot's indicatrix.[1]

Construction of Projections

In any projection only certain parallels and meridians can be true, that is, the same length as on the globe of corresponding scale. We usually lay out projections by these true distances. If a meridian is true, every degree on it will be 69.1 miles in the scale of the map. If a parallel is true, a degree of longitude on it will be 69.1 cos φ miles, where φ is the latitude (see Table 15.7). If a parallel or meridian is divided truly, it is divided equally also; but if it is divided equally, it does not necessarily mean that it is divided truly. *Here and in the following discussion the earth is regarded as a perfect sphere of equal volume unless stated otherwise.*

We often have to lay out projections along a curving line. This can be done by using a flexible scale or with a divider if we employ chord distances. Nothing is gained by laying out very short distances repeatedly with the divider, as this may

[1] O. S. Adams, "General Theory of Polyconic Projections," U.S. Coast and Geodetic Survey Spec. Pub. 57, 1934, pp. 153–163; Arthur H. Robinson, "Elements of Cartography," John Wiley & Sons, Inc., New York, 1959, pp. 324–329.

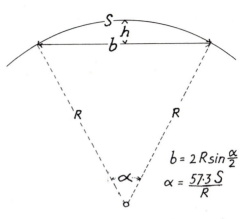

Fig. 17.3. In constructing projections we often have to lay out arc distances. This can be done most accurately by measuring off the chord distances.

lead to cumulative error. Rather, we calculate the chord distance for longer stretches, as in Fig. 17.3, where the formula and a table of chord distances are given.

TABLE 17.4. CHORD DISTANCES AND HEIGHT OF CHORDS (EXPRESSED IN PERCENTAGE OF ARC DISTANCES)

Degrees $\alpha°$	Difference between arc distance and chord distance $\left(\dfrac{S}{b} - 1\right)100$	Height of chord $\dfrac{100h}{b}$
5	0.03	1.1
10	0.12	2.2
15	0.31	3.3
20	0.55	4.4
25	0.81	5.5
30	1.16	6.6
35	1.58	7.7
40	2.05	8.8
45	2.61	9.9
50	3.25	11.1
55	3.96	12.2
60	4.72	13.4
65	5.60	14.6
70	6.56	15.8
75	7.50	17.0
80	8.60	18.2
85	9.85	19.5
90	11.2	20.7

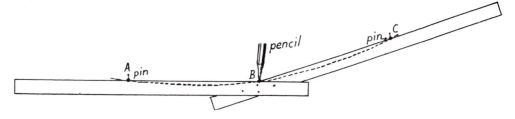

Fig. 17.5. A circular arc can be drawn through three given points by moving the cardboard strips along the two pins in A and C.

Most projections are symmetrical on the central meridian and on the equator. It is often sufficient to construct one quarter only, take a piece of tracing paper and copy that quarter on it with pencil, then reverse it for the other quarters, copying its back side and using it in the same way we use carbon paper.

Parallels and meridians often are flat circular arcs with very long radii. Figure 17.5 shows how to draw these arcs with two strips of cardboard.

Classification of Projections

The usual classification of projections refers to the idea of projecting the globe upon a developable surface, such as a cylinder, a cone, or a plane (see Fig. 17.6) which can be flattened out to make a map. Actually many projections did not originate

Fig. 17.6. Most map projections are related to a cylinder, cone, or plane. The projections actually used are usually modified from the original geometrical conception.

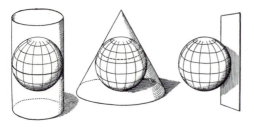

from this idea and most frequently projections are not geometrically projected. Yet it is still convenient to keep this grouping in a modified form, and in the balance of this chapter and in the next chapter we will discuss the systems in the following order:

1. Projections with horizontal parallels (cylindrical and others)
2. Conic group of projections
3. Azimuthal and related projections
4. Other unrelated projections

In the following pages, we will speak of a globe rather than the earth. It facilitates understanding to assume that the reduction to scale has already been accomplished.

Projection with Horizontal Parallels

Horizontal parallels make it easy to compare latitudes, an important controlling factor of our climate. Most lettering on the map is horizontal, which gives it an orderly appearance. The meridians can be either vertical or curved. The maps with vertical meridians are often called "cylindrical," although the actual projection of a globe upon a cylinder is almost never used. It is a part of our civilization to prefer horizontal-vertical lines; thus cylindrical projections are popular in spite of the lack of convergence of the meridians.[2]

Cylindrical Projections. *Equirectangular Projection.* This is the simplest projection composed of an evenly spaced network of horizontal parallels and vertical meridians. As our first step we select a parallel near the center of our area. We lay this parallel off horizontally and divide it truly; that is, the meridians will be cos φ distance from each other. We draw them as vertical lines, divide them true to scale, and draw the horizontal parallels (see Fig. 17.7). Thus the scale is true on all meridians and on the central parallel. In the Northern Hemisphere, the northern parallels are too long and the southern parallels are too short. East, West, North, and South are true; all other directions are changed.

This projection is good for city maps or estate maps because of its simplicity. It makes a fair world map, as seen in Fig. 17.8. The map is not conformal or equal-area, but the polar regions are less exaggerated than in the Mercator projection. If the standard parallel is the equator, we have a grid made up of even squares (see Fig. 18.23).

Mercator Projection. Mercator drew his 1569 world map in this projection. It has vertical meridians and horizontal parallels. The meridians are placed truly on the equator (see Fig. 17.9); the parallels are placed so that for any small area the scale along parallels and along meridians is the same as on a globe. This is a *conformal*

[2] For the study of cylindrical projections, see C. M. Miller, Notes on Cylindrical World Map Projections, *Geog. Rev.*, vol. 35, 1943, pp. 424–431. Miller presents some of his own projections, which are now often used.

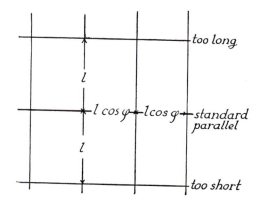

Fig. 17.7. In the equirectangular projection the scale is true along the standard parallel and along all meridians. The rest of the parallels are either too short or too long.

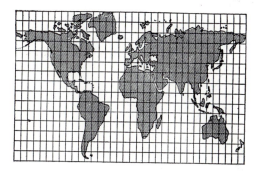

Fig. 17.8. Map of the world in equirectangular projection. The standard parallels are at 48° lat. (*Modified from U.S. Coast and Geodetic Survey Spec. Pub. 68.*)

Fig. 17.9. The Mercator projection shows a great increase in scale at high latitudes. Only the equator is true to scale. (*From Finch and Trewartha, "Elements of Geography."*)

TABLE 17.10. MERCATOR PROJECTION TABLE (DISTANCES OF PARALLELS FROM THE EQUATOR IN MINUTES OF LONGITUDE ON THE EQUATOR TAKING ACCOUNT OF THE EARTH'S ELLIPTICITY AS $\frac{1}{294}$)

Degrees	y	Degrees	y
1	59.596	41	2686.280
2	119.210	42	2766.089
3	178.862	43	2847.171
4	238.568	44	2929.594
5	298.348	45	3013.427
6	358.222	46	3098.747
7	418.206	47	3185.634
8	478.321	48	3274.173
9	538.585	49	3364.456
10	599.019	50	3456.581
11	659.641	51	3550.654
12	720.472	52	3646.787
13	781.532	53	3745.105
14	842.842	54	3845.738
15	904.422	55	3948.830
16	966.296	56	4054.537
17	1028.483	57	4163.027
18	1091.007	58	4274.485
19	1153.893	59	4389.113
20	1217.161	60	4507.133
21	1280.835	61	4628.789
22	1344.945	62	4754.350
23	1409.513	63	4884.117
24	1474.566	64	5018.419
25	1540.134	65	5157.629
26	1606.243	66	5302.164
27	1672.923	67	5452.493
28	1740.206	68	5609.149
29	1808.122	69	5772.739
30	1876.706	70	5943.955
31	1945.992	71	6123.602
32	2016.015	72	6312.610
33	2086.814	73	6512.071
34	2158.428	74	6723.275
35	2230.898	75	6947.761
36	2304.267	76	7187.387
37	2378.581	77	7444.428
38	2453.888	78	7721.700
39	2530.238	79	8022.758
40	2607.683	80	8352.176

projection. For example, at lat 60°, a degree of latitude is twice as long as a degree of longitude. On the globe the meridians converge toward the poles. In the Mercator projection the meridians are parallel. Thus a degree of longitude is increased toward the poles by $1/(\cos \varphi) = \sec \varphi$. In order to have the right proportion, a degree of latitude has to be increased also with the secant of the latitude.

For any parallel the distance y from the equator is the sum of secants of latitudes.

$$y = \sec 1' + \sec 3' + \cdots + \sec \varphi$$

A table for y is given in 17.10.

Although the Mercator projection has very good shapes for small regions, the scale is enormously exaggerated toward the poles. The poles cannot be shown at all; they are of infinite distance. It is not advisable to carry this projection north of lat 75°. Because of the great distortion, the usual kind of scales cannot be used in this projection. Figure 17.11 shows a scale for the Mercator projection.

This is the only projection in which all loxodromes (lines of constant compass directions) are shown as straight lines; therefore most charts for sea navigation are in this projection. For scale the graduated sides of the chart can be used, as a nautical mile is 1' of latitude. Directions can be read from the frequent compass roses on the chart.

In spite of its extreme variation in scale, the Mercator projection is quite popular for world maps of all kinds. The construction is easy and the projection fills a page well. But perhaps the chief reason for its popularity is its exaggeration of Northern latitudes. A cartographer who has to letter in "Netherlands," or "Copenhagen," will certainly appreciate this. It is advisable to inset the two polar maps on every Mercator world map.

A "transverse" form of the Mercator pro-

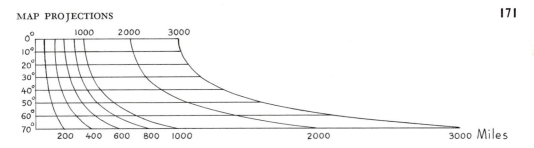

Fig. 17.11. Scale for a world map in Mercator projection. The curves are sinusoids.

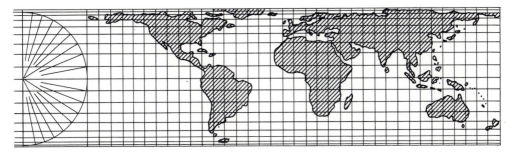

Fig. 17.12. Cylindrical equal-area projection of the world. (*From Finch and Trewartha, "Elements of Geography."*)

jection, used for the Universal Transverse Mercator grid, will be discussed later. Tables for an oblique development can be bought from the U.S. Coast and Geodetic Survey.

Other Cylindrical Projections. By varying the distances between the parallels, any number of cylindrical projections can be designed.

In the *cylindrical equal-area projection* the parallels are projected with horizontal rays upon a tangent cylinder, as shown in Fig. 17.12. It is rarely used in this form because of the extreme vertical compression of the polar regions, but in an oblique development it makes a good map for the British Commonwealth.

Gall's projection can be understood from Fig. 17.13. It exaggerates the Northern lands less than Mercator's and is often used in British atlases. The poles are lines the same length as the equator.

Projections with Horizontal Parallels and Curved Meridians. In this group of projections, the meridians converge in different

Fig. 17.13. Gall's cylindrical projection derived from a cylinder cutting the globe at 45°. The parallels are projected from the antipodal point on the equator to any meridian.

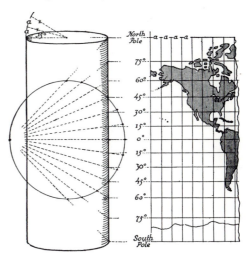

Fig. 17.14. Sinusoidal projection is good in the tropical regions but distorts the high latitudes. (*From Finch and Trewartha, "Elements of Geography."*)

TABLE 17.16. MOLLWEIDE PROJECTION (DISTANCE FROM THE EQUATOR TO THE POLE = 1)

Latitude	Distance of parallels from the equator	Latitude	Distance of parallels from the equator
0°	0.000	50°	0.651
5	0.069	55	0.708
10	0.137	60	0.762
15	0.205	65	0.814
20	0.272	70	0.862
25	0.339	75	0.906
30	0.404	80	0.945
35	0.468	85	0.978
40	0.531	90	1.000
45	0.592		

ways. They are used chiefly for equal-area world maps. Their main liability is great angular distortion in the peripheral portions.

The Sinusoidal Projection. We lay out a horizontal equator and a vertical central meridian and divide them both truly. We draw horizontal parallels and divide each of these truly,

$$1° \text{ long} = 1° \text{ lat cos } \varphi$$

for the meridians. The meridians are sine (or rather cosine) curves. The projection is sometimes called the Sanson-Mercator projection, after its early users. This is an equal-area projection rarely applied to world maps. Its central part is often used for maps of Africa, South America, and

Australia. It is also good for smaller countries, particularly if they extend in a North-South direction.

The Mollweide (Homalographic) Projection. We lay out a horizontal equator and divide it equally. We draw a central meridian half the length of the equator and divide it according to Table 17.16. We draw horizontal parallels and a central circle going through the poles. This will be the central hemisphere. We divide each parallel equally within the circle. The divisions are connected and form the meridians. The same divisions, if laid out on the parallels outside the circle, will make the outer meridians. The meridians are ellipses.

Fig. 17.15. The Mollweide projection, if centered on America, distorts Europe and Asia. (*From Finch and Trewartha, "Elements of Geography."*)

Fig. 17.17. Goode's interrupted homolosine projection combines equality of areas with little distortion in shape. (*From Finch and Trewartha, "Elements of Geography."*)

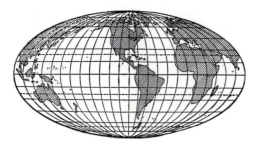

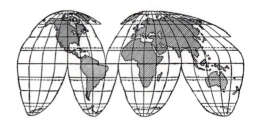

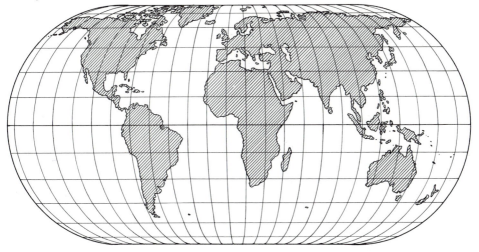

Fig. 17.18. The Eckert IV projection. The poles are represented by lines half the length of the equator.

The projection is equal-area since the distance of the parallels is calculated to make it so. If we compare the map with a globe of equal surface, only the two 40°40′ parallels are true to scale.

The projection is popular, particularly in Europe, for world maps, centered on the prime meridian. It is not as suitable with America centered on the prime meridian as Asia is cut in two. The projection makes fair hemisphere maps.

Goode's Interrupted Projections. We draw a horizontal equator and divide it equally. We draw a vertical central meridian for each continent. With these we draw the central portions of the sinusoidal or the Mollweide (or Eckert, Aitoff, etc.) projections, interrupting the globe in the oceanic areas. In the *homolosine* projection of Fig. 17.17, the tropical zone is in the sinusoidal, the rest in the Mollweide projection.

These projections combine good shape with equality of area and are common in atlases. They are less good when showing world relationships.

The Eckert Projections. These projections are similar to the sinusoidal and to the Mollweide, but the poles are lines half as long as the equator. In the Eckert IV projection the meridians are ellipses and the parallels are so placed as to make the projection equal-area (see Table 17.19). In the Eckert VI projection the meridians are sinusoids. These projections give better shapes than the Mollweide projection in the Northern areas and are becoming increasingly popular.

TABLE 17.19. SPACING OF PARALLELS IN THE ECKERT IV PROJECTION *

Latitude	Distance of parallels from equator	Latitude	Distance of parallels from equator
0°	0	50°	0.718
5	0.078	55	0.775
10	0.155	60	8.827
15	0.232	65	0.874
20	0.308	70	0.915
25	0.382	75	0.950
30	0.454	80	0.976
35	0.525	85	0.994
40	0.592	90	1
45	0.657		

* After A. Robinson.

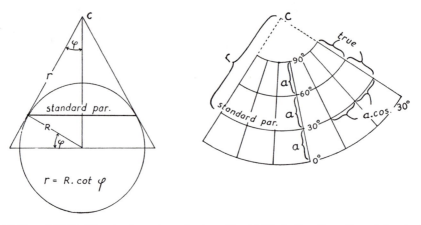

Figs. 17.20 and 17.21. The simple conic projection: Fig. 17.20, finding the radius of the standard parallel; Fig. 17.21, layout of the projection.

Several variations of this group of projections have been proposed, including the Bogg's Eumorphic projection, the Putnin's (Riga) Elliptical, Craster's Parabolic projection.

Exercise 17.1

In pencil draw the parallels 30°, 40°, 50°, and 60° and meridians 10°W, 0°, 10°E, corresponding to a globe with an equator 72 in. in circumference, in the following projections. Add your calculations to each projection.

a. Equirectangular with central parallel 50°
b. Mercator
c. Sinusoidal
d. Mollweide
e. Eckert IV

The Conic Group of Projections

In the basic construction of the conic projection, we project the globe upon a tangent cone, open up the cone along one of its elements, and lay it out flat.* The

projection is not used in this form; but all conic projections are modifications of the basic construction. All of them have circular parallels and straight, evenly placed, radiating meridians. Conic projections are used particularly for maps in the middle latitudes; they have remarkably little scale error.

The construction of all conic projections starts with finding the radius of the *stand ard parallel*, where the cone is tangent to the globe. From Fig. 17.20

$$r = R \cot \varphi$$

where φ is the latitude, and R the radius of the globe. Table 17.22 gives the values of r in fractions of the radius of the globe.

Simple Conic Projection. We choose the standard parallel in the center of our map area and draw it as a circle with a radius

$$r = R \cot \varphi$$

---

* NOTE TO THE INSTRUCTOR: This can easily be demonstrated with the help of a globe and a cone of transparent paper.

TABLE 17.22. RADII OF STANDARD PARALLELS
IN THE CONIC PROJECTION

$$r = R \cot \phi$$

The earth's radius: $R = 1$

0°	∞	50°	0.8391
5	11.430	55	0.7002
10	5.671	60	0.5774
15	3.732	65	0.4663
20	2.747	70	0.3640
25	2.145	75	0.2679
30	1.732	80	0.1763
35	1.428	85	0.0875
40	1.192	90	0.0000
45	1.000		

$$\frac{r_a}{a} = \frac{d}{b-a}$$

$$r_a = a \frac{d}{b-a}$$

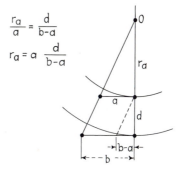

Fig. 17.23. Finding the center O for the conic projection with two standard parallels; a and b are the true lengths of the parallels or parts (say 10°) of the parallels; d is the true distance between the standard parallels.

as in Fig. 17.21. We divide the standard parallel truly,

$$1° \text{ long} = 1° \text{ lat} \cos \varphi$$

and connect the divisions with C for the meridians. We divide one meridian truly, and parallel circular arcs will form the parallels. The pole will be a circular arc, at its true surface distance from the standard parallel. Scale is true along all meridians and along the standard parallel; everywhere else the scale is too large. For this reason it is not used as often as the following modification.

Conic Projection with Two Standard Parallels. Instead of one standard parallel we use two, one in the upper part of the map and one in the lower part, so that roughly two-thirds of the map is between them. The standard parallels are circles spaced their true surface distance apart, and their centers and radii can be found according to Fig. 17.23. All other parallels are circles at their true distances. We divide the standard parallels truly and the meridians are straight lines through the division point. As a check on the construction, they have to meet at one point. Scale is

slightly too short on the central parallels and too long on the marginal ones. On a map of the United States the greatest scale error is less than 2 per cent. The projection is sometimes, not quite correctly, called the "secant conic projection."

Conic Equal-area Projection with Two Standard Parallels, or Albers Projection. We may modify the previous projection by placing the parallels so that the zones between the parallels have the same area as on the globe. This projection was calculated by H. C. Albers of Gotha in 1805.[3]

This projection combines small scale error with equality of area. It is one of the best projections for a map of the United States and at present it is used for most government maps. The greatest scale error is 1.25 per cent. The projection is used also in the U.S.S.R. and Europe.

The Lambert Conformal Conic Projection with Two Standard Parallels. If we place the parallels so that for every small

[3] The calculation involved and tables for it can be found in the U.S. Coast and Geodetic Survey Spec. Pub. 130, 1927.

TABLE 17.24. TABLE FOR THE CONSTRUCTION OF A MAP OF THE UNITED STATES *

Latitude	Radius of parallel	Spacings of parallels
	Meters	Meters
20°	10 253 177	
21	10 145 579	107 598
22	10 037 540	108 039
23	9 929 080	108 460
24	9 820 218	108 862
		109 249
25	9 710 969	
26	9 601 361	109 608
27	9 401 409	109 952
28	9 381 139	110 270
29	9 270 576	110 563
29°30′	9 215 188	110 838
30	9 159 738	
31	9 048 648	111 090
32	8 937 337	111 311
33	8 825 827	111 510
34	8 714 150	111 677
		111 822
35	8 602 328	
36	8 490 392	111 936
37	8 378 377	112 015
38	8 266 312	112 065
39	8 154 228	112 084
		112 065
40	8 042 163	
41	7 930 152	112 011
42	7 818 231	111 921
43	7 706 444	111 787
44	7 594 828	111 616
		111 402
45	7 483 426	
45°30′	7 427 822	111 138
46	7 372 288	
47	7 261 459	110 829
48	7 150 987	110 472
49	7 040 925	110 062
		109 592
50	6 931 333	
51	6 822 264	109 069
52	6 713 780	108 484

* Using Albers Equal-area Projection with two standard parallels at 25° and 45°. (From Deetz and Adams, "Elements of Map Projection.")

TABLE 17.24. TABLE FOR THE CONSTRUCTION OF A MAP OF THE UNITED STATES (CONTINUED)

Longitude from central meridian	Chords on latitude 25°	Chords on latitude 45°
	Meters	Meters
1°	102 184.68	78 745.13
5	510 866.82	393 682.00
25	2 547 270	1 962 966
30		2 352 568

area the scale along the parallels is the same as along the meridians, we have a conformal projection, having true directions at every point. The calculations for this projection were made by J. H. Lambert (1728–1777) in Germany. Lambert designed a number of other projections and did much to formulate the science of map projection.[4]

This projection is used for air navigation charts as it shows directions well over fairly large areas. Maximum scale error is slightly larger than in the Albers projection, but on a map of the United States it does not exceed 2¼ per cent.

All the conic projections mentioned before are easy to construct and have small scale error. They are particularly good for countries having their largest dimension in an East-West direction which does not require a great departure from the standard parallels. They are not good for world maps or large continents. They are often used in atlases. If a large map is prepared, any part of the map can be framed into a page, as any meridian can be central.

The following projections are further modifications of the conic system because they have curved meridians.

[4] Tables for the projection are contained in U.S. Coast and Geodetic Survey Spec. Pub. 52 and 68.

Bonne Projection. We draw a vertical central meridian and cross it with a selected standard parallel with a radius of

$$r = R \cot \varphi$$

We divide the central meridian truly and draw circles through the divisions for parallels. Thus far the construction is identical with the simple conic. The difference is in dividing *each* parallel truly, and curves drawn through the division points delineate the meridians.

The projection is equal-area. Shapes are good near the central meridian; distortion increases progressively to the left and right. The pole is a point, but not the same as the center of the parallels. If the standard parallel is the equator, the projection becomes the same as the sinusoidal.

The Bonne projection was commonly used for countries in the mid-latitudes. Most of the topographic sets of maps of France, the Netherlands, Switzerland, etc., were made in this projection. One network is drawn for the whole country and is then cut into individual sheets. Thus the parallels and meridians on a sheet may all curve in one direction. The Bonne projection is often used for the continents of Europe, Asia, and North America. With 15° lat as standard, it makes a fair world map. The projection is named after Rigobert Bonne (1727–1795), but it was actually used earlier.

The Polyconic Projection. We draw a vertical central meridian and divide it truly. The parallels are nonconcentric circles, the radii of which are $r = R \cot \varphi$ as every parallel is regarded as standard. We divide off the parallels truly for the meridians. The equator will be a horizontal line,

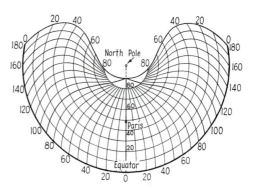

Fig. 17.25. Northern hemisphere in Bonne projection centered on Paris. Note that the pole is not identical with the center of the parallels. (*From Deetz and Adams, "Elements of Map Projection."*)

and the poles are points at their true surface distance from the equator. The projection seems to envelope the globe in a great number of cones; hence the name "polyconic."

The projection is neither equal-area nor conformal. Scale error is small near the central meridian but increasingly great away from it. The scale error is less than 1 per cent within 560 miles of the central meridian.

The polyconic projection is particularly suitable for countries such as Chile, which extend in a North-South direction. Its chief use is for topographic sheets. Tables are available for the horizontal and vertical coordinate for each intersection for various scales.[5]

The projection was devised by Ferdinand Hassler, the first director of the U.S. Coast Survey, in 1820. It became so popular with

[5] U.S. Geological Survey Bulletin 809; U.S. Coast and Geodetic Survey Spec. Pub. 5 and 57. Prof. John Leighley, Extended Uses of Polyconic Projection Tables, *Annals Assn. Am. Geog.*, 1956, pp. 150–173. This article gives many useful suggestions.

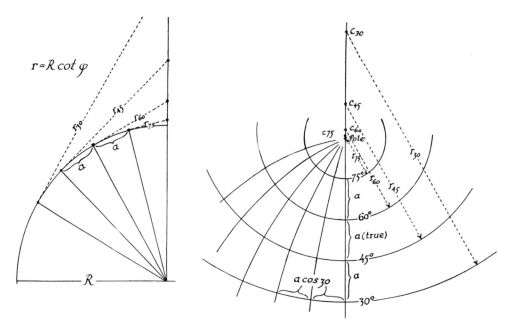

$$r = R \cot \varphi$$

Fig. 17.26. The polyconic projection has a truly divided central meridian, with nonconcentric circular parallels. The parallels are truly divided for the meridians.

our government that not only our topographic sheets are in this projection, but it was used also for maps of the whole United States in spite of its 6 per cent maximum scale error. At present the Albers projection is preferred.

Polyhedric Projections. Many European countries have their topographic sheets in the polyhedric projection. In these projections the small quadrangles of the globe are projected upon plane trapezoids with straight sides. There are many modifica-

tions. The central meridian or the sides can be made true to scale. Within the limits of a topographic sheet, the differences are microscopic.

Exercise 17.2

Draw parallels 30°, 40°, 50°, and 60° and meridians 10°W, 0°, 10°E, on the scale of a globe 72 in. in circumference in (1) the simple conic, standard parallel 50°; (2) conic with parallels 40° and 60° as standard; (3) Bonne projection; standard parallel 40°; (4) polyconic projection. Add your calculations to each sheet.

18

Azimuthal Projections, Grid Systems

Azimuthal Projections

In these projections we project a part of the globe upon a plane from some eye point. The various terms used are best understood from Fig. 18.1. Different eye points result in different projections. Some projections are modifications of this process, but all azimuthal projections have the following in common:

1. All great circles, passing through the center, appear as radiating straight lines and have correct bearings or azimuths; hence the name "azimuthal." For example, on a polar projection all meridians radiate from the pole.
2. All points equally distant from the center of the projection on the globe are equally distant also on the map. A circle connecting these points is called a "horizon circle." In the polar case the horizon circles are the parallels.
3. Places equally distant from the center have equal distortion. There is no distortion at the center.

4. All azimuthal projections differ from each other only in the radii of the horizon circles; thus they can be transformed easily into one another. Radial scales can best be obtained from the polar projections.

In the cylindrical and conic projections discussed above, we usually work with projections in which the axis of these solids corresponds to the axis of the globe. Azimuthal projections, however, are common in all positions, *equatorial, polar,* and *oblique,* as in Fig. 18.1. The plane upon which we project need not be tangent. Projection upon a parallel plane would differ only in scale.

Fig. 18.1. Azimuthal projections are drawn from any point of view. (*C. S. Hammond & Company.*)

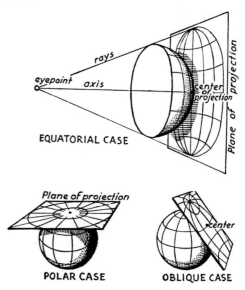

179

POLAR CASE

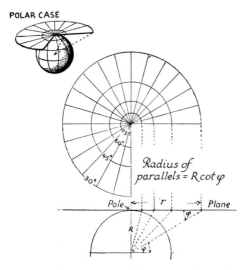

Fig. 18.2. Construction of the gnomonic projection, polar case.

Gnomonic Projection. In this projection the eye point is in the center of the globe. The construction of the polar and equatorial cases can best be understood from Figs. 18.2 and 18.3. Here, as in the other following constructions, we use the principles of descriptive geometry. We draw a side view in which the parallels show at true angles. Underneath we use a plane view, in which the equatorial plane is

Fig. 18.3. Construction of the gnomonic projection, equatorial case.

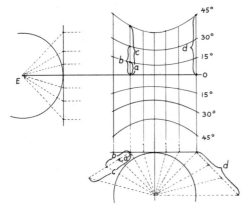

turned flush with the paper to obtain the meridians. In both, the plane of projection is perpendicular to the paper. At the bottom of Fig. 18.3, we turned the meridional planes down flush with the paper to obtain *a b c d*. The construction of the oblique projection is difficult.[1]

Scale is enormously exaggerated away from the center and it is not practical to show more than 45° from the center. The projection has one great virtue—this is the only projection in which *all great circles are straight lines.*

The U.S. Hydrographic Office has published gnomonic charts for all oceans. The navigator lays out longer voyages or flights with a straightedge, but he transfers this line upon a Mercator chart since he has to navigate with a compass (see Fig. 15.11). Gnomonic charts are also used for radio and seismic work, because the waves travel in more or less great-circle directions. A photograph of the stars is in gnomonic projection (unless curved plates or special lenses are used).

Orthographic Projection. In this projection the surface of a globe is projected with parallel rays upon a perpendicular plane. In other words, the eye point is at infinity. The constructions of the equatorial and the polar cases are shown on the side and the bottom of Fig. 18.4. The important view, however, is the oblique, because it looks like a globe. The parallels are ellipses, the large axes of which can be measured off from the side view; the small axes can be projected across from the side view with horizontal rays. The meridians are also ellipses and can be projected from the polar

[1] See E. Raisz, "General Cartography," McGraw-Hill Book Company, Inc., 1948, p. 81.

view. The largest area that can be shown in the orthographic projection is a hemisphere.

The radial scale decreases rapidly toward the peripheries and only the central area and all circles around the center of the projection are true to scale. Although distortion is very bad toward the peripheries, we do not see it so because we imagine that we see a globe and not a map. This projection can be focused on a region which will appear large while the peripheral regions are in the background. It is also often used in art and illustration. The popular but ill-defined term *global* usually refers to a map in this projection or to a photograph of the globe.

The perspective (globe-view or photographic) projection is the picture of a globe from a finite distance. The farther we are from the globe, the more similar the picture will be to an orthographic projection. Obviously we cannot see a complete hemisphere. The construction of the projection is more difficult than for the orthographic projection. It is easy, however, to photograph a globe, which explains the popularity of this projection with artists. It has the advantage that we can focus upon a relatively small area by bringing the camera close and we do not have to show too much of the rest of the world. Many of Richard Edes Harrison's maps are in this or in the orthographic projection (see Fig. 1.6).

Orthoapsidal Projections. For these projections we draw a parallel-meridian network on any suitable solid other than a sphere and then make an orthographic projection of that. The solids usually used are ellipsoids, toroids, hyperboloids, trifoliums, quadrifoliums, etc. The advantage of this

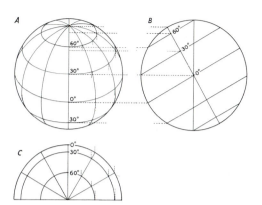

Fig. 18.4. Construction of the oblique orthographic projection with the axis of the globe tilted at 60°. Note the visual quality of the projection.

is that the whole earth is visible, not merely a hemisphere. The *armadillo projection* (see Fig. 18.5) has more land in proportion to sea than any other world map. The construction of the orthoapsidal projection is similar to the orthographics.[2]

Stereographic Projection. The eye point here is the point antipodal to the center. The projection has a unique quality—*any circle drawn on the globe will show as a circle on the map also.* Thus all parallels and meridians are arcs; therefore they can be drawn with a compass easily and exactly, as shown in Fig. 18.7. Parallels and the meridians intersect at right angles and the projection is conformal.

The oblique case is important because it is easy to draw and can be used for transformation into the other more difficult oblique azimuthal projections. Construction is shown in Fig. 18.8. At present it is used for star maps, for plotting charts, and

---

[2] E. Raisz, Orthoapsidal World Maps, *Geog. Rev.*, vol. 33, 1943, pp. 132–134.

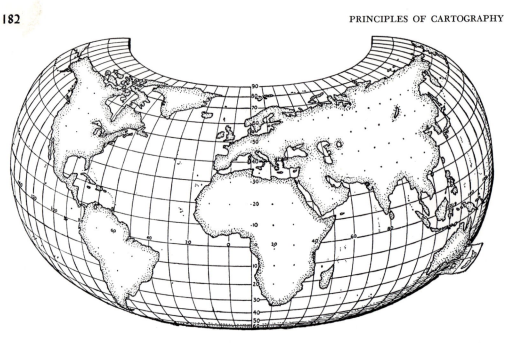

Fig. 18.5. The armadillo projection got its name from the little Mexican animal that coils itself into a sphere if frightened. The construction of the map is similar to the orthographic projections. (*Geog. Rev.*, 1943, p. 133.)

for various maps used in geophysics, as it is easy to solve problems in spherical geometry on it. In the past almost all Eastern

Fig. 18.6. In the stereographic projection the scale is increasing from the center toward the peripheries. (*From Finch and Trewartha, "Elements of Geography."*)

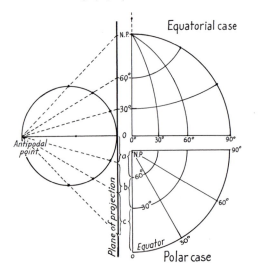

and Western Hemisphere maps were drawn on this projection in spite of considerable scale error.[3]

Azimuthal Equidistant Projection. This projection can best be understood from the polar case. We draw radiating straight meridians, and the parallels are equally spaced circles. We cannot arrive at this by projecting the globe upon a plane from any single eye point; yet this is still an azimuthal projection. This is the most frequently used projection for the Arctic and the Antarctic, and its most famous application is the emblem of the United Nations. This is the only projection in which *all points are in true distance and true global direction from the center*. The point antip-

[3] For formulas and tables for construction of this projection as calculated by J. A. Barnes, see Arthur Robinson, "Elements of Cartography," John Wiley & Sons, Inc., 1953, app. F.; U.S. Coast and Geodetic Survey Spec. Pub. 57.

odal to the center becomes a circle with a diameter of the earth's circumference.

The equatorial case is rarely used, but the oblique case is important. The map can be centered on any city, and every place in the world will be in correct distance and direction. Airlines like to use such maps and they are also used for radio and seismic work.

Construction of the oblique equidistant projection can best be done by transforming it from the stereographic. First we construct an oblique stereographic projection of the same tilt. Then we make a paper or plastic scale upon which we draw a stereographic radial scale. This we get from a meridian of the polar projection of a hemisphere of the same size, as in Fig. 18.6. Then we

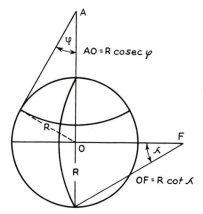

Fig. 18.7. Construction of the parallels and the meridians in the equatorial case of the stereographic projection.

draw an equidistant scale upon the same paper scale with red ink, starting from the same 0°. This scale can be larger or smaller

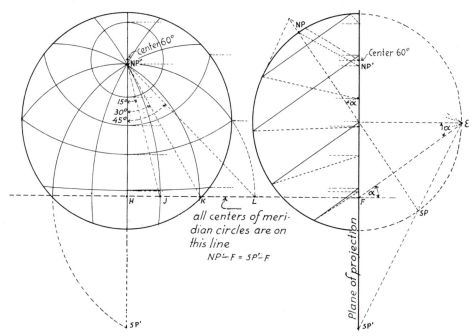

Fig. 18.8. Construction of the oblique stereographic projection. On the right is a side view of a tilted globe, halved by the plane of projection. E is the eye point. Each point is projected from the eye point upon the plane of projection and then carried over to the map on the left. H is the horizontal parallel and locus of the centers of meridians. South of H the parallels curve the other way.

Fig. 18.9. In the azimuthal equidistant projection, every point is at its true distance and in the right direction from the center. (*C. S. Hammond & Company.*)

as desired. We pin down the zero point very exactly on the center of the oblique stereographic projection and turn it very exactly to each intersection. We read off the radial distance in degrees as precisely as we can and prick the same distance along the same radius, according to the equidistant scale. The stereographic projection

Fig. 18.10. In the azimuthal equidistant or equal-area projections, the antipodal points to any point will be on a line passing through the center of projection, 180° apart.

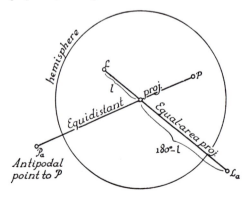

cannot be drawn much larger than a hemisphere, but the equidistant projection can show the whole world. We can get the antipodal point to each intersection according to the method shown in Fig. 18.10.

Azimuthal Equal-area Projection. Let us draw again radiating meridians for a polar projection. We will now vary the distance between parallel circles so that the area of each zone between two parallels is the same as on a globe. The construction is shown in Fig. 18.11. The oblique and equatorial cases can be transformed from a stereographic projection in the same way as the equidistant, but using a scale derived from the polar case as in Fig. 18.11. The antipodal points are found by laying out 180° − 1, as in Fig. 18.10.[4] The equatorial case is common for hemispheres. The oblique case makes excellent continent maps and is often used to show land and water hemispheres. The polar case can be extended to make a world map. The projection was designed by J. H. Lambert in 1772.

The Aitoff and the Hammer Projections. If we take any circular Eastern or Western Hemisphere map, we may double the horizontal distance from the central meridian of every intersection and thus transform the circle into an ellipse. This we can use for a world map similar to the Mollweide, but the parallels will be curved.

This was done to the equidistant projection by Aitoff in 1889, and to the azimuthal equal-area projection by Hammer in 1892. There was some confusion about the authorship of these projections which was cleared up in Karl-heinz Wagner's

[4] The U.S. Coast and Geodetic Survey Spec. Pub. 68 has tables for construction of this projection tilted 40°.

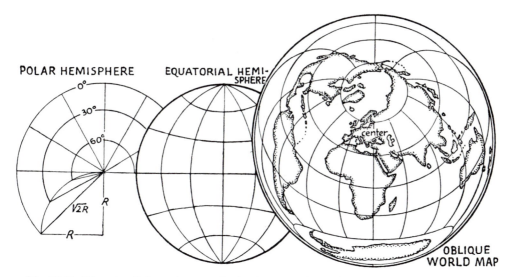

Fig. 18.11. The azimuthal equal-area projection is commonly used for continents and world maps centered on the North Pole. The oblique map is centered on the center point of all lands. (*C. S. Hammond & Company.*)

"Kartographische Netzenentwürfe," and the whole story was interestingly told by John B. Leighly.[5]

William Briesemeister used an oblique azimuthal equal-area projection map and extended it into an ellipse 1.75 times wider than high.[6] This map is used for the "Atlas of Diseases" of the American Geographical Society, for it shows intercontinental relationships very well.

Transverse Projections. These projections are related to a globe that is turned 90° from its usual orientation. Some authors call the oblique projections "transverse" or "meridional," causing confusion. Any projection can be transformed into a "transverse" orientation. The simplest way to construct transverse projections is to draw a stereographic equatorial projection and

[5] J. B. Leighly, Aitoff and Hammer, *Geog. Rev.*, vol. 45, 1955, pp. 246–249.
[6] William Briesemeister, A New Oblique Equal-area Projection, *Geog. Rev.*, vol. 43, 1953, pp. 260–261.

superimpose the polar case on it in another color. The latitude and longitude of each intersection can be read in the other orientation, and the same latitudes and longitudes can be applied to any projection.[7]

The Transverse Mercator Projection. If we relate a Mercator projection to a cylinder with a North-South axis, we may

[7] Tables are also available as in U.S. Coast and Geodetic Survey Spec. Pub. 67.

Fig. 18.12. The Hammer projection is the same as the Lambert azimuthal equal-area projection, but the horizontal distances from the central meridian are doubled. (*From Finch and Trewartha, "Elements of Geography."*)

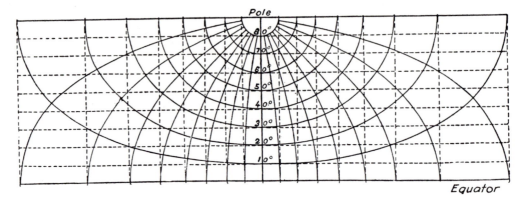

Fig. 18.13. Northern hemisphere in the transverse Mercator projection can be related to a cylinder tangent along a meridian. Near the central meridian it makes a good conformal projection. The related equatorial projection is shown by dotted lines. Note the great distortion away from the central meridian. (*After Deetz.*)

imagine it related to a cylinder with an East-West axis. Thus the central meridian will be divided truly, and the equator will be divided according to the increasing scale. As we could not show the poles in the upright case, so we cannot show the antipodal meridian in the transverse case. The projection is also conformal, but the loxodromes are curved.[8]

The part of the projection close to the

[8] Tables of the azimuth and radial distance of each intersection are given in U.S. Coast and Geodetic Survey Spec. Pub. 67.

Fig. 18.14. Transformation of one spherical coordinate system into another.

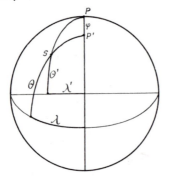

central meridian combines conformality with small scale error. It is used often as base for rectangular grids for Great Britain and for many of our states which extend in a North-South direction. The most important application is, however, the Universal Transverse Mercator Grid (see Fig. 18.20).

Transformation of One Spherical Coordinate System into Another. Having a globe with the usual system of parallels and meridians converging, we could select any other axis and draw a new system of parallels and meridians. Any intersection of the new system can be expressed by its latitude and longitude in the old system. In Fig. 18.14 the new pole is shifted $\varphi°$ from the old pole P to the new pole P'. From the solution of the spherical triangle $PP'S$ for any intersection S in the new system,

$$\sin \theta' = \sin \theta \cos \varphi - \sin \varphi \cos \theta \cos \lambda$$

$$\sin \lambda' = \frac{\sin \lambda \cos \theta}{\cos \theta'}$$

in which θ is the latitude of S in the new system. λ is the longitude in the new system, while θ' and λ' are the coordinates of S related to the old system.[9]

Other Projections

Many commonly used projections do not fit in any of the previous classes. To design new projections is a fascinating problem, and hundreds of networks have been presented. One may think that all possibilities are exhausted by now, yet year after year some new projections appear. Some examples are presented below but many good ones had to be omitted for lack of space.

The Globular Projection. This projection is often used for hemispheres, as it is most easy to construct. Divide the central meridian, the equator, and the circumference equally and draw circles through the intersection points, as in Fig. 18.15. The projection is not conformal or equal-area, but distortions are small.

Star Projections. In these projections the Northern Hemisphere is in a polar projection and the Southern continents are appendages to it, as in the emblem of the Association of American Geographers (see Fig. 18.16).

Heterohedral Projections. In these projections the globe is projected upon a system of polygons, such as triangles, squares, pentagons, hexagons, octagons, etc., and these surfaces are opened up and laid out flat to make a world map. This was tried by Leonardo da Vinci and by Albrecht Dürer in the Renaissance. "World Maps and Globes," by Irving Fisher and O. M.

[9] Tables for every 5° of tilt are given in Hammer's "Über die geog. wichtigsten Kartenprojectionen," Stuttgart, 1889; in somewhat different form in Karl-heinz Wagner, "Kartographische Netzenentwürfe," Leipzig, 1949; also in the U.S. Coast and Geodetic Survey Spec. Pub. 67.

Fig. 18.15. In the globular projection, the equator, the central meridian, and the peripheral circle are divided into even lengths.

Miller,[10] gives many examples of such projections. Some of them, like Fisher's "polygnomonic maps," can be fitted into a near globe. Cahill's butterfly projection (see Fig. 18.17) is an example where the octahedral sides are modified to give less scale error.

[10] Essential Books, 1944.

Fig. 18.16. Star-shaped maps are usually centered on the North Pole or on the center of the land hemisphere near Nantes, France.

Fig. 18.17. B. J. S. Cahill's butterfly projection.

Interruption of Projections. A globe is a continuous surface; a map is not; thus every map is interrupted. Most world maps are interrupted along a meridian, usually in the Pacific Ocean. Figures 18.8 and 18.11 are discontinuous along a point, the antipodal point, extended into a large circle. Usually we call a map "interrupted" only if more than one meridian is involved, as in Fig. 18.17. Interruption gives better shapes and has small scale error, but is less good to show world relationships.

Identification of Projections. We should always name the projection on our maps. Often, however, this is not done and we may need to identify projections. The key to projections given in Table 18.18 will help to identify the most common projections.

Choice of Projection. When Richard E. Harrison chose the oblique orthographic projection for his famous maps in *Fortune* (also see Fig. 1.6), he made an intelligent choice. He considered not only the areas to be shown and the purpose of the maps, but also the mentality of the busy and independent American businessman, who wants to understand geographical problems quickly and is not concerned about cartographic traditions such as North orientation. Much of the success of a map depends on our choice of a projection to fit the user.

For large-scale maps there is little problem. Almost every projection with some truly divided parallels and meridians will do. In medium-scale maps, the conic group is the most common choice. For tropical countries and continents, the sinusoidal has advantages. Asia, Europe, and North America are often shown in the Bonne, azimuthal equal-area, or Mollweide projections. O. M. Miller's bipolar oblique conic conformal projection is especially designed for both Americas.

The real problem comes with world maps. Here the purpose of the map is considered. When the late B. J. S. Cahill published his butterfly map (see Fig. 18.17), with its excellent shapes and small scale error in his "A World Map to End World Maps," he was convinced that all problems had been solved. Yet the map is hardly ever used. With world maps we usually want to show intercontinental relationships rather than to use them for navigation. Thus many will prefer a continuous Mollweide, an Eckert, or even a cylindrical projection.

In simple or diagrammatic maps, we like horizontal parallels. Horizontal lettering is easy to read and the parallel-meridian network can be omitted and indicated only on the sides. Strictly speaking, for statistical maps, particularly for dot maps, we have to use equal-area projections. Yet many statistical maps of the United States are on the polyconic projection which has considerable scale error. For maps of the United States, the Albert conic equal-area projection is preferable with its maximum scale error of 1.25 per cent. Certainly no dot map should be drawn in the Mercator projection.

These are just examples. Each map pre-

TABLE 18.18. KEY TO PROJECTIONS *

	Parallels	Meridians	Projection	Merit	Use
1	Horizontal, spaced equally at true distances	Vertical, spaced equally, true on standard par.	RECTANGULAR	Easy to construct	City maps, less exact maps
2	Horizontal, spaced closer near equator	Vertical, spaced equally, true on equator	MERCATOR	Conformal Straight loxodromes	Charts, world maps
3	Horizontal, spaced equally at true distances	Sine curves, spaced equally, true on each par.	SINUSOIDAL	Equal-area	Tropics
4	Horizontal, spaced closer near poles	Ellipses, spaced equally.	MOLLWEIDE	Equal-area	World maps, hemispheres
5	Horizontal, spaced closer near poles (poles are lines half length of equator	Ellipses, spaced equally. Sine curves, spaced equally.	ECKERT IV VI	Equal-area Equal-area	World maps
6	Concentric circles, spaced equally	Radiating straight lines, spaced equally, true on one or two standard par.	CONIC	Small distortion	Middle latitudes, series maps
7	Concentric circles, spaced closer at N and S ends	Radiating straight lines, spaced equally, true on one or two standard par.	ALBERS	Equal-area	Maps of U.S.
8	Concentric circles, spaced wider at N and S ends	Radiating straight lines, spaced equally, true on one or two standard par.	LAMBERT'S CONFORMAL CONIC	Conformal	Aeronautical charts
9	Concentric circles, spaced equally at true distances	Curves, spaced equally, true on each par.	BONNE	Equal-area	Middle latitudes
10	Nonconcentric circles, spaced true on central meridian	Curves, spaced equally, true on each par.	POLYCONIC	Tables available	Topographic sheets
11	Nonconcentric circles, spaced equally on periphery and central meridian	Circles, spaced equally, true on equator	GLOBULAR	Easy to construct	Hemispheres
12	Nonconcentric circles, spaced closer near center	Circles, spaced closer near center	STEREO-GRAPHIC	Conformal	Oblique transformations
13	Ellipses, spaced closer near periphery	Ellipses, spaced closer near periphery	ORTHO-GRAPHIC	Visual	Hemispheres, continents
14	Curves, spaced closer near poles on central meridian	Curves, spaced closer near sides of equator	AZIMUTHAL EQUAL-AREA	Equal-area	Hemispheres, continents

* Polar interrupted, and rare projections are not included.

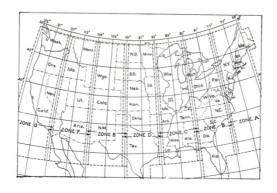

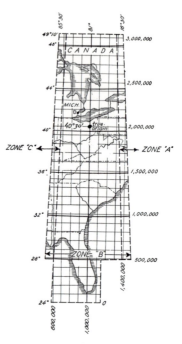

Fig. 18.19. Progressive military grid of the United States. In Zone B (right) the coordinates of point D 865.9–2172.1 fix it within 100 yd.

sents a problem of its own, and it is hoped this chapter will help the cartographer in making the proper selection.

Exercise 18.1

Azimuthal Projections. Draw one of the following maps in the specified projections, with parallels and meridians at every 10°. Ink in the map, parallels, and meridians with hairlines, the equator with a heavier line, and the Arctic Circle and the tropics with dashed hairlines. Color land yellow, and for oceans paint only a narrow strip of blue along the shores.

a. Draw an outline map of the world as projected from the center of the globe upon a tangent cube. Open up the cube, and lay it down flat to form a cross consisting of two polar and four equatorial projections.

b. Draw a map of the hemisphere in oblique stereographic projection, centered on 60°N and 100°W. Make the diameter of the hemisphere 6 in.

c. Draw a hemisphere on oblique orthographic projection centered on N 60° lat and W 100° long; make the diameter of the hemisphere 6 in.

d. Draw a world map by extending a globular hemisphere laterally as in the Hammer projection. Diameter of the hemisphere should be 4 in.

e. Draw a North polar projection in the Lambert azimuthal equal-area network as far as 60°S.

f. Construct a star or butterfly projection of your own.

g. Draw an orthoapsidal world map of your own design.

Square-grid Systems

In World War I, it was found that giving orders by latitude and longitude to designate exact locations and directions was most cumbersome, as the length of a degree of longitude varies. The Allied command therefore drew up a map of the Western Front in the Lambert Conformal Conic projection. On this map a square kilometer grid was drawn centered on Paris. This grid, at some distance from the Paris meridian, cut the parallels and meridians at an angle, and it further complicated matters by introducing the "grid north" added to true and magnetic north. Yet such was the advantage of easy reference to grid locations and grid azimuths that at present every military map has some kind of grid on it. Almost every general staff designed its own grid in meters, yards, or feet, using different spheroids and different projections, so that by World War II dozens of grids were used, often overlapping each other and causing endless confusion.

The grid system for progressive maps of the United States divided the country into seven North-South sections 9° wide. A polyconic map was drawn for each section and over this was drawn a 1,000-yd-square grid, starting at the intersection of the central meridian with 40°30′N. In giving references, however, a "false origin" 2 million yd south and 1 million yd west was chosen to avoid negative numbers (see Fig. 18.19). This grid was later extended to cover the whole world, but at present it has been superseded by the Universal Transverse Mercator grid.

State Grids. Most surveyors prefer to work on a plane surface, not taking into account the earth's curvature. The U.S. Coast and Geodetic Survey has designed a map for each state either on the Transverse Mercator or on the Lambert Conformal Conic projection. Larger states are divided in sections 150 miles wide; this makes the scale error less than 1:10,000, which is accurate enough for most surveys. On these maps a grid a thousand feet square is drawn, and

a false origin assigned. In the official lists, every triangulated or leveled point is expressed also in state-grid coordinates in feet, and the local surveyors can attach their measurements with ease.

Universal Transverse Mercator Grid. This grid is now standard for all of our military maps for any part of the globe. The world is divided into 60 North-South zones each 6° wide, and a Transverse Mercator projection is drawn for each zone. On these maps a square kilometer grid is drawn, starting with the intersection of the equator and the central meridian. False origin is 500 km west of the central meridian on the equator for the Northern Hemisphere and 10,000 km below this for the Southern Hemisphere (see Fig. 18.20).

Fig. 18.20. In the Universal Transverse Mercator grid the world is divided into sixty 6° zones and each zone into 8° latitude belts. The polar regions have their own stereographic grid.

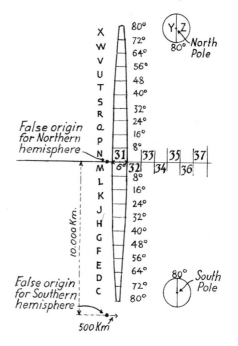

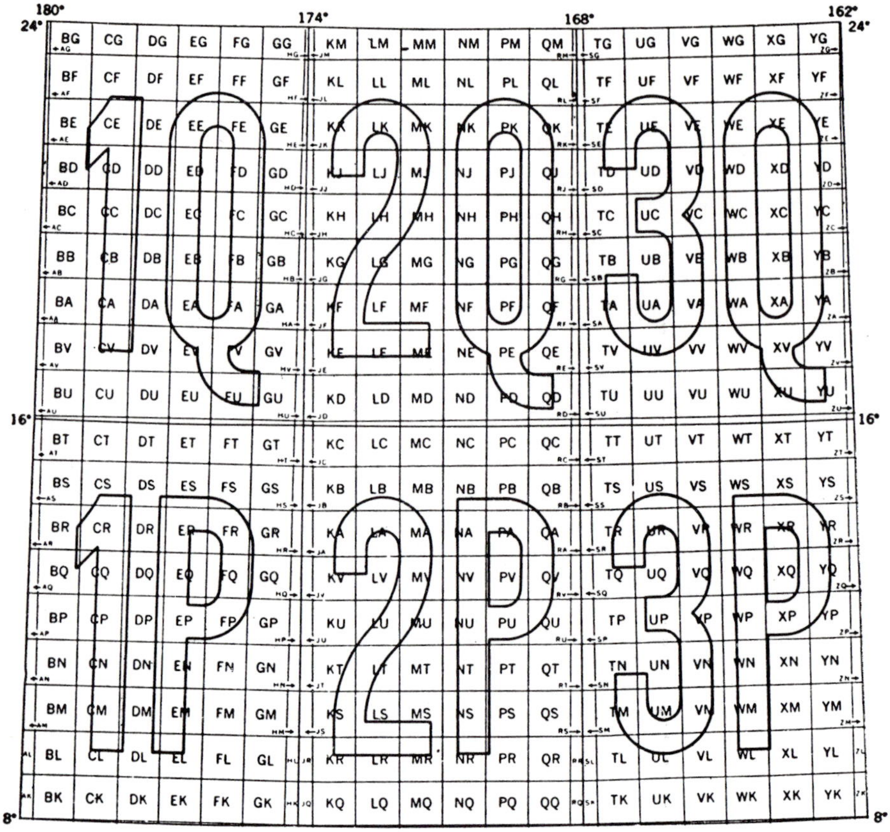

Fig. 18.21. In the Universal Transverse Mercator grid each zone is divided into 100-km squares and designated by two letters.

References are given in three stages. First we number the zones from 1 to 60, starting with 180°; second we give them letters from C to X, according to Fig. 18.20 (each letter including 8° latitude, ca. 550 miles). We use this reference only if we want to define our point in world relationship; in limited operations this reference is usually omitted.

A second reference is given by two letters which define 100-km squares. The scheme of lettering is shown in Fig. 18.21. This scheme of lettering repeats itself for every 18° of latitude and a little over 24° of longitude.

The third phase of reference is by numbers giving meters from the Southwest corner (within the 100-km squares), always going right and up, as in the bottom of Fig. 18.22. Points are usually defined within 100 m, which is sufficient for easy identification of most objects.

The Universal Polar Stereographic Grid. The polar areas from 80° to the pole are drawn on the stereographic projection, as this is a conformal projection which makes calculations easier. This is divided into 100-km squares starting from 0° + 180° and 90°E—90°W meridians. From there on, reference is given in the same manner as before.

The British system is still found on many maps. It uses 5 by 5 = 25 letter squares. First one large letter will define 500-km squares prepared for geographical units rather than systematic world coverage. A second letter will define 100-km squares. Within these, the definition is by numbers, the first half of which is easting, the second half northing. Thus point HV7442 is defined to within 100 m.

The World Geographic Reference (GEO-REF) System. While the Universal Transverse Mercator system (UTM) is well adapted to limited operations and is marked on our topographic and large-scale Army maps, it is not shown on most of our aeronautical charts. For long-range operations of planes, rockets, and missiles, spherical coordinates are still the best, and the Air Force returned to our latitude-longitude system of degrees and minutes, but with a simplified scheme of reference.

The world is divided in twenty-four 1-hr or 15° bands of longitude lettered A to Z (with I and O omitted), starting at 180°. Similarly twelve 15° belts of latitude lettered A to M start at the South Pole. Thus we have 15° quadrangles totaling 288, and identified by two letters. As before, we read to the right and up. Within each quadrangle, single degrees are marked with letters A to Q, starting at the lower left corner. Thus four letters like B G M Q in Fig. 18.23 will identify a 1° quadrangle, 60 nautical miles square. Each degree quadrangle is divided into 60' again, starting from the lower left, and referred to by numbers. The minute quadrangles can be further subdivided by tenths or hundredths by additional figures similar to the UTM

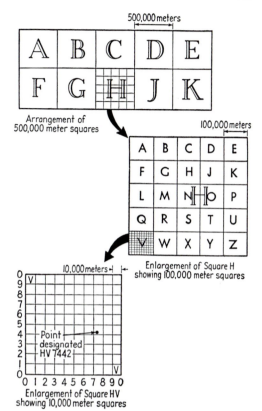

Fig. 18.22. The British and most other grid systems use a 25-letter system for 500- and 100-km squares; within the 100-km square the position is given by numbers.

Fig. 18.23. The GEOREF system is used for air navigation and global reference. It uses degrees and minutes but records them by letters and numbers starting from the 180° meridian and the South Pole.

system. The divisions, however, are not in kilometers, but in hundredths of minutes. The system was made effective in 1951 and is described in Air Force Regulation 96-6.

As most of these grids are indicated on every topographic map, a cartographer has to know their meaning. On a single map we may have a Universal Transverse Mercator grid in solid lines; the older polyconic grid indicated by "ticks" (short lines) on the sides; and the state grid similarly indicated. The parallel and meridian system is also a grid; so is the section and township system. Thus we may have five grid systems on the same maps. We may be in the overlapping area of two grid zones, which makes matters even less simple.

19

Thematic (Statistical) Maps

The maps discussed up to now have been general maps showing the total picture, although selectively. We can, however, pick out a single topic, such as geology, rainfall, number of people, etc., and relate the entire map to that. These maps are also called "topical," "single-factor," "special-purpose," or "distribution" maps. They can be *qualitative*, such as maps of soil types or of political adherence, etc.; or *quantitative*, such as maps of temperature or density of population—these we call "statistical maps."

Nonquantitative Maps. These maps show the distribution of such features as soil, vegetation, geology, dominant religions, etc. We cover, for instance, the area of Silurian rocks with one color, the area of Devonian rocks with another color, and thus produce a "color-patch" or *chorochromatic* map. We choose the colors carefully, not too clashing, yet having sufficient distinction.

Similar features should have a similar color. For instance, we show all Tertiary formations with various tints of yellow and all Mesozoics with tints of green—irrespective of the actual color of the rock—as we want to show only the age of the formation according to an international code. But we will choose a dark green for boreal forest and a medium green for deciduous forest, light green for grass, and earthy brown for plowed fields; here we show features of typical colors. To make certain that colors are well identified, we often place index letters and numbers in each patch, remembering that 1 man in 25 is color blind to a certain degree (see Fig. 19.1). If we cannot

Fig. 19.1. Color-patch (chorochromatic) map showing the geology near Bend, Oregon. Here the colors have been replaced by patterns. Index numbers are added for easier identification. (*From E. L. Baldwin, "Geology of Oregon."*)

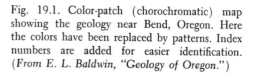
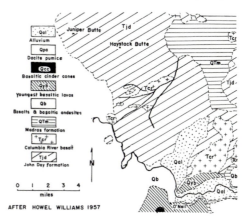

195

use colors, we may use various shades of gray or various patterns. We make the patterns symbolic, if possible (see land-use patterns in Fig. 23.2).

Often we have to show *transitional distribution*. To mix colors is dangerous. For instance, if we merge a yellow area into a blue area, the green in between looks like an independent color rather than a transition. If the colors are of different darkness, however, they show better transition. Fading out colors is difficult; it can be done by brush or crayon work reproduced by halftone process or mechanical vignetting (see Fig. 13.5).

Another method is *interdigitation*, as shown in Fig. 19.2. This method is often used, although it is obviously not quite exact. It is acceptable only if we regard it as a symbol for mixed distribution, as it is quite obvious that it does not mean actual location.

Statistical Maps

A contour map, a rainfall map, or a population map shows variation in amount or

Fig. 19.2. Interdigitation is a common but inexact way to show mixed or transitional values.

Christians
Moslems
Native re-
ligions

quantity. Many devices are used to present these, such as dots, choropleths, isopleths, superimposed diagrams, etc. Many statistical relations are complex and to show them well is a challenge to the imagination of a cartographer.

Dot Maps. This is the most obvious method of showing distribution. We draw even-sized dots for a given quantity such as 1,000 people, 100 acres of wheat, etc. Dot maps are best for rural distribution—to show agricultural products, for example—but not so good on small-scale maps for manufacturing concentrated in cities. Dots may be used, however, on large-scale city maps to show some urban distributions, such as newspaper or consumer distribution.

First the statistics have to be assembled. Then we have to draw a base map, about twice publication size, with the statistical divisions such as counties, townships, etc. The "Minor Civil Divisions" maps of the Census Bureau are most useful.

Size and Value of Dots. Our next concern is to select the value and size of the dots. As a general rule we make our choice so that the dots *coalesce* (just touch each other) at the *densest distribution*. The dots should be neither too large nor too small; the most popular size seems to be 30 to 40 dots to an inch, about 1,000 dots per square in. (publication size) if they just touch each other. If we make our drawing twice publication size, we may use a No. 3 Payzant, a No. 6 Leroy, a No. 2 Wrico, or a 0.7 Pelican Graphos pen. Now we figure what value to give each dot. We select an area of somewhat less than the very densest distribution, as this may be some exceptional locality. We measure the area of the selected land division and calculate how many dots it could take (see Chap. 8). We divide the number of statistical units in the area by the number of dots and the next round number will give us the dot scale. For instance, Florida's highly productive Alachua County, with 50,000 head of cattle is 0.5 sq in. on the map. Choosing 1,000 dots per square inch, this can be covered with $1,000 \times 0.5$, or 500 normal-sized dots. Thus we have 50,000/500, or 100 cattle for each dot.

Dasymetric method. In the foregoing, we assumed that the county, province, or other area is evenly inhabited throughout. If this is not the case, we use only the dense areas, with allowance for the less dense. For instance, in the province of Camagüey in Cuba, the swampy Southwest and the forested mountainous part have relatively few cattle. Thus we reduce the area of the province by half, for which we allowed four-fifths of all the cattle, and this relationship is what we use to find the dot scale. This is a very rough method, but much better than distributing the cattle evenly throughout the province. This is the *dasymetric* (roughly measured) method. The formula is

$$\text{Density} = \frac{N - N_1}{A - A_1}$$

in which $N - N_1$ is the number of units in the dense area after we subtract the number in the less dense area, and $A - A_1$ is the area of the dense part discounting the less dense area. Thus if there are 1 million head of cattle on the 10,000 sq miles of Cama-

$$\frac{1,000,000 - 200,000}{10,000 - 5,000} = \frac{800,000}{5,000}$$

$$= 160 \text{ cattle per sq mile}$$

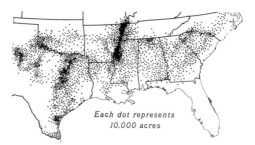

Each dot represents
10,000 acres

Fig. 19.3. On a dot map, the size and value of dots are selected so that the dots coalesce at the densest distribution. (*From "Cotton Production," by the U.S. Bureau of Agricultural Economics.*)

güey, the density in the dense part will be which we use for selecting the dot scale.

Placing of Dots. In Fig. 19.4 we see two maps. On the left-hand one the dots were evenly distributed for each county; on the right, the number of dots in each county was kept, but distributed to take into account the relief and land utilization. The right-hand map has a personal element in it and depends a great deal on the geographical intelligence of the cartographer, yet it gives a much truer picture than the left-hand map.

Strictly speaking, every dot should be in the pivotal point of the distribution it represents (see Centrograms, Chap. 21). This would lead, however, to endless calculations, and the best we can do is to study the geography of the region for proper placement within the unit area. For instance, the dots for orange production in Florida will not be on the swampy lowlands but on the well-drained uplands of the county. The finer the statistical subdivisions and the smaller the scale, the less difficult is the placement of dots.

Dot maps are often *differentiated*. For instance, if we want to separate dairy from

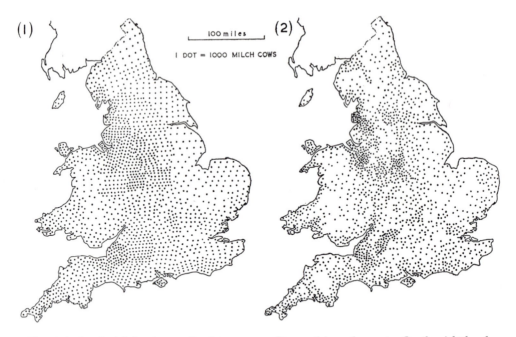

Fig. 19.4. On the left-hand map, the dots were evenly spaced in each county. On the right-hand map, the same numbers of dots were placed, taking relief, soil, and land use into consideration. (*From F. J. Monkhouse and H. R. Wilkinson, "Maps and Diagrams," Metheun & Company Ltd., London, 1952.*)

beef cattle, we may use dots for dairy and small circles for beef cattle. The circle is lighter in appearance, as beef cattle have less value than the milch cows. Sometimes the dots are replaced by initial letters of

Fig. 19.5. Choropleth maps show values for each political division or any other areal unit. They show sharp changes at boundaries where the change may be transitional. (*By Leslie Hewes and A. C. Schmieding, Geog. Rev. vol. 46, 1956.*)

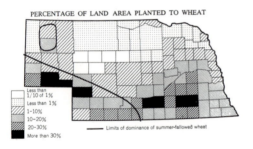

various products, as in Fig. 23.12, which shows percentages of world production.

Percentage or per-mille dot maps are often used for comparisons. If we make one map in which every dot represents 1 per-mille of the United States wheat production and another of corn production, and so forth, the regional specialization of each part of the country becomes apparent.[1]

Choropleth Maps. Statistical data are given by civic divisions such as states, counties, townships, etc. The simplest way to prepare a statistical map is to color or tint the civil divisions darker or lighter in proportion to the density of distribution. The unit is not necessarily a political division;

[1] J. Ross Mackay, Percentage Dot Maps, *Econ. Geog.*, 1953, pp. 263–266.

it can be any shape—even squares or circles can be used, as long as we have the data. The choropleth map is a kind of color-patch or chorochromatic map in which the tints show densities. For the application of proportional tints see Chap. 13. We can easily make 10 distinct shades with dots, dashes, and lines. The boundary lines between the patches should be light, dashed, dotted, or omitted altogether. Heavy lines would give the impression of darkness where the divisions are small.

Choropleth maps are simple to construct but not exactly true. For instance, if we show farm population by counties, there can be a sudden change at the boundary of the county, while in reality the change is likely to be gradual. The smaller the land division, the truer the map will be. On the other hand, if the units are too minute and map scale too small, the patterns are difficult to recognize. If possible, choropleth maps should be replaced by dots, isopleths, or superimposed diagrams.

Choropleth maps cannot have many place names, rivers, or symbols, particularly in the darker portions. City names may be boxed in.

Isopleths. These are lines which connect places of equal value. We have already studied one kind of isopleth, the contour lines which connect places of equal elevation. If we connect places of equal temperature, we have isotherms; barometric pressure is shown by isobars; rainfall by isohyets; magnetic declination by isogons; depth under sea level by isobaths, etc.

Isopleths have the same inherent virtues and deficiencies as contour lines. They give and demand exact information. They do not show what happens between isopleth intervals. Like contour lines, isopleths have

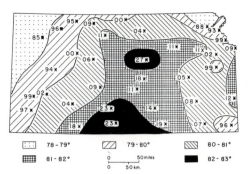

Fig. 19.6. Isopleth map showing July mean temperatures in Kansas. Numbers refer to full degrees and tenths of degrees; 11 means 81.1°. (*From D. I. Blumenstock, The Reliability Factor in the Drawing of Isarithms, Annals Assoc. Am. Geog., vol. 43, 1953, pp. 289–304.*)

a datum plane, usually the zero value. Like contour lines, isopleths are not easy to visualize. Layer tinting is even more helpful, as in such cases we have no rivers to indicate the direction down or up. On most isopleth maps, the zones between the lines are shaded so that the higher the value, the darker the shade. For methods of shading, see Chap. 13.

When Are Isopleths Used? If we have a simple distribution, such as number of cattle or acres of wheat in the United States, we can best see the pattern of distribution from a dot map. This gives us a true picture of the distribution, but we cannot tell the number of cattle per square mile at a given place unless we count the dots and divide them by the area—a rather cumbersome process. Thus the economist or sociologist who is interested in numerical values prefers isopleths on which he can observe the values directly. If we have ratios such as yield per acre, number of doctors, percentage of parochial schools, etc., we also use isopleths or sometimes choropleths. Isopleths are always used for *continuous distribution* such as rainfall and temperature, where we have values rather than numbers of units.

Terminology. Much has been written about the two different concepts of isopleths. They may be so called for *continuous distributions,* such as contour lines, where every point has to fall in between the lines above and below; others use the term also to show *discontinous distributions,* such as density of population, where there may be spots with no population between even the highest of the isopleths. They have been given different names, too, such as "isopleths," "isarithms," "isolines," "isograms." In this author's opinion, no such difference exists. Isopleths can be used *only* for continuous values. If they are used for population, for instance, the distribution of population is generalized to such an extent that it is regarded as a continuous variable. If we have truly discontinuous factors, such as the distribution of industries, where large empty areas may be adjacent to the highest concentrations, we must use a color-patch map (see Fig. 8.12).

Fig. 19.7. In drawing of isopleths over scattered spot numbers, the lines are in proportionate distance between numbers.

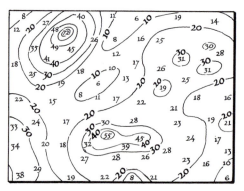

Preparation for Isopleth Maps. Before we can make an isopleth map, we need a base map well covered with spot numbers indicating the extent of our distribution at the spot. We have various ways of obtaining data. Spot observations provide data for rainfall or temperature; for social and economic problems the statistical yearbooks of the U.S. Census Bureau may give the answers. These are usually arranged in minor civil divisions, such as counties and townships. The "Statesman's Yearbook" and the various publications of the UN will help for foreign countries. Sometimes we have to do our own sampling—in market surveys, for example. Various methods of sampling have been worked out in detail by sociologists.[2]

Isopleth intervals should be close enough to show the pattern of the data compiled, but they should not be as close as contour lines would be drawn over the same map. Spot numbers may be much less reliable than spot heights obtained by a surveyor's transit, and we may want to show the general pattern rather than every local variation. Thus we consider the following:

1. Scale of map
2. Reliability of data
3. Density of spot numbers
4. Nature of the distribution
5. Purpose of map

If the distribution is rather even we use even line intervals like 0, 10, 20, 30, 40, etc. If the numbers are low over large areas with relatively small extent of high values, as in a map of population, we use a logarithmic scale: 0, 2, 4, 8, 16, 32, 64, etc. A

[2] M. J. Hagood and D. O. Price, "Statistics for Sociologists," New York, 1952.

common compromise is 0, 1, 2, 5, 10, 20, 50, 100, 200, etc. This is less even, but it has round numbers.

The drawing of isopleths is similar to laying out contour lines and is best understood from Figs. 7.2 and 19.7. We always place the isopleth in proportionate distance between two numbers. For instance, the line 10 will be twice as near to 12 as to 6. Often we have to choose between various solutions. For instance, in the upper right we may connect the three areas of the 30 isopleth, or we may keep them separate. The best way to come to a decision is to study the region on topographic maps and to keep in mind the nature of distribution.

Often our data is not quite reliable. For example, we may have a single spot of exceptionally high temperature. This can be due to local conditions. It would be misleading to introduce this on a general climatic map without explanation.[3]

Making true and expressive isopleth maps is something of an art and requires the best geographical knowledge of the region. It is helpful to imagine every spot number as a spot height and regard it as if the values would form an undulating surface upon which we have to draw contour lines. It may actually help to stick into the map pegs of proper height at each spot number to visualize the distribution. An oblique photograph of this may show an effective statistical model.

[3] If a relatively flat distribution becomes at places very steep as, for instance, 2, 5, 10, 30, we draw the 20 line somewhat nearer to 30 than to 10 as the line graph of transition will be rather a continuous curve than a sequence of straight lines. P. W. Porter, Putting the Isopleth in Its Place, *Proc. Minn. Acad. Sci.*, vols. 25 and 26, 1957 and 1958, pp. 372–384.

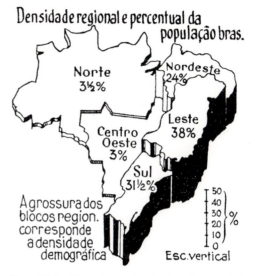

Fig. 19.8. Three-dimensional statistical model showing the density of population of the major divisions of Brazil. (*By Prof. João Soukup, São Paulo, Brazil.*)

Statistical Models

On Fig. 19.8, Professor Soukup of São Paulo laid out vertically the density of population of the regions of Brazil upon a map giving a vivid illustration. Here he transformed a choropleth map into a statistical model. Similarly, on Fig. 19.9 isopleths were offset vertically upon an oblique view on the map. In this model, parallel profiles were added to enhance its visual impres-

Fig. 19.9. Rainfall of Oregon shown by an isometric block diagram.

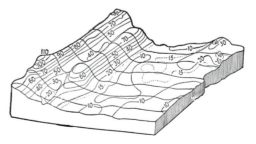

sion. We can produce a book illustration by isometric construction as in the samples mentioned or prepare an actual model and photograph it obliquely. The construction of isometric blocks is discussed in Chap. 22 and construction of relief models in Chap. 25. If it is possible, we orient the models so that the low values should be in the foreground. The preparation of statistical models is good practice for cartographers and makes excellent display material.

Exercise 19.1.

Draw a statistical map. Show the distribution of colleges and universities in your state, indicating student population. Subdivide symbols for men and women and mark type of school. Select your own system.

20

Diagrams

We are informed that the Mississippi-Missouri is close to 3,900 miles long, while the Rio Grande is 1,900 miles, the Ohio-Allegheny River 1,300 miles, and the Hudson 310 miles. Even such simple facts are more easily comprehended and remembered when seen on a graph, as in Fig. 20.1, than from merely reading the numbers. In presenting more complicated relationships, diagrams are indispensable; indeed, modern science and technology could not have developed without graphics. The terms *diagrams* and *graphs* are both used, and larger ones are also called *charts*. Many textbooks deal with graphics, as shown in the bibliography. Here we present only those which are common as geographic literature, and particularly those which are commonly superimposed on statistical maps.

Bar Graphs. These are used for simple comparison of quantities (see Figs. 20.1 and 20.2). They are drawn wide or narrow, close or separated, vertical or horizontal, as we choose. Vertical bars are remembered better than horizontal bars. Bars are often subdivided and may show percentages. The exact value in numbers is usually printed on each bar. Horizontal bars should be labeled above and not along the side. Vertical bars often have to be labeled vertically along the side or obliquely above. Bars in general should start at zero value. If we show only the upper part of a set of bars, we may give the impression of greater variation than exists. "Mutilation" should be indicated by broken bars.

Unit Graphs. A bar may be divided into a row of equal countable units, such as squares or circles or figurettes. In recent years, quite an industry has grown up producing pictorial unit graphs, using rows of men, bales, sacks, autos, etc., advertised as pictographs. They do not show more than an ordinary bar graph, and serve only as kind of sugar-coating on the dry statistics. The term can rightfully be applied to any other kind of pictorial diagram also.

Line Graphs. These show the relation of one variable to another. The most common is the variation of a quantity, like production, in time. There are, however, many other kinds. For example, a profile

Fig. 20.1. Relative values of simple quantities are better visualized and remembered if shown graphically.

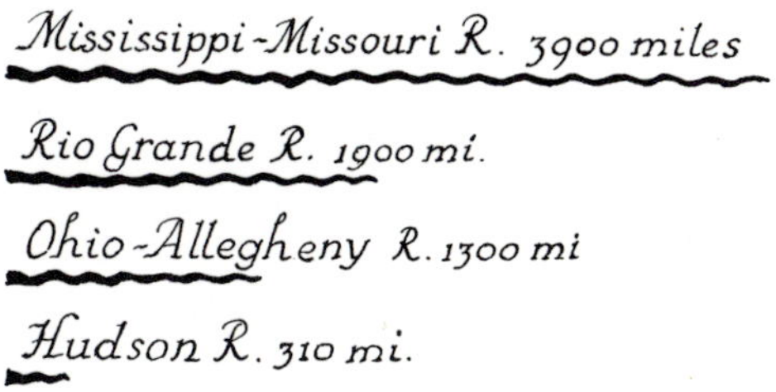

Mississippi-Missouri R. 3900 miles

Rio Grande R. 1900 mi.

Ohio-Allegheny R. 1300 mi

Hudson R. 310 mi.

FARM REAL ESTATE

AVERAGE VALUE PER ACRE IN THE U. S.

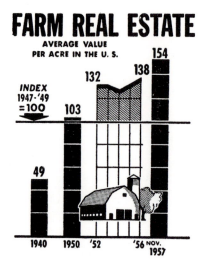

Fig. 20.2. Bar graphs serve for simple comparison of quantities. This one is divided in countable units and a part is shown as a line graph. (*Courtesy of Fiscal Information Service.*)

made on a topographic sheet is also a line graph, as it shows the variation of altitude with horizontal distance.

We have to figure the vertical scale carefully. Graphs with very steep lines which look jagged are difficult to comprehend, and those which are too flat are ineffective. As a rule of thumb, the average inclination of the lines should be about 45°. Subdivided line graphs are better if the smaller or the least variable quantities are at the bottom. We may place several line graphs

Fig. 20.3. Pictorial unit graphs are used generally for visualization of quanities. (*From "Our Cities," National Resources Committee, 1937.*)

EACH TELEPHONE REPRESENTS 1 TELEPHONE
EACH DISC REPRESENTS 10 DOLLARS
EACH AUTOMOBILE REPRESENTS 2 AUTOMOBILES

on the same coordinate for comparison, but too many may become confusing. It helps to use different types of lines.

Logarithmic Line Graphs. Such graphs show the rate of change rather than the actual change, as in Fig. 20.5. Stock quotations are often presented this way. In geography we frequently use them to compare the rate of change of population, trade, income, etc., of different-sized countries. For these we use the commercially available semilogarithmic paper, where one of the coordinates has even divisions. Logarithmic paper, with both coordinates logarithmic in units, is rarely used in geography.

Generalization. If the line graph is based on spot observations, we plot these as dots on two coordinates called a *scatter diagram.* If we want to get the general trend expressed by a line graph, we do not connect every point but we use either a *median line* (a smooth curve with an equal number of dots above and below) or a *central curve* (a curve on which the distance of the

Fig. 20.4. Line graphs are used for showing the change of one variable against the other, which is most often time. (*Courtesy of Junior Scholastic and Graphic Syndicate.*)

PAYROLLS

IN MANUFACTURING

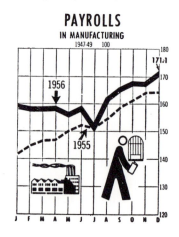

points is also considered). For the median line, we divide the field into equal vertical strips and then we draw our curve through the median point of each strip. For the central curve, a fast solution is shown in Fig. 20.6, but mathematically it is not quite precise. The amount of correlation between the variables can be expressed by a line of regression, which is discussed later in this chapter.

If the variable is *continuous* but our data is not, we are nearer the truth if we use a smooth curve rather than a jagged one, even if the intermediate values are not known. Figure 20.7 shows how to do this with tangents at the known points.

Radial Graphs. These can be truly radial when the rays represent actual directions, as in a *wind rose.* More often, however, we use radial graphs for cyclic variation, such as yearly rainfall and temperature. Figure 20.8 shows a radial rainfall-and-temperature curve for Havana, Cuba. The asset of this kind of curve is that January

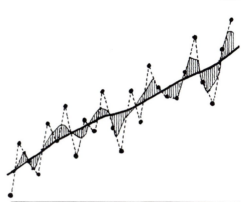

Fig. 20.6. A quick method of generalizing a line graph from scattered dots consists in connecting successive dots and halving the connecting lines. The heavy line is derived from generalizing the line produced by the first halving.

comes back to January; the liability is that the eye has to follow radial distances, which are not as familiar as vertical distances.

Radial coordinates are used also in *star diagrams,* showing, for example, the composition of rocks in which the eight fundamental chemicals are arranged so as to form a characteristic eight-sided star [1] or to show

[1] H. Kemp, "Handbook of Rocks," 6th ed., D. Van Nostrand Company, Inc., New York, 1940.

Fig. 20.5. We bought $2 worth of one item and $5 worth of another. Both items doubled their value in a certain time and later went back to half of the original value. The diagram on the left shows the actual values involved, while the semilogarithmic one on the right discloses that the rate of change was identical.

Fig. 20.7. If the variation is continuous, we are nearer the truth if we make a continuous curve.

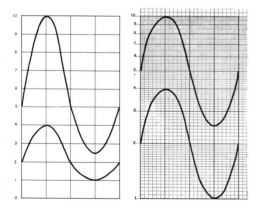

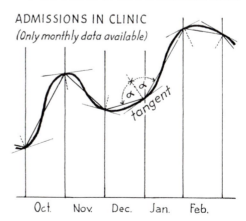

ADMISSIONS IN CLINIC
(Only monthly data available)

tangent

Oct.   Nov.   Dec.   Jan.   Feb.

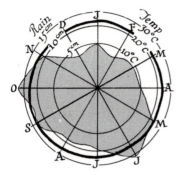

Fig. 20.8. The heavy line represents mean monthly temperatures and the shaded area the mean rainfall for Havana. The divisions on the spokes correspond to square roots of values.

the power potential of nations, as in Fig. 20.9. The value of star diagrams is that each rock, nation, or other unit has a characteristic shape. We have to assign a reasonable scale for each arm. If we use the area of the star for comparison, we have to use a square-root scale; otherwise a nation which should be shown as having twice the amount of all the factors would actu-

Fig. 20.9. Star diagram showing power potential of nations. The axes are divided by square-root scale to make the areas of the start comparable. The above star is typical of an underdeveloped country.

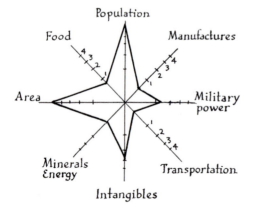

ally appear to be four times larger, and so on.

Triangular Diagrams. Let us say we have a soil which consists of 60 per cent sand, 30 per cent clay, and 10 per cent lime. We can designate this soil by its location within a triangle, as in Fig. 20.10, and compare it with other soil samples. We draw on all three sides a line at the proper percentage parallel to the side on the left. The three lines must meet in a point, as in Fig. 20.10. Similar diagrams may show occupations of people in various regions, occupations divided between agriculture, industry, and others. Anything which can be divided into three factors and expressed in percentages can be brought to a point in a triangular diagram.

Climatograms. The climate of a region can be expressed by simple rainfall and temperature curves, by radial coordinates as in Fig. 20.8, or by a line graph in which rainfall and temperature are the two axes. A more elaborate climatogram used for pre-

Fig. 20.10. Three variables expressed in percentages are best shown on triangular graphs. The earth sample A is of 60 per cent sand, 30 per cent clay, and 10 per cent lime.

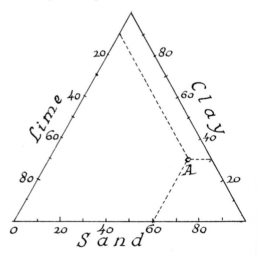

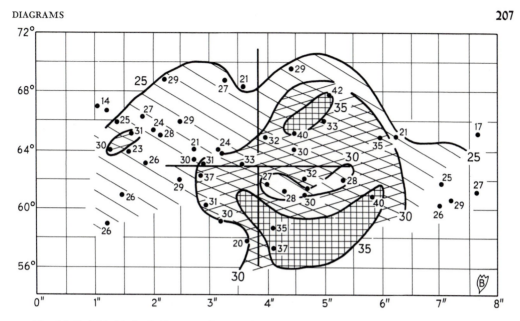

Fig. 20.11. This kind of climatograph is used in forecasting crops. Numbers represent the yield of wheat in Indiana, placed according to spring rainfall and temperature of each year from 1887 to 1939. Darker shading means higher yields. Best yields are in the years of a wet and cool spring. (*By S. S. Visher.*)

diction of crops from the weather in the growing season is shown in Fig. 20.12.

Moisture Utilization Diagrams. These were introduced by William Thornthwaite. They show the yearly precipitation and how much water is required during the growing season to replace evaporation and transpiration by plants. Transpiration depends on temperature, the amount of sunlight, and the crop or vegetation cover.[2] In his newer diagrams, Thornthwaite takes into account the loss of heat produced by evaporation and transpiration. These diagrams are used in appraising the agricultural possibilities of regions and for planning irrigation.

Two-dimensional Graphs. In these diagrams the areas of geometrical figures are proportionate to the values. Any shape can

[2] W. C. Thornthwaite, An Approach toward a Rational Classification of Climate, *Geog. Rev.*, vol. 38, 1948, pp. 55–94.

be used, but the most common are circles, squares, and triangles. To have the areas in correct proportions, the diameters of the circles or the sides of the squares, triangles,

Fig. 20.12. Moisture utilization diagram by W. C. Thornthwaite. The dash-dot line shows monthly precipitation; the central cone shows water need during growing season. In this case, the need is not quite met by rainfall.

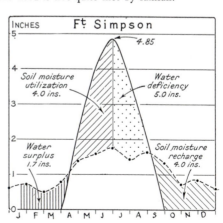

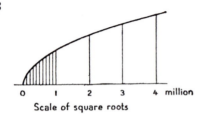

Scale of square roots

Fig. 20.13. This type of scale is used for finding the radii of pie graphs.

etc., have to be proportionate to the *square root* of the values. If the side of a square is doubled, the area is quadrupled; thus if we want double the area of a square, its side will be $\sqrt{2}$, or 1.42, times larger. On Fig. 20.13, we can scale off the square roots of quantities along the vertical component.

Pie graphs (also called circular, coin, or sector graphs) are the most common two-dimensional graphs (see Fig. 20.20). They are often subdivided into percentages. Guides on transparent paper in which a circle is divided radially into 100 parts can be bought in engineering-supply stores. We center these exactly over our circle and prick through the percentages. Pie graphs are not very easy to label but they are compact and easily drawn.

Square or oblong graphs are somewhat easier to subdivide and label, but it is more difficult, for instance, to show 1 per cent in a square than in a circle. When some of

the subdivisions are very small, we may use *triangular area graphs*. They are used for subdividing, as in Fig. 20.14. We can place the small items near the apex, as in Fig. 20.15. Circles and rectangles can be viewed obliquely, as in Fig. 20.14.

Three-dimensional Graphs. These are used when the quantities for comparison vary greatly, such as the population of cities or production of various industries. Figure 20.16 shows that the graphic comparison of 1 to 100 requires a great deal of space when shown by a bar graph, much less by a circular graph, and least by a three-dimensional graph. If the comparison were between 1 and 1,000, the bar graph would be impractical. We may use pictures of small spheres, blocks, boxes, barrels, or sacks. Theoretically, the diameter of the spheres or the sides of the boxes should be proportionate to the cubic root of the quantities. A study by Robert Williams disclosed that the eye is very deficient in correct appraisal of the size of these pictures of solids.[3] Actually the value of the solids comes nearer in most people's estimation if their sides are made proportionate to the square root, not the

[3] Robert Williams, "Statistical Symbols for a Map," Yale University Press, New Haven, Conn., March, 1956. 115 pp.

Fig. 20.14. Complex subdivisions are best shown on two-dimensional graphs.

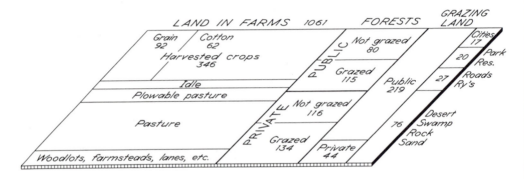

cubic root, of the values. For this reason, the countable block pillars are preferable to spheres.

Block Pillars (Block Piles). For the unit, we use a picture of a small isometric block. In Fig. 20.17, the unit block is a cube, but if placed on maps it is preferable to make the unit block half as high as it is wide. Usually a pillar is not more than 10 blocks high; for larger values we group them together. These blocks are more easily labeled, subdivided, and counted than the spheres. They are usually superimposed on maps. By using fractions of the unit block, quantities related as 1 to 10,000 can be drawn (see Fig. 23.13).[4] They can be drawn in the forms of buildings, tanks, silos, boxes, or sacks if pictorial rendering is desired.

Volumetric Graphs. In case of three variables, we may use isometric drawing with one axis vertical and the two others representing two directions in a horizontal plane. Usually the axes are drawn 120° apart, as shown in Fig. 20.18. This is actually a statistical model, as discussed in the previous chapter.

Amount of Correlation

(*By Prof. Philip W. Porter, University of Minnesota*)

The Line of Regression. The study of the similarities and dissimilarities in the occurence of things on the earth's surface lies at the very heart of geography. One frequently encounters two phenomena which seem to vary together in a systematic fashion. It may be noted, for example, that high and low densities of rural population

[4] Erwin Raisz, The Block-pile System of Statistical Maps, *Econ. Geog.*, vol. 15, 1939, pp. 185–188.

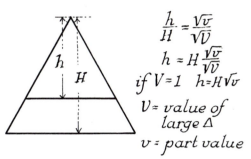

$$\frac{h}{H} = \frac{\sqrt[3]{v}}{\sqrt[3]{V}}$$

$$h = H\frac{\sqrt[3]{v}}{\sqrt[3]{V}}$$

$$if \ V = 1 \quad h = H\sqrt[3]{v}$$

$V =$ *value of large* Δ

$v =$ *part value*

Fig. 20.15. Triangular area diagrams are good when one value is very large in comparison with the others.

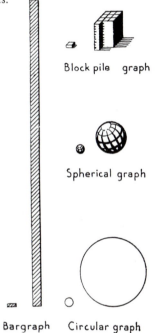

Block pile   graph

Spherical graph

Bargraph   Circular graph

Fig. 20.16. Different types of graphs showing the comparison of two quantities related as 1 to 100.

Fig. 20.17. Block-pillar systems of comparison of quantities. The blocks are 10 units high. Fractional pillars are always in the foreground.

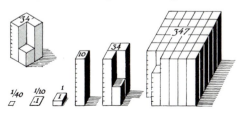

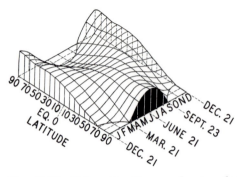

Fig. 20.18. Volumetric diagram showing the relation of three variables, sunlight varying with latitude and seasons. (*Modified after William M. Davis.*)

vary spatially with greater and lesser amounts of rainfall, that the heights of a particular species of tree correlate with the length of the frost-free season, that the amount of pedestrian traffic appears to be related to land values in the central business districts of cities, and so forth. For the geographer who is not content with such subjective phrases as a strong correlation, a moderately strong association, or an apparent similarity, statisticians have a particularly useful method which makes it possible to quantify the description of the covariance of two variables. This is known as the *line of regression*. Stated simply, it tells us: Given this much X, one can expect that much Y. Turning to Fig. 20.19, we may say: Given a reading on the X scale of 50 ($5,000 per frontage foot), one can expect a value on the Y scale of 31.82 (3,182 pedestrians per hour) in the business districts of a city.

On the scatter diagram in Fig. 20.19, it is obvious that any number of lines can be drawn which will divide the distribution into two equal parts; yet it is easy to see that only one line can be drawn which will

minimize the distance the dots would have to travel to reach the line. This line is known as the *line of least squares*, which means that the sum of the squares of the distances of the observations from this line is less than that obtained by using any other line. The formula which describes the graphic relationship of the two variables is expressed

$$Y = a + bX$$

The symbol a tells us the value of Y when the value of X is zero. In other words, it gives the point of origin of the regression line. It is commonly known as the Y *intercept*. The letter b is a coefficient which tells how much Y increases with each unit increase of X. It describes the *slope* of the regression line.

The regression of Y on X is found by calculating the summations of X, Y, XY, and X², and the means of X and Y.

(N = 10)	X	Y	XY	X²
	21	9	189	441
	23	6	138	529
	26	12	312	676
	32	13	416	1,024
	35	21	735	1,225
	39	21	819	1,521
	41	31	1,271	1,681
	55	35	1,925	3,025
	57	45	2,565	3,249
	71	43	3,053	5,041
	400	236	11,423	18,412

These totals and means are then substituted in the following formulas, the first of which gives the value of a, the second of which gives the value of b.

$$a = \overline{Y} - b\overline{X}$$
$$b = \frac{N\Sigma XY - (\Sigma X)(\Sigma Y)}{N\Sigma X^2 - (\Sigma X)^2}$$

where

X is the independent variable (i.e., property value)

$\overline{X}$ is the mean value of the independent variables

$\overline{Y}$ is the dependent variable (i.e., pedestrian traffic)

Y is the mean value of the dependent variables

N is the number of pairs

Σ is the summation sign

Solving first for b, we get

$$\frac{(10)\ (11{,}423) - (400)\ (236)}{(10)\ 18{,}412) - (400)^2}$$

$$= \frac{114{,}230 - 94{,}400}{184{,}120 - 16{,}000} = \frac{1{,}983}{2{,}412} = 0.8221$$

Substituting 0.8221 for b in the first formula, we solve for a:

$$23.6 - (40)(0.8221) = 23.6 - 32.884 = -9.284$$

We may now express the line of regression:

$$Y = -9.284 + 0.8221X$$

Returning to the scatter diagram, we must find two points in order to draw the line of regression. One of these can be the Y intercept; where X is 0, Y is −9.284. Then taking a value of 50 on the X axis, we find Y as follows:

$$Y = -9.284 + (0.8221)(50) =$$
$$-9.284 + 41.105 = 31.821$$

This point is found, and a line is drawn from it through the Y intercept at −9.284. Using X as a standard, we have found the line which minimizes the Y variables. That is, we have minimized the sum of the *vertical* distances (squared) which the dots must travel along Y to reach the line.[5]

This line might now be used as a means of predicting the pedestrian traffic for areas

[5] There is, of course, another line, the regression of X on Y, which minimizes the sum of the *horizontal* distances (squared) which the dots must travel along X to reach the line.

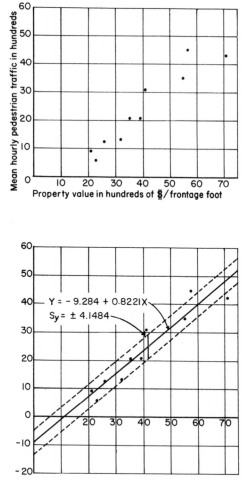

Fig. 20.19. Scatter diagram showing observations plotted against two coordinates (above). In the diagram below, Y is the "line of regression" and S_y is the "standard error of estimate." (*By Philip W. Porter.*)

whose property values are known. The geographer might use this regression as a yardstick against which to map and measure the differences between actual pedestrian traffic and the pattern which would be expected in terms of the distribution of property values.[6]

[6] This idea was used in the maps constructed by Robinson and Bryson in their article in *Annals Assn. Am. Geog.*, December, 1957, pp. 378–391.

It is important to be able to say how well a regression line represents the scatter diagram on which it is based. For the regression line computed above, we may describe the confidence we have in the line by saying that 95 per cent of all Y values fall within a range ± 8.3 from the regression line. For this we find the *standard error of estimate*, using the formula

$$S_y = \sqrt{\frac{\Sigma(d^2)}{N}}$$

where S_y = the standard error of estimate
   N = the numbers of cases
   d = the deviation of each *actual Y* value from the Y value theoretically expected when read from the regression line.

This is a measure of the degree of scatter about the line of regression.

Y (actual)	Y (expected)	d	d^2
9	7.980	1.020	1.0404
6	9.624	3.624	13.1334
12	12.091	0.091	0.0083
13	17.023	4.023	16.1845
21	19.490	1.510	2.2801
21	22.778	1.778	3.1613
31	24.422	6.578	43.2701
35	35.932	0.932	0.8686
45	37.576	7.424	55.1158
43	49.085	6.085	37.0272
			172.0897

$$S_y = \sqrt{\frac{172.0897}{10}}$$

$$S_y = \sqrt{17.20897}$$

$$S_y = \quad 4.1484$$

The dashed lines on either side of the regression line are laid off 4.1484 vertical units on Y. This is one standard error of estimate and will enclose 68 per cent of all observations. Plus and minus two standard errors of estimate (± 8.2968) will enclose 95 per cent of all cases.

Many limitations and refinements attend the use of statistical techniques such as these. For example, many scatter diagrams are best described by a curved regression line; and this is likely to be especially true of the variables with which geographers commonly deal. If handled intelligently, the line of regression has great merit and utility. The interested student is advised to consult any standard text on statistics for greater detail on correlation theory.[7]

Statistical Maps with Superimposed Diagrams

When do we use a map with superimposed diagrams in preference to color patches, dots, or isopleths? Often we do not have sufficiently detailed data. For instance, if tractor exports are given only by countries, this may best be shown by superimposed bar graphs. Sometimes there is too great a difference between high and low values. For instance, mining of metals in the United States is shown by superimposed block pillars in Fig. 23.13, where the value of the largest producer is 2,000 times that of the smallest. Perhaps the variable is subdivided, as it is in the graphs in Fig. 20.20. There can be variations in the variable itself, such as in the logarithmic line graph showing population changes in Missouri (see Fig. 20.21) or in a map with a number of rainfall and temperature curves

[7] Also useful in this regard is H. H. McCarty et al., "The Measurement of Association in Industrial Geography," Department of Publications, State University of Iowa, Iowa City, 1956. (For the Department of Geography.)

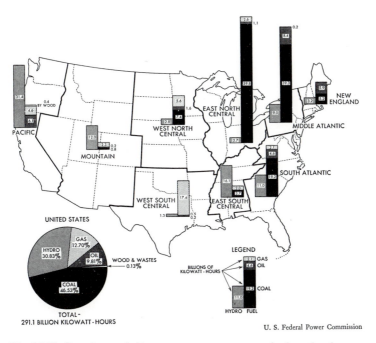

Fig. 20.20. Superimposed diagrams upon maps are used when the data are generalized or subdivided. the above map shows the sources of electric energy in the United States in 1949. (*From Woytinsky, "World Population and Production," Twentieth Century Fund.*)

superimposed. Diagrams can easily be labeled, while neither dots nor isopleths or choropleths allow much lettering. Sometimes the variable is concentrated in certain localities. For instance, the exports and imports of New York, Philadelphia, and Boston can be shown by horizontal bar graphs extending out in the sea.

The cartographer's task in applying superimposed diagrams is to find the right scale and use the available space well, but not to crowd it too much, so that all the labels are readable. The best way is to draw all the diagrams on a separate paper. This will give an idea of how large the map needs to be to contain them. It is not a bad practice to draw the base map and to cut out the diagrams and experiment with their spacing. The center of pie graphs and the base of blocks should be in exact loca-

Fig. 20.21. Population map by F. D. Stilgenbauer. The circles show the population of the black urban areas. (*Denoyer-Geppert Company,* 1953.)

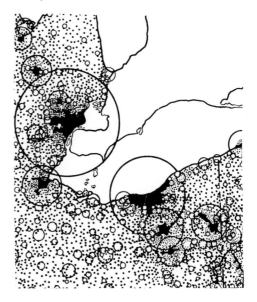

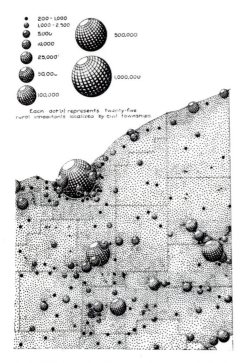

Fig. 20.22. Density of population map. Spherical symbols stand out clearly, but they are not easily commensurable. (*Map by Guy-Harold Smith, Geog. Rev., 1928.*)

tion; if they are not, arrows should point to the spot.

Diagrams can be superimposed upon simple color-patch or isopleth maps too, if not too much will be covered up. For instance, a tint may show the extent of various types of forest and the production of paper will be shown by bar graphs.

Population Maps. While rural population is more or less evenly distributed, city population is highly concentrated and may vary from a few thousand to several million. The most successful method is to show rural population by dots and urban population by transparent circles or by

small spheres or block piles. Figures 20.21 and 20.22 show the two methods. The transparent circles have an advantage in that they do not blot out the rural population underneath. On the other hand, block pillars or spheres show up more impressively. Only on very small-scale maps can population be generalized to such an extent that it can be shown by isopleths.

Diagrams and statistical maps are always a challenge to a cartographer. The aim is to present statistical facts so that their size and significance should be easily understood and the picture easily remembered. The map itself is only the background for the statistics and its content should not compete with the dots, tints, and diagrams.

Exercise 20.1

To prepare an economic map, each student may select a state, a small foreign country, or any region desired. Compile all the necessary data. The central map should show land-use categories with colors and patterns. Mineral production should be shown by value. Show main transportation lines. Cities may be indicated with block pillars proportional to the amount of manufacturing. The map should be surrounded with diagrams and cartograms showing imports and exports, rainfall, leading products, etc. Increase or decrease of production through the last decades may also be included, together with anything else for which there is pertinent data. Colors may be used freely.

Exercise 20.2 *

Select a commodity such as coal, rubber, wheat, etc. Show world production, production of key areas, transportation, and consumption.

* NOTE TO THE INSTRUCTOR: Use as an alternative exercise.

21

Cartograms

A *cartogram* may be defined as a *diagrammatic map*. Every map is abstracted to some degree; where the line is drawn between map and cartogram depends on the point of view. Some foreign authors call every single-factor map, such as a rainfall map, a cartogram. Here we do not go so far, but limit ourselves to highly abstracted maps where actual outlines or locations are distorted, or to maps which express motion, such as traffic-flow maps, or such mathematical concepts as centrograms.

Fig. 21.1. Blackboard sketch of the main gaps and historic trade routes of Central Europe. The use of colored chalks greatly improves the effectiveness of sketches.

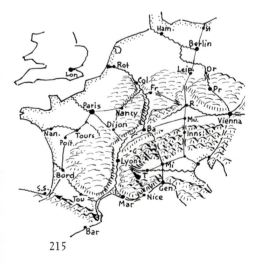

215

Blackboard Maps

Good teachers often sketch simple maps on the blackboard. Students understand them better than wall maps with all their complexities, and they are fascinated by the performance. Figure 21.1 represents the main gateways of Central Europe. All detail is omitted, the lines are simplified, and the mountains strongly symbolized. Whole continents may be drawn with the help of guide lines forming squares. Our task is simpler if we use for the model of the layout a projection with horizontal parallels, such as the sinusoidal or the Mollweide. The United States fits two squares fairly well, as in Fig. 21.2.

Fig. 21.2. Blackboard outline maps can be easily drawn by any teacher. The horizontal guide lines are parallels but the vertical ones are not exactly meridians.

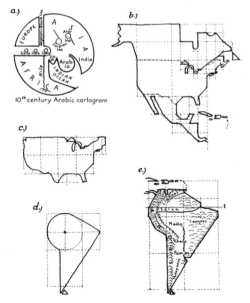

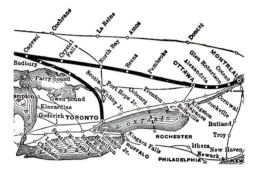

Fig. 21.3. Timetable maps are used chiefly for listing the stations and showing connections. The lines are simplified and the land is distorted.

Timetable Cartograms

Transportation companies like to use simple cartograms to show their stations and transfer points. The lines are straightened and simplified and stations are arranged like beads on a string, with emphasis on transfer points. All that most travelers want to know is, for instance, that they have to transfer at North Bay if they want to go to Cochrane and that the previous

station is Brent. The intricacies of all the curves and the detail of the landscape would be for many passengers more confusing than helpful. Most city transportation systems have such cartograms displayed.

Value-by-area Cartograms

Figure 21.4 clarifies this idea. All states are transformed into squares, oblongs, triangles, etc., the area of which is proportionate to population. As far as possible, they are placed in the position they have on the map. First we tabulate the values and select a geometrical figure which resembles the main body of the country. Then we divide it proportionately, always going from the large divisions to the small divisions. For example, in Fig. 21.4 the large oblong represents the population of the United States without New England or Florida. Then we slice off the portion for the Atlantic states minus New England and Florida. This can be calculated according to formula in Fig. 20.4. We may lay out

Fig. 21.4. Value-by-area cartogram showing the population of the United States in 1958.

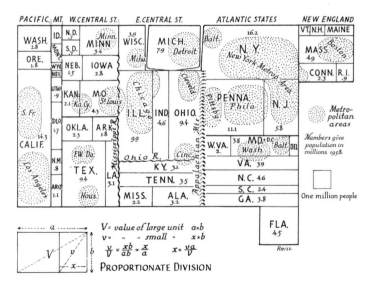

the cartogram so that it corresponds exactly to a scale. For instance, if the United States (minus New England and Florida) has 150 million people, the cartogram can be made 15 in. wide.

Next we slice off the South Atlantic states from the rectangle of the Atlantic states; this leaves us the North Atlantic states. From these, we cut off New York State, leaving in Pennsylvania and New Jersey; these we also divide proportionately.

Another approach is to work on graph paper divided into 0.1-in. squares. We make a 0.1-in. square to equal 100,000 people, for instance, and count squares for each division. We will find, however, that this takes longer than the other method, though it is useful in laying out small units.

Value-by-area cartograms can be made for population, wealth, number of college students, hydroelectric potential, or any statistical distribution. They often are laid out in an oblique view in an isometric scale for correlation. Figure 22.14 shows the world laid out flat according to population and the per capita income is superimposed as a third dimension.

Value-by-area cartograms are important. Our socioeconomic overview of the world will be more realistic if we think of the relative importance of its parts in the proportions of a population cartogram rather than in the proportions of a map. The results are often quite surprising. Such cartograms can make certain problems startlingly clear, such as the comparison of two cartograms of the United States, one showing taxation, the other showing representation in the Senate.

Exercise 21.1

Prepare two value-by-area cartograms of the states of the United States, one showing Federal taxation, the other showing representation in the Senate.

Figure 21.5 shows a cartogram where the area of the states is shown proportionate to retail sales. Instead of using quadrangles,

Fig. 21.5. Value-by-area cartogram where the shapes are distorted to conform in area to value of retail sales.

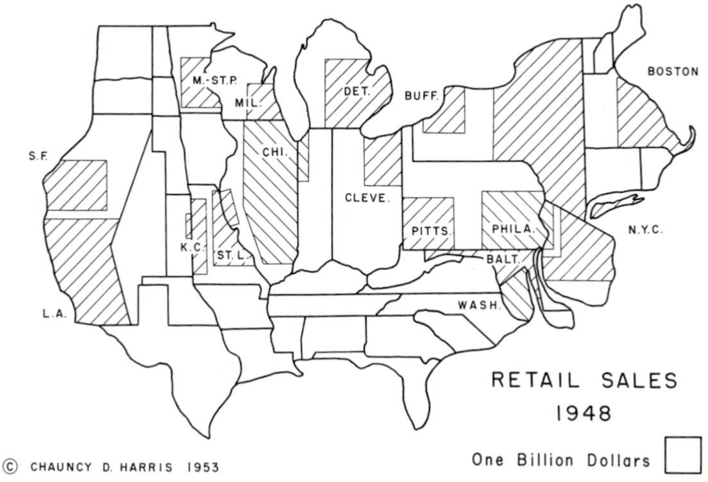

© CHAUNCY D. HARRIS 1953

RETAIL SALES
1948

One Billion Dollars

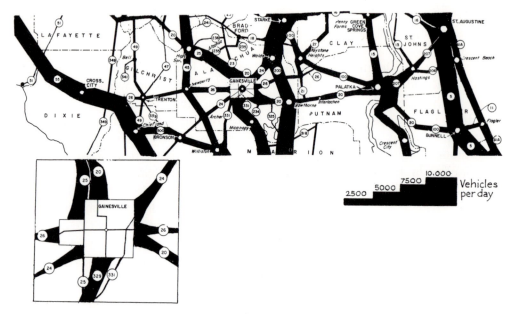

Fig. 21.6. Part of a traffic-flow map of Florida. Suburban traffic of Gainesville is in inset. City traffic is not shown. (*Modified from map of Florida State Road Department.*)

the author distorted the size of the states but more or less retained their familiar shapes. This makes place recognition easier, but comparison more difficult.

Traffic-flow Maps

In these the width of the traffic lines is proportionate to the number of trains, passengers, cars, ships, etc. Lines are usually greatly generalized; often we use composite lines, with different colors for coming traffic and going traffic. The chief problem

in drawing them is congestion of lines at the larger cities, so that quite often cities are circled and the circle is drawn separately on a larger scale. Sometimes we have to show very great differences in traffic. For example, the line with the heaviest traffic carries 100 trains a day. We show this with a line $\frac{1}{4}$ in. thick. If we want to show a line with only a single daily train, the line could be $\frac{1}{400}$-in. thick, which is too thin. One solution is to use dashes and dots, as shown in Fig. 21.7.

Fig. 21.7. By using dashes and dots instead of a solid line we can quadruple our range of scale on traffic-flow lines.

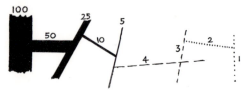

Migration Maps

In our mobile society most people do not live in the city or even in the state of their birth. The internal and external migration of our people is a major concern of geography. The simplest kind of migration map is the birth-residence-remainder map, such as that prepared by W. C. Thornthwaite for every decade. The thickness of arrows leading from one state to another is proportionate to the number of migrants. Data for these maps are published frequently by the Census Bureau. The map may show movement both ways, or emigration may be subtracted from immigration, so that only the remainder is shown.

The greatest mass movement of people at the present time is between home and work. This is an important study for urban geographers and a sample of such a cartogram is shown in Fig. 21.8.

Isochrones

In most history books there are maps showing how far one was able to travel in

Fig. 21.8. Cartogram showing commuting from house to work place. It is the same as a highly generalized traffic-flow map in which the lines are straightened. (*From Rudolf Hoffman, Germany, Salzgitter area.*)

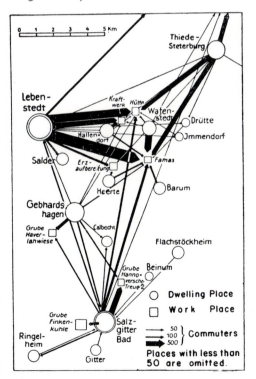

the past in one day, two days, and so on. It would be far more difficult to prepare isochrones for today, as we may fly in one day to La Paz, but it would take many days to get to a nearby mine in the Bolivian Andes. Isochrones may show any factor which spreads in time, such as the growth of a city or the spread of the Dutch elm disease.

Accessibility Maps

If we draw a map showing every railway and navigable river and motorable road with lines 20 miles wide in the scale of the map, we roughly show the areas to and from which heavy goods can be transported without the excessive cost of back packing or pack animals. Europe and most of the United States would be completely covered but other places would show great interline areas. Inaccessibility is rapidly vanishing, and a comparison of maps of today with those of only a decade ago indicates

Fig. 21.9. Isochronic map showing the rates of travel from New York City in 1800. (*Paullin and Wright, "Atlas of Historical Geography of the United States."*)

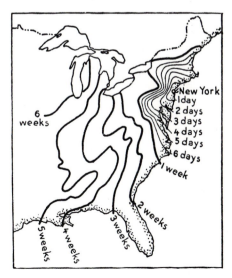

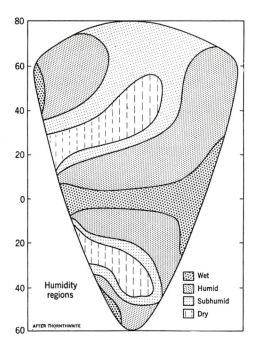

Fig. 21.10. The four major humidity regions are shown broadly generalized on the outline of an imaginary flat continent. (*From Finch, Trewartha, Robinson, and Hammond, "Elements of Geography," McGraw-Hill Book Company, Inc., 1957.*)

one of the most far-reaching changes of our age.

Idealized Geographic Concepts

To understand some geograpical concepts, it is best to show how they would work under ideal conditions without incidental complications. Figure 21.10 shows a hypothetical flat continent in its broadest generalization. Other cartograms may show the ideal distribution of cities, the zones of a typical city, etc. Many cartograms are used in climatology to explain such phenomena as the planetary wind system, cyclones, monsoons, etc. These cartograms offer a fertile field for a geographer's imagination.

Centrograms

These show the centers of some distribution in an area. The most common are centrograms of population of a country, but centrograms of wealth, education, and other subjects can be made in similar fashion.

The simplest kind of center is the *median point*. If, for instance, we divide the population of the United States into two equal parts with a meridian and do the same with a parallel, the intersection will be the median point. From that point, there will be the same number of people in the North and South, East and West, Northeast and Southwest (but not in the Northeast and Northwest). Any number of people may migrate from the Carolinas to Florida without affecting the median point. The use of a parallel for division is not quite correct; strictly speaking, a great circle should be used.

The technique for finding the pivotal point was described by John Fraser Hart.[1] Draw a series of equidistant lines enclosing tiers numbered as in Fig. 21.11. Multiply each tier value by the tier number to obtain the tier moment upon the baseline at the bottom. Add all the tier moments. Divide the sum of tier moments by the total population, and this will give the tier at which the *AB* fulcrum is located. Do the same with a set of vertical lines and the intersection will give the pivotal point.

In figuring the median points of the United States, we divided it into halves. It may just as well be divided into quarters, tenths, or any other parts, and thus we obtain *quartilides, sextilides, decilides*, etc.

World Centers. The median point of the

[1] *Econ. Geog.*, vol. 30, 1954, pp. 48–59.

world population in 1940 was calculated by the author. If we use a parallel and a meridian, this point was at Lahore, Pakistan.[2] This center more closely approaches our geographic way of thinking than that obtained by using a great circle instead of a parallel. The great circle is geometrically more correct, but would bring this center to Lake Balkhash in the U.S.S.R. The center of gravity of the world's land area was obtained by taking a light globe and covering all land with a sheet of lead. The globe was immersed in water and the bottom point was near Varna on the Black Sea. All other cultural and economic centers such as centers of wealth, education, etc., clustered around France.

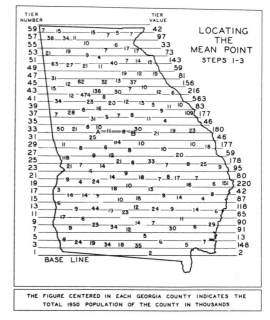

THE FIGURE CENTERED IN EACH GEORGIA COUNTY INDICATES THE TOTAL 1950 POPULATION OF THE COUNTY IN THOUSANDS

[2] Erwin Raisz, Our Lopsided Earth, Jour. Geog., vol. 43, 1944, pp. 81–91.

Fig. 21.11. Finding the center of gravity or mean point of population. (By John Fraser Hart.)

Fig. 21.12. The unevenness of distribution of land on the Earth is shown by this cartogram. (From Woytinsky, "World Population and Production.")

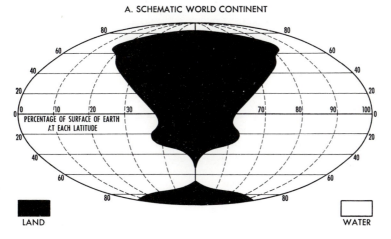

A. SCHEMATIC WORLD CONTINENT

PERCENTAGE OF SURFACE OF EARTH AT EACH LATITUDE

LAND

WATER

22

Science Maps

Almost every science uses maps and diagrams. Obviously the earth sciences rely most heavily on maps but the social sciences, particularly history, are using maps at an ever-increasing rate. In these days when we are overwhelmed with the magnitude of scientific literature, maps and diagrams facilitate greatly the comprehension of the material.

The Presentation of Maps and Graphs

Most scientific maps and graphs are presented in three ways: by lantern slides, in books or periodicals, or in theses. Each of these has different requirements. Slides must be very simple, the lettering needs to be large, but color may be used. For scientific periodicals, only black-and-white maps are used, but they can be more complex than those for slides; fold-in maps are sometimes permissible, but rarely encouraged. Thesis maps are usually photostats or contact prints, which can be large and folded; as only a few copies are involved they can be individually colored by hand.

We endeavor to make maps and diagrams which can be used in all three ways. They should be simple enough for slides and usable without color, but planned so

that coloring would enhance their appearance. Often this is not possible. A full-page map in a periodical is about 5 by 8 in. Lines and lettering adapted to such a map would be difficult to read if the map were made into a slide (see Chap. 5). A simpler version with larger letters may be necessary for this.

Avoiding Common Mistakes. Editors of magazines usually complain that maps are too large, lettering is too small, and cellotones are too fine to be produced on a magazine page. Some suggestions to avoid difficulties follow:

1. Do not make the manuscript more than twice publication size.

   Use cellotones which do not have more than 80 to 100 lines per inch in reduced size. Most lists of tints have accompanying charts showing how the tint will look when reduced to various sizes. One way to check is to have photostats made in publication size and inspect them.

   Avoid halftones. Halftones may look well on the original drawing, but the engraver would have to use a halftone screen in reproduction, which dulls letters and lines and increases costs.

2. Avoid unnecessary borders. A border is effective on the manuscript, but framed in between text, particularly on a slide, it is often not needed and it reduces the scale. Diagrams and cartograms in particular are more effective without borders. Maps which show parallels and meridians, however, require at least a neatline.

3. Avoid incorrect tints. Special tints are available for forest, grass, bush, swamp, etc., and advantage should be taken of them. If tones graded for darkness are required, take the recommendations of the

cellotone companies. For example, parallel lines of the same weight but going in two different directions can hardly be differentiated when reduced.

4. Use colors on slides and thesis maps. Although a map or diagram should be able to convey information clearly without colors for microfilming, there is no reason why colors should not be added to slides and to thesis maps. Most slides used today can be reproduced by Kodachrome or a similar process. Coloring, however, should be light, particularly over cellotones.

5. Do not cover lettering with lines or cellotones. Lettering has the right of way over everything else. Cut out the cellotone neatly over the lettering. Sometimes touches of white opaque paint applied with a fine brush will clean up around the letters.

6. Spread the lettering of large regions. It takes less effort to letter ROCKY MOUNTAINS compactly in a straight line, but it is far better to spread the letters along the crest. Names of rivers are not spread, but follow the curvature of the river.

7. Letter horizontal names parallel to the parallels. To make all lettering in the same direction is easier but it reveals the lack of geographic sense of the map maker.

8. Name symbols and tones in the legend. Too often we find that symbols and tones are numbered in the legend; the numbers are then explained in the caption underneath. This means that the reader must first look up the tint number in the legend, then find what it means in the caption. The cartographer has, in this case, put his own convenience before the reader's. As a general rule, put yourself in the place of the reader; do unto him as you would be done by.

Geography

Is not every map geographical? Indeed it is, but here we are concerned primarily with maps which accompany papers in geography or which are prepared in geography departments.

A paper in regional analysis will usually start with a *location map*, showing the relation of the region to a larger, generally known segment of the world. The location map should show political boundaries, as well as landform regions and transportation lines. The area of the study is usually squared in.

Next comes a *general map* of the study area. This should contain all place names mentioned and should be handled as a simplified topographic map reproducible without color. Then come the various *special maps*, such as climatic, geologic, economic, population, and social-composition, using dots, isopleths, choropleths, and diagrams, according to the principles discussed in previous chapters.

Geology

There is a close historical connection between geology and geography in America. Our topographic maps are made by the Geological Survey and many of our early geographers were also geologists. Departments of geology and geography are still combined in many universities. In the thirties geographers tried to emancipate themselves and orient geography toward economics and the social sciences. This went so far that many young geographers never even had a course in geology. At present the pendulum is swinging back, particularly since aerial photographs have disclosed the profound influence of geological fundamentals upon the land and its use.

The usual geologic maps show surface formations by age, using colors, tints, and index letters (see Fig. 19.1). Geologic mapping of the world is well advanced, advancing simultaneously or even preceding topo-

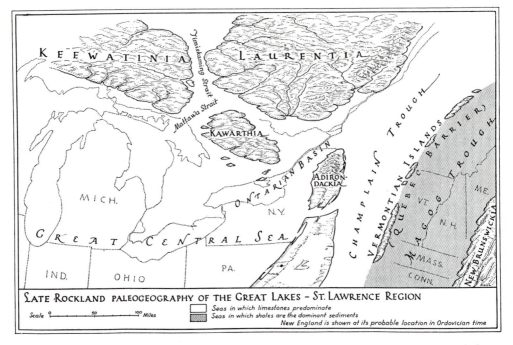

Fig. 22.1. Paleographic map. Note the difference between the shorelines of emergence and the shorelines of submergence. The dotted outlines, modern states, are pulled apart along the line of the future Green Mountains. (*After G. Marshall Kay.*)

graphic mapping, especially in areas where minerals are found. Almost all our states have 1:500,000 or 1:1,000,000 geologic maps, and the 1:2,500,000 geologic map of the entire United States is one of our basic maps. Equally important is the tectonic map of the country, showing folds, faults, trends, and the depth of key formations. The detailed 1:62,500 folios and the more recent and even more detailed geologic maps cover only a small part of the country. A number of photogeologic maps are now in progress on 1:24,000 scale.

All of Europe is mapped in great detail. The "Carte Géologique Internationale d'Europe," 1:1,500,000, started in the 1880s, was one of the first international maps. Canada, Australia, the U.S.S.R.,

Mexico, and most of Asia and Africa are mapped in some detail.

Paleogeographic maps try to reconstruct the conditions of the past geological ages. Obviously they are highly speculative, as we do not know how far the various formations extended before being eroded.[1] Although we do not know the exact shorelines of the past, we do have a fairly good idea about the nature of the land forms. Mountains, plateaus, and lava were probably sculptured as they are at present, and shorelines of emergence differed from shorelines of submergence in the Cambrian period as they do now. However, in the early ages we may assume a faster, badland type of ero-

[1] A. I. Layorsen, "Paleogeologic Maps," W. H. Freeman, 1960.

sion on lands unprotected by grass or forest. The term *paleomorphic map* would better express the nature of these maps.

Geomorphological maps show the various surface features. Michigan has a remarkably detailed map of its moraines, till plain, drumlins, and other glacial features. Belgium has been mapped by Mme. M. A. Lefèvre, and Denmark has a superb geomorphological atlas. The "Glacial Map of North America," the "Glacial Map of Canada," the map of the "Eolian Deposits of the United States," and similar maps of Germany, Poland, Sweden, and the U.S.S.R. all deserve mention.

Remarkable attempts in detailed large-scale morphologic mapping were made by R. Lucerna in Prague, by Hans Annaheim of Basel with about 80, and by M. Klimaszewski of Poland with 150 different symbols. As the work is done by students and volunteers, the progress is slow, but the educational value, particularly for the makers of such maps, is very great. A different approach for small scale is represented by the landform maps discussed in Chap. 8.

Block Diagrams

Block diagrams show a block cut out from the earth's crust and viewed obliquely from above. The relation between the surface features and the geology is in plain view.

Isometric blocks are drawn on one vertical and two chosen axes. The three axes are usually 120° apart. All parallel lines remain parallel; thus all vertical lines remain vertical. Distances are true on lines parallel to the axes; all other distances and directions are distorted. These blocks are used widely in mining geology to visualize subsurface structures, ore bodies, fault problems, and wells. Isometric blocks are accurate and measurable but they seem to grow wider in the background. The surface is drawn as shown in Fig. 22.3. First we draw a network of squares on the map. Then we draw the same network isometrically distorted, and transform the map on this. Then we copy the highest contour line first on an overlay; to copy the next contour we move the overlay paper one contour interval lower, and so on, until we obtain the impression of relief, as shown in the figure. Rivers,

Fig. 22.2. Preparation of isometric block diagrams. The block on right is produced by drawing the highest contour first and moving the paper repeatedly toward the bottom for copying each contour of the central figure.

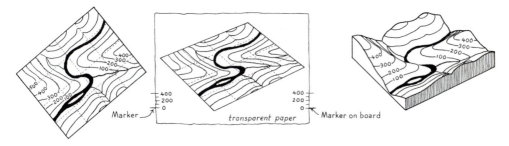

Fig. 22.3. Stages in the preparation of block diagrams, drawn in one-point perspective.

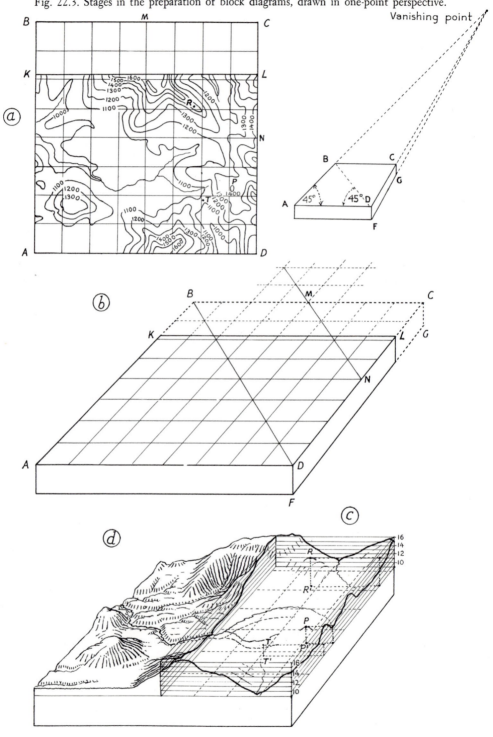

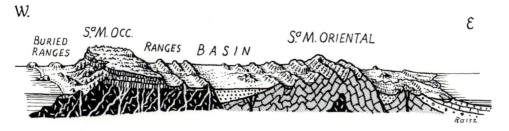

W.

BURIED RANGES   S.° M. OCC.   RANGES   B A S I N   S.° M. ORIENTAL   E

Raisz

Fig. 22.4. Panoramic section across Mexico. These sections are one of the best devices for teaching, particularly if colored for vegetation and cultivation.

roads, and other details have to be adjusted to this relief.

One-point Perspective. This also is not quite a true perspective, as there is no eye point from which a block could be seen exactly in this manner. The construction can be understood from Fig. 22.3. All parallel lines converge at a vanishing point on the horizon line. We draw a network of squares on the map; we construct a block that looks like a square according to the figure, even if we will not use the whole square. On this we draw a corresponding network. The surface of the block represents a chosen level, perhaps the largest plain and not necessarily the lowest contour line. We draw a railing of guide lines around the block, somewhat like those around a boxing ring; on this railing, we draw the profiles. The headwaters of rivers have to be raised and peaks elevated to the proper height before we can draw the landscape.

Panoramic Sections. These are elongated block diagrams. The frontal section shows geology and soil, and the surface is only a narrow strip. First we draw the front profile, then the back profile, and this will be enough to draw the landscape. Some vertical exaggeration is permissible. Panoramic sections are very effective tools in teaching the interrelation of structure, soil, vegetation, cultivation, etc. Several 10- to 15-ft-

long sections have been drawn by the author and by his students in class with pastels on wrapping paper in remarkably short time. The medium-dark surface and natural color of the wrapping paper make highlighting possible with white and light-colored pastels. Spraying with a liquid plastic prevents smudging.

Two-point Perspective. This is a true perspective as used by architects and artists. We use it for more elaborate block diagrams. Two vanishing points are chosen; the block is laid out on the horizontal and projected upon a vertical plane (see Fig. 22.5). The network for transferring the map is usually made by successive halving and diagonals. If the vanishing points are

Fig. 22.5. The exact method of constructing perspective blocks. The horizontal layout is in the foreground. Its projection upon a vertical plane is above. All horizontal parallel lines meet in vanishing points on the horizon.

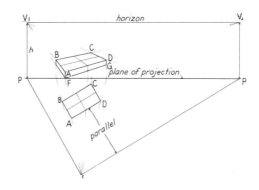

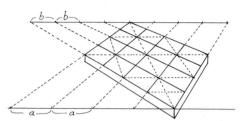

Fig. 22.6. Two-point or true perspective blocks are drawn by subsequent halving and drawing diagonals on the map and also on the block. Even divisions can be projected from horizontal (or vertical) lines.

far away, we can make use of the rule that even divisions on any horizontal (or vertical) line on the block remain even (see Fig. 22.6). Along all other lines they are

distorted. After the grid is drawn, the procedure is the same as with the one-point perspective.

Dufour Diagrams. We can transform a contour map into a block diagram by a long (ca. 6-ft) rod which slides along a groove or along the side of a table, held close to it by a long rubber band. The other end of the rod has a nail for a pointer; in the center is a pencil. The map is placed edgewise at the other end, as in Fig. 22.7, and the paper is placed in the center. If we move the pointer along the rectangular borders of the map, the pencil will draw a rhomboid. We move the pointer along the lowest contour; then we move the paper

Fig. 22.7. Dufour diagrams can be drawn fast and effectively. Their scale can be made smaller (above) or larger (below) than the original map.

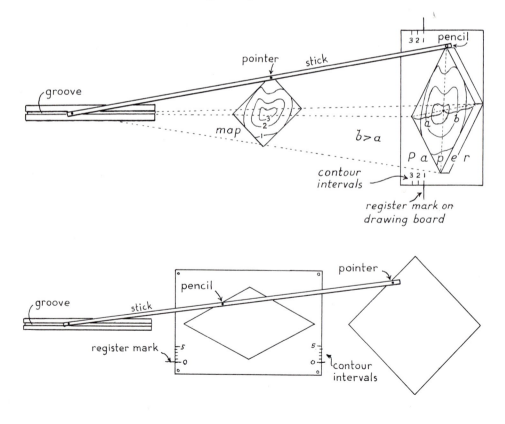

down one contour interval, draw the contour, and repeat this with all successive contours. The result will be a serviceable block diagram done in a very short time and with reasonable accuracy. By hand, we add rivers, roads, and houses. Some hachure lines and shading will make it more realistic. If we want to make the block diagram larger than the map, we exchange the pointer for the pencil, as shown in the lower part of the figure. The bottom of the block should be away from the map because this way the curvilinear distortion inherent in this construction will make a more natural view as it enlarges the foreground. The Dufour system does not work well with too elongated maps. It is also important that the map be set near a 45° angle; otherwise it will look distorted.

Exercise 22.1

To prepare a Dufour diagram; take a contour map 10 in. square. The best procedure is to take about a 3-in.-square area on a topographic sheet and enlarge it three times. Prepare a Dufour pantograph and draw the block diagram as described. Some of the students should enlarge it and some reduce it. Add geological section, fully labeled and colored. Some landform hachuring, plastic shading, and light crayon coloring will help. Place names are best above the diagram with arrows pointing to their locations.

Crisscross Sections. We draw geological sections in two directions and dovetail them into each other by notching them halfway. We can place this framework over the map and we will have a kind of relief model showing underground structure. An oblique photograph of it will serve as a good illustration. There is no difficulty in constructing the same view by isometric drawing.

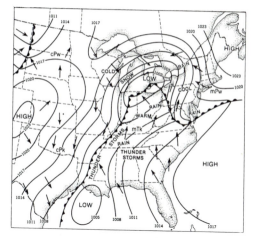

Fig. 22.8. Daily weather maps based on air-mass analysis. (*From Finch, Trewartha, Robinson, and Hammond, "Elements of Geography," McGraw-Hill Book Company, Inc., 1957.*)

Meteorology and Climatology

Weather and climate vary both horizontally and vertically. Thus their cartographic problems are often difficult.

Weather Maps. There are perhaps more weather maps printed every day than any single other kind of map. Daily weather maps are issued at Washington and at regional centers. They show temperature, precipitation, wind, pressure, fronts, and station data. Each station has about 20 different elements of weather recorded by a clever symbol system. These elements are teletyped so that they can be drawn on outline maps in a very short time all over the country. International organizations help to draw daily weather maps of the entire Northern Hemisphere, including information such as high-altitude conditions and jet streams, as well as electrical conditions.

Climatic Maps. Climate is average weather for a period of years. We draw maps and diagrams of rainfall, temperature, growing

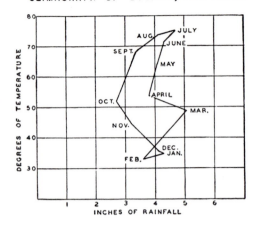

CLIMAGRAPH OF EUBANK, KENTUCKY

Fig. 22.9. A monthly rainfall-temperature clima-tograph will have a characteristic shape for every locality.

season, sunshine, wind, paths of storms, and climatograms showing the combination of several elements (see Figs. 19.6, 19.7, 19.9, 20.8, 20.11, 20.12, 20.18, 21.10). The literature is large and many excellent sources are available.[2] For abroad, the volumes of "Handbuch d. Klimatologie" by Köppen and Geiger will give much information. The Navy's "Marine Climatic Atlas of the World" is soon to be completed for all the oceans. A set of maps showing the world's climate for every month classified by rain-fall and temperature was prepared at Clark University and published by the Quarter-master General to help in the selection of proper clothing for our soldiers. A similar set of maps of the Northern and of the Southern Hemispheres was published by the Aeronautical Chart Service.

Astronomy

Celestial globes and star maps have been made since antiquity. A most important atlas is the "Bonn Durchmusterung of 1855," which records the stars with such exactness that it is possible to ascertain the proper motion of stars during the last century. Its descendants, "Bonner Stern-karten," show stars to the ninth magni-tude.[3] For most of the work, however, pho-tographs prepared by Palomar, Harvard, Bloemfontein, and a great number of other observatories are used. The usual star maps showing constellations use the stereo-graphic projection, which, as a conformal projection, can be easily calculated and does not distort the shapes.

Earth magnetism is recorded every few years by the U.S. Coast and Geodetic Sur-vey.[4] Separate charts show declination, hor-izontal intensity, vertical dip, rate of change, etc. Both polar and equatorial projections are used. Particularly revealing are oblique maps centered on the magnetic poles, as the divergence from radial or con-centric lines gives us some clue about the magnetic forces of the earth. Large-scale magnetic maps were made in the United States and Canada by airborne magne-tometers, and some important iron deposits were thus located.

Oceanography

A fathometer, using echo sounding, gives a continuous profile of the sea bottom. In the last decades we have learned much about submarine topography such as moun-

[2] S. S. Visher, "Climatic Atlas of the United States," Harvard University Press, Cambridge, Mass., 1954, 403 pp.; C. W. Thornthwaite, "Atlas of Climatic Types in the United States, 1900–1939," Department of Agriculture, Misc. Publ. 421, 1942; M. K. Thomas, "Climatological Atlas of Canada," National Research Council, Ottawa, 1953.

[3] Published by Ferdinand Dummler, Bonn, 1950, 1951, etc.
[4] S. A. Deel and H. H. Howe, "U.S. Magnetic Tables and Magnetic Charts," U.S. Coast and Geodetic Survey Spec. Pub. 667, 1948, 137 pp.

tain chains, trenches, rift valleys, sea mounts, submarine canyons, and our whole conception of the earth's tectonic history has had to be revised. The new Hydrographic Office charts give us most of the information. The "Carte Générale Bathymetrique des Océans," published by the International Bureau of Hydrographics in Monaco, has a set of colored charts of the world which is constantly revised, but it is not able to keep pace with the rapid accumulation of data. Maps of the configuration of the bottom of the North and the South Atlantic Oceans was drawn in the style of landform maps by Bruce C. Heezen and Marie Tharp at the Lamont Geological Observatory, Palisades, N.J. The colored and shaded oceanic hemispheres painted by Kenneth Fagg for *Life* are superb pieces of art.

For winds, currents, fog, storms, ice, etc., the "Monthly Pilot Charts" of the Hydrographic Office are jointly prepared with the Weather Bureau and the U.S. Coast and Geodetic Survey. Separate charts are issued for each ocean. The "Ice Atlas," "Atlas of Sea and Swell," and "Current Charts and Bottom Sediment Charts" of the Hydrographic Office give us a wealth of information. The "Morskoi Atlas" of the U.S.S.R., which was recently published, is the largest and richest atlas of oceanography, and came as a surprise from a nonmaritime nation. The cartographic devices of oceanography are complex and, in many respects, similar to those of climatology.

Soil Maps

The mapping of our most valuable and also most abused natural resource is rather recent. The U.S. Department of Agriculture published the "Soil Atlas of the United States," and a list of large-scale county maps can be obtained from the same source. For the most part, the maps use simple colored patches, but some of them are combined with land-classification letters, numbers, or patterns. Soil maps of foreign countries are found in their national atlases. The soil maps of Africa by Marbut and Schantz [5] and the 1:1,000,000 soil maps of the U.S.S.R. are the result of prodigious work. The Soil Conservation Service has mapped large areas showing land use, erosion, slope, and soil types.

Vegetation Maps. If we look at the earth from a great height, the most noticeable feature is its vegetation and cultivation pattern. Mapping the rich tapestry of the earth's surface is relatively recent. Air photos give results quickly and exactly but have a limited use without ground study.

Small-scale maps of countries and continents are common, with perhaps a dozen types of forest, grasslands, etc., indicated. Medium- and large-scale maps are made of only a very few areas. The usual vegetation map shows color patches, patterns, or index letters—often all three. Index letters usually indicate plant associations, while the numbers indicate height, age, quality, etc. (see Geostenography in Chap. 23). Vegetation is often done by sampling, as the same associations often cover very large areas. A. W. Küchler, of the University of Kansas, has written several articles on mapping vegetation, and Prof. Henry Gaussen of Toulouse produced remarkably detailed vegetation maps of France which, in an intensely cultivated country, show land use as well.

[5] H. L. Schantz and C. F. Marbut, The Vegetation and Soils of Africa, Am. Geog. Soc. Research Series No. 13, New York, 1923.

Fig. 22.10. Light vegetation patterns can be combined with heavier landform symbols. (*From Moldenke, "Plants of the Bible."*)

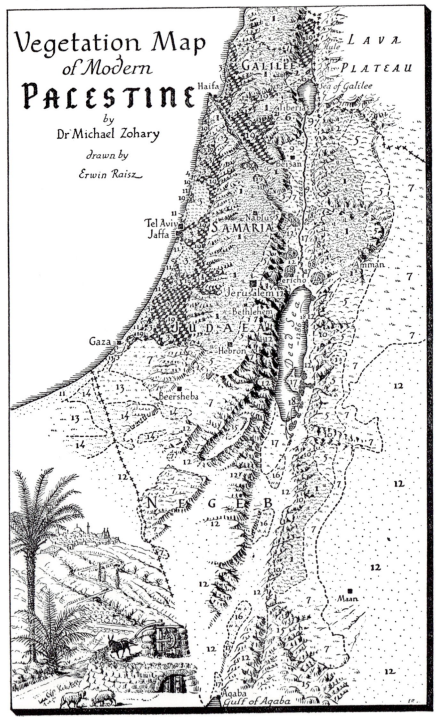

Maps of natural vegetation usually show what the vegetation would be if man abandoned the region. It is not quite true that natural-vegetation maps show the status before man appeared on the scene, because men burned the forest and grass for tens of thousands of years. In the present climatic period, the plant cover never has been quite natural.

Animal-distribution maps show either the prevailing association of mammals, birds, fish, and insects in a certain area or the distribution of a single species, genus, or family. They are handled as vegetation maps are, but, as animals are more mobile, the maps are even more complex. Maps which correlate plants, animals, climate, relief, etc., offer interesting cartographic problems.

Historical Maps

Every event occurs in time and space, and historians from the earliest times relied heavily on maps. Indeed, most medieval maps were historical, as they presented the Roman world. Historical maps are common jobs for cartographers. The difference between geographical and historical maps is chiefly in attitude. Most history maps are political in emphasis. They show battlefields, routes of armies, and other factors of political consequence. Yet the physical environment is important too. Mountains and swamps, for instance, limited the use of horse-drawn carts and impeded the supply of armies. We sometimes carry our present ideas of well-surveyed boundaries too far into the past. Almost every map of the Roman Empire shows a very definite boundary line in the Sahara Desert. In reality, Roman control was most uncertain beyond the coastal fortresses. We have to discriminate also between claims and settled possessions. The usual map of the eighteenth century, dividing North America between Spain, Great Britain, France, and Russia, would have certainly been derided by the midcontinental Indians.

History books have fewer statistical and economic maps because of lack of data.

Fig. 22.11. Historical events or economic products can be labeled effectively if boxed in on maps.

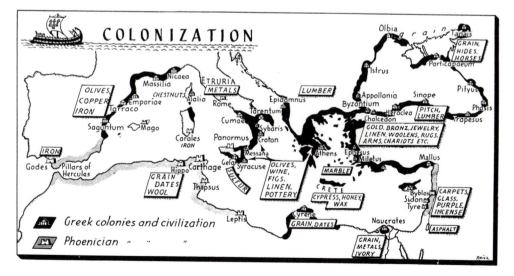

M   Mediterranean
N   Nordic
D   Dinaric
ɑ   Armenoid
A   Alpine
m   Mongoloid
n   Negro
ɒ   Negrito

Fig. 22.12. Racial composition can best be shown on maps by index letters. (*After E. Hooton.*)

Yet Paullin and Wright's "Atlas of the Historical Geography of the United States" and H. C. Darby's "Historical Geography of England before 1800," published by the Cambridge University Press, have a number of maps showing population densities and racial, religious, and economic conditions of the past.

Historical maps are a challenge to the cartographer. They can be made dynamic and symbolic with arrows, routes, barricades, and gateways, and they can show the

Fig. 22.13. Age-group diagram for England. The 2044 age pyramid is calculated by assuming the decline of the birth rate.

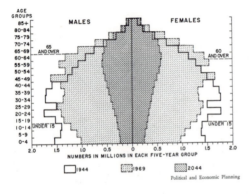

Political and Economic Planning

ever-changing drama of history on the world-wide stage.

Maps of Races, Languages, Religions, and Occupations

These maps have two common problems. First of all, these distributions are likely to be mixed. Second, these maps have little meaning without considering the density of the population. The usual color-patch maps of dominant distributions may give imperfect information.

One way to show mixed distribution is with grouped initial letters. For example, in Fig. 22.12 the first and largest letter will show the dominant race; the others will range in accordance with their importance. Where exact statistics are available, we can use pie graphs. The size of the circles is proportionate to the population, the sections to languages, religions, etc. The fact that poor cartography can affect the destiny of nations is demonstrated by the Treaty of Trianon in 1918, when large tracts of Transylvania were given to Rumania, even where the Magyars were numerically dominant. The Magyars lived in valleys and cities, while the Rumanian shepherds roamed over the wide ranges. On the available color-patch maps, the cities hardly showed up against the overwhelming color of the sparsely inhabited mountain areas.

Social Maps

The various other social compositions, such as wealth, health, and education, are best shown on maps by various superimposed diagrams, such as divided pie graphs, block piles, or star diagrams. Each case is a problem in itself, and cartograms and diagrams should be used creatively. Age-group pyramids have distinctive shapes for each

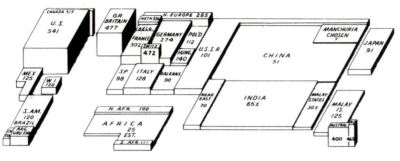

Fig. 22.14. The area of the bases of the various countries is proportionate to their population; height represents per capita incomes.

country (see Fig. 23.13). Educational pyramids, showing the number of students in each grade, are similar in construction.

Figure 22.14 shows a population-by-area cartogram, drawn in isometric perspective, on which the average incomes are placed as the vertical dimension. Similarly, per capita wealth, death rate, literacy, etc., can be shown. It would be a mistake, however, to show the total wealth as a vertical component; by this procedure, rich Denmark would appear to be poorer than India.

Fig. 22.15. By the use of a perspective projection, the scale of the foreground can be made larger. (*K. D. Metcalf, "A Graphic Summary," from Harvard University Library.*)

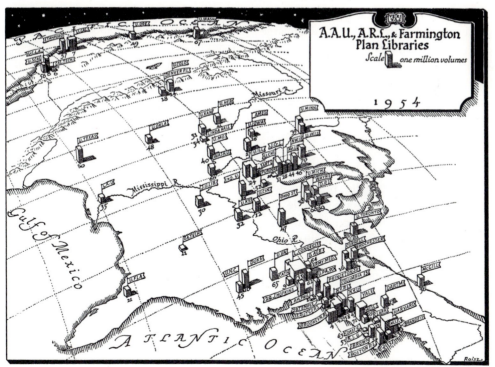

23

Land-use and Economic Maps

Our country has detailed maps of topography, geology, soil, population, etc., yet there is no land-use map available of the country as a whole in similar detail. Several states issued more or less generalized land-use maps, but only a few detailed large-scale maps are available. This is remarkable, as land use is one of the most important distributions. In most European countries land use is indicated to a certain extent on topographic sheets. Our maps rarely go beyond showing a green overprint for forests.

Land-use Maps

The British solved the problem with little expense. They have a 6-in./mile map of the whole country upon which every field is shown. These maps were distributed among interested citizens, students, scouts, and teachers who filled in one of the following letters:

F—Forest
M—Meadow
H—Heath and rough pasture
G—Gardens and orchards

236

W—Wasteland, cities, yards, cemeteries, etc.
P—Water

From these, 1 in./mile and 10 miles/in. reductions were made which portray, in vivid colors, the use of British land. Much credit is due for this work to Prof. L. Dudley Stamp.

We in the United States do not have 6-in./mile maps but we do have airplane photographs of almost the whole country. On these photographs the above categories are fairly visible even without going into the fields. It would be a most valuable educational project to send our students, scouts, and interested citizens into the field to annotate the photographs with interesting detail. The method of marking is discussed later in this chapter under Geostenography.

It is hardly to be expected that this large country could be completely surveyed for land use by volunteer work alone. Yet it would not only benefit land planning and business in general, but it would arouse interest in geography and provide excellent training for cartographers. For the less inhabited areas, it would make little additional work for surveyors to add to every topographic sheet a small land-use map. The reason that this has not yet been done is that land use is rapidly changing in some parts of the country. A young forest will certainly grow old during the usage of the map, but any land-use map, even if dated, is better than no land-use map.

Our greatest hope is in the various city, state, and regional planning offices, where much has already been accomplished. The Tennessee Valley Authority carried through a land-use survey of the valley with the most elaborate fractional code index system. As an example, III 3 $\dfrac{IN233}{4 \quad 4132114}$ tells all about the land and needs three pages of legend. Such detail is not recommended. A simpler index with less data is described in the author's "General Cartography," giving type, composition, value, upkeep, slope, drainage, soil, and stoniness for each land unit and a simpler index for buildings, roads, and structures.[1]

The Soil Conservation Service prepared a number of detailed maps of various parts of the country where different patterns show (1) cultivated land, (2) idle land, (3) pastureland, and (4) woodland. The main emphasis, however, is on soil erosion. The present stress of the Department of Agriculture is on Land Capability Maps, in which the land is classified into eight categories, of which only four are suitable for cultivation. Such maps are made even

[1] E. Raisz, "General Cartography," McGraw-Hill Book Company, Inc., New York, 1948, pp. 284–287.

for individual farms. A list of available maps can be obtained from the Department of Agriculture.

The International Geographical Union has a committee on land-use maps of the world, founded by Samuel Van Valkenburg. At every congress of the I.G.U. we learn of the immense amount of work done, not only in the United States and Europe but also in tropical countries. Its categories are shown as follows:

LAND-USE CATEGORIES AND COLORS OF THE I.G.U. COMMITTEE OF LAND USE

1. Settlements and associated nonagricultural lands (dark and light red)
2. Horticulture (deep purple)
3. Tree and other perennial crops (light purple)
4. Cropland
   a. Continual and rotation cropping (dark brown)
   b. Land rotation (light brown)
5. Improved permanent pasture (managed or enclosed) (light green)
6. Unimproved grazing land
   a. Used (orange)
   b. Not used (yellow)
7. Woodlands
   a. Dense (dark green)
   b. Open (medium green)

Fig. 23.1. The land-use map of England was made by volunteer workers who marked every field in the 6-in./mile map of Great Britain with one of seven letters.

Fig. 23.2. The U.S. Soil Conservation Service maps show land-use patterns and indicate erosion, slope, and soil types by index letters and numbers. The originals are in three colors.

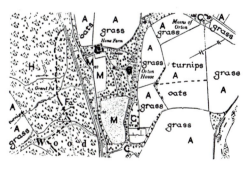

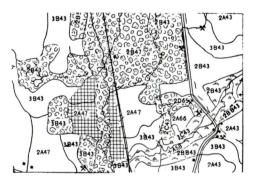

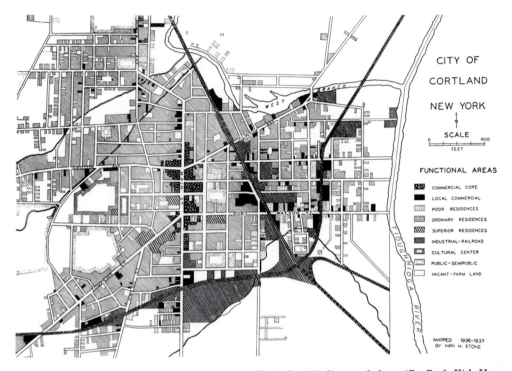

Fig. 23.3. Functional city maps can be made effectively with lines and dots. (*By Prof. Kirk H. Stone.*)

   c. Scrub (olive green)
   d. Swamp forests (blue green)
   e. Cut-over or burnt-over forest areas (green stipple)
   f. Forest with subsidiary cultivation (green with brown dots)
8. Swamps and marshes (fresh- and saltwater, nonforested) (blue)
9. Unproductive land (gray)

Functional city maps show urban land use and are fundamental for city planning and basic maps for urban geography. The various categories are shown by colors or tints. If tints are used, we use the darkest tones for the more valuable tracts, somewhat in this order:

Business centers, first, second, and third class
Industrial areas, warehouses
Public buildings, schools, hospitals
Apartments, first, second, and third class
Residences, first, second, and third class
Railways
Parking lots
Parks, playgrounds, stadiums, airports, cemeteries
Empty lots
Farmland
Water areas

City engineers usually have maps on large enough scale, 1:5,000 or 1:10,000, which can be annotated as described on the following pages, either on foot or from a slow-moving auto. Zoning maps are available for most cities but the actual land use may be quite different. Air photos help very much to make a preliminary plan, but cannot replace house-to-house surveys.

The street maps of the U.S. Bureau of Public Roads show types of pavement, business districts, and other data, and can be used as base maps, or for general information. The fire insurance maps of the Sanborn Map Company, in Pelham, New York, have remarkable detail, showing almost every house in our cities.

Land-use Profiles. It is part of human nature that we can now more easily understand a profile than a map. A land-use profile is only along one line but there it can show geology, soil, land slope, and a picture of land use in a single figure. This is the same idea as the panoramic sections described in the previous chapter, but with greater emphasis on land use. The addition of such profiles adds much to the value of a paper on regional analysis.

Geostenography

Our automobiles, trains, and planes cross the country with ever-increasing velocity. Fields, forests, and buildings pass by so fast that our ordinary methods of taking notes are too involved to keep pace. The method here described is designed not only for making notes on a sheet of paper rapidly, but also for marking maps and airplane photographs. The proposed codes of letters and symbols are suggestions only. They can

be redesigned and augmented to suit special regions and special interests.

Figure 23.5 of an imaginary region will best explain the procedure. We assume that we travel by car or bus and no detailed map is at hand. We use a good-sized notebook or pad of typing paper and mark at the bottom the time, place, mileage, and barometric altitude. We do this at the bottom of every page, but we note mileage and altitude at every important point. We draw a vertical line in the center of the paper and this will be our route line. If we have two men, one working to the left and the other to the right side, the route line can be nearer to one side of the page. If the road is sinuous, we may indicate this with a wavy line, but we mark every major change in direction with an arrow and reassume the new direction at the center of the page. We indicate the North frequently with short thick arrows. For this we need an auto compass, or if the sun is shining we may mark the direction of the shadow instead. An auto road map may help us in general orientation. On a very twisting road we make notes only when we travel in our main direction; otherwise on a hairpin turn we may easily mark a right-

Fig. 23.4. Land-use profiles are most useful tools in understanding the geography of a region.

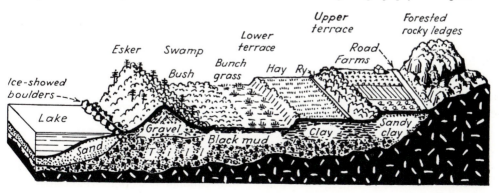

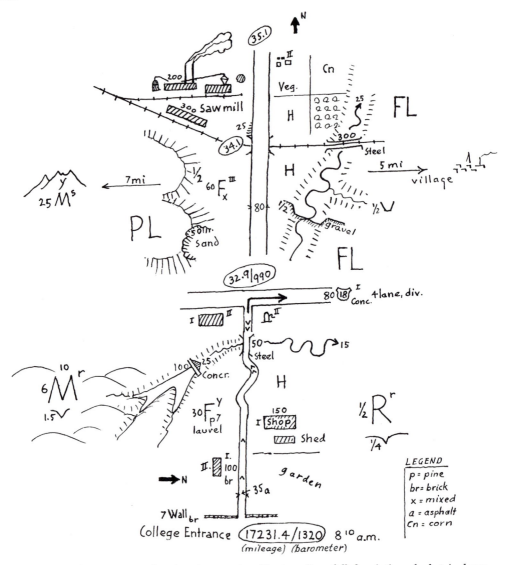

Fig. 23.5. Geostenogram of an imaginary region. Try to write a full description of what is shown here, after studying the next pages.

side feature on the left side. We make special effort to note distant features, always indicating their estimated distance and direction.

We try to make the stenograms in roughly the same scale, but one place may need many more notations than another. Frequent indications of mileage will help us to place the features properly on the map.

In half a day's auto trip we may expect to make 20 to 40 pages of notes. Whenever possible they should be reedited and inked in the same evening while the memory is fresh. If possible they should be converted into panoramic sections (elongated block diagrams) on much smaller scale. If our purpose is to make a land-use map of the region we crisscross at moderate speed and stop frequently.

We indicate land forms, vegetation, cultivation, buildings, and structures with sketches, symbols, letters, and numbers. The basic arrangement is the same for each with minor differences. The main type is indicated by the central capital letter or sketch, and the markings around it are given in the subsequent figures.

Land forms are indicated by both sketches and letter symbols supporting one another. One need not be an artist to draw simple outlines illustrating the nature of the hills. How he divides his time between letter symbols and sketches is up to the individual. Generally in mountains we sketch more; in lowlands we use more letter symbols.

The nature of letter symbols can be best understood from Fig. 23.6. T is the main landform type written with a large capital letter. Other symbols are as follows:

FL—Flat
R—Rolling land
M—Matureland (most of the land in slope)
KN—Knob
AL—Alluvium

Fig. 23.6. Land forms are indicated by letters, symbols, and simple sketches.

(*a*) Full notation according to *a* is rarely taken. Unimportant or unknown factors are omitted.

(*b*) Means 600-ft high, rounded mature land composed of granite with narrow valleys spaced 2,000 ft apart on the average.

(*c*) A plateau of rolling upland with 50-ft relief occupying seven-tenths of the area. The rest consists of steep-walled but flat-bottomed valleys 200 ft deep and cut in horizontally bedded sediments.

(*d*) A gneissic peneplain of 300-ft relief. Knolly peaks cover six-tenths of the area. The broad valleys are sand-covered. The trend of gneiss is indicated by lines (without trend lines it would be granite).

(*e*) A valley 200 ft wide at the bottom and 400 ft wide at the top, cut into gravel. The valley has sharp edges and steep banks 150 ft high.

(*f*) Rolling land with 150-ft upland relief. The valleys are narrow and 25 ft deep.

(*g*) Ridge 300 to 400 ft high and sharp-edged. On top is a flat plateau of gently dipping sedimentary rocks.

(*h*) Maturely dissected sharp-crested mountain, 2,500 ft high, 10 miles away.

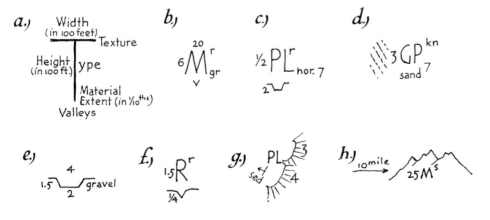

RI—Ridge
U—Upland
GP—Crystalline (granitic) peneplain
V—Valley

It is better to spell out such features as sand, fans, karst, sinks, dunes, and mesa. Transitional or mixed forms can be marked with two letters; *MU* means a matureland with some relatively flat uplands, for example.

The *h* in front of the main letter means *height* in hundreds of feet (or meters). For a single hill or ridge, we note our best estimate. In the case of intricately dissected slopes, we may note the total relative relief or we may write it as 10–4–2 for the depth of the primary, secondary, and tertiary dissection.

The *width* of a feature is marked on top of the main letter. We mark it only if it has some significance, such as the width of a mesa or of an elongated ridge. In maturelands, we may indicate the average distance between crests in this manner (see Fig. 23.6).

Texture is marked on the upper right and is designated by letters:

s—Sharp
r—Rounded
m—Medium dissection (usually omitted)
f—Fine dissection
ff—Very fine (badland) dissection
kn—Knobby
pt—Pitted

We may mark the spacing of rivulets as 2 *reg*, meaning regularly spaced rills 200 ft apart.

Specialty on lower right usually denotes geological composition:

ls—Limestone
ss—Sand
g—Gravel

la—Lava
al—Alluvium
hor—Varied horizontal sediments

Inclined sediments are marked *sed*, and dip is indicated with an arrow, as in Fig. 23.6g.

Extent, also marked in the lower right corner, means how many tenths of the total area are covered by the designated type. For instance, U_7 means that the flat upland occupies about seven-tenths of the area—the rest is valley.

The V at the bottom means the nature of the valley. A narrow V means narrow valleys; a wider $\vee$ denotes wider valleys. Flat-bottomed valleys can be shown by the actual profile of the valley, as in Fig. 23.6e. Asymmetry may also be indicated by the shape of the profile.

Rivers are drawn in the proper direction as we go along. If the stream is meandering or braided, we indicate this character by the type of line. We note the width of the river in feet and the direction of flow. We usually write out remarks such as "dry," "swift," or "black." It is perfectly feasible to design a stenogram for a river such as

$$20 \; \mathbf{R}^{\,5}_{\,200\;gr}$$ which would mean a 20-ft-wide river flowing with a velocity of 5 knots in a floodplain of gravel 200 ft wide. As rivers change rapidly with the seasons, such detail has only relative value.

Lakes may be sketched to show shape, and we write the nature of the shore, indicating whether it is swampy, sandy, stony, or rocky. We indicate swamps and marshes with short horizontal lines.

Vegetation is marked in a manner similar to land forms. The main type of vegetation is indicated with small-sized capital letters:

F—Forest
B—Bush

GR—Grassland
SA—Savanna
H—Herbaceous plants

As the types can be mixed or transitional, we may use several letters. *FB* means a woodland of forest and bush. More unusual types should be listed in the legend. Figure 23.7 explains the system. In sketching *forest*, we indicate the average top level for height. If two or more top levels are present, we can use several numbers. Diameters of the large timber trees are indicated in inches. Value is graded from I to V, but we give this information only for commercial timber. Age is noted only if the forest is young or very old, but we omit age if the forest is of average age. We indicate the kinds of trees by letters in order of their significance. We also list these tree types in the legend. Undergrowth is indicated by a number showing its density graded from 1 to 10. If there is a significant species, we may name it. We cannot ex-

pect, however, to make a detailed vegetation map from a moving automobile.

Cultivation is shown by rough drawing indicating the extent and pattern of the fields on the stenogram and naming the type of crop:

P—Pastureland
H—Hayfield
CN—Corn
CT—Cotton
TO—Tobacco
TR—Truck farming
O—Orchard
PL—Plowed field with crop unknown

More unusual types should be listed in the legend. Sometimes we add on the upper right the value or grade of the crop classified I—V. For instance *PI* means pastureland with improved grass. *Orchards* are handled like forests; 40 O_{ap}^{II} is an apple orchard with 40-ft-high trees in good condition.

Buildings are symbolized by small blocks

Fig. 23.7. Letter symbols for vegetation types.

(a) Model for spot notation. In case of mixed distributions several such models can be used as in c, marking the dominant type first. A legend for types and species should be attached to the stenogram.

(b) Slash pine forest with trees of high value, 80 ft high, and 20 to 30 in. thick. The forest is one-seventh of full density. The undergrowth is half of full density.

(c) Bush of mixed (x) plants, 15 ft high, from which 50-ft-high and 20-in.-thick palms are sticking out, comprising three-tenths of fully dense palm forest. Note the pictorial symbol for palms.

(d) Herbaceous plants 3 ft high, covering two-thirds of the area, apparently unused (u). The remaining third is covered by clumps of young mixed forest 20 ft high.

(e) Parana pines 100 ft high, of half of full density with 3-ft grass underneath. The use of sketches instead of letter symbols enlivens the stenogram.

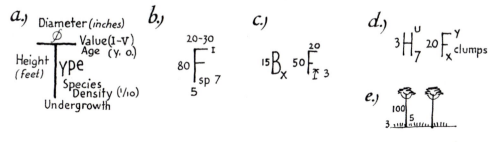

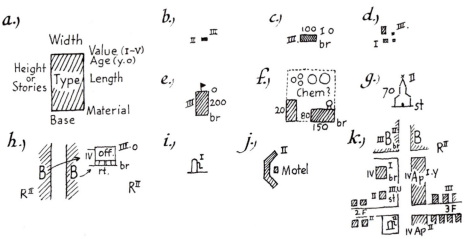

Fig. 23.8. Letter symbols for buildings.

(a) Model. Dimensions are in feet. Base can mean basement or ground floor, if significant. If no
type is indicated, the house is residential.
(b) Two-story third-rate frame house of medium age, used for residence. Factors typical for a
region may not be marked.
(c) Large three-story residence once first-class, but old. Built of brick.
(d) Farm compound of less than medium grade with one-story frame house and large barn.
(e) Three-story old brick schoolhouse about 200 ft long.
(f) Industrial compound of good appearance with 80-ft-high, 150-ft-wide main factory. No
chimney but tank on top. Twenty-foot-high storehouse and four round reservoirs for liquids.
Looks like a chemical plant.
(g) Stone church of medium age, 70 ft high, in good condition.
(h) Our route crosses through the busines section B consisting of old third-rate four-story brick
buildings with retail stores on the first floor and offices above. Second-class residences on the
side streets.
(i) First-class filling station.
(j) Motel, second class.
(k) City streets with business houses, apartments, old mansions (one unused, u), two-family
houses, and three-family row houses.

roughly proportionate to size. If we can
indicate a building's actual shape, so much
the better. Its characteristics are indicated in
Fig. 23.8. Driving through a city we have to
work fast, so we may not frequently repeat
the average characteristics, but we concentrate on special features. The average characteristics should be noted in a general
description. Thus in Fig. 23.8b we did not
indicate that it is a frame house used for residence and of medium age, as this is typical
for the area. In the heart of the city, as

in Fig. 23.8h, all we were able to note was
that we were driving through a business
section and as far as we could see to left
and right, there were second-grade residences. One should not expect to survey a
functional city map by driving through the
city once. Yet this could be done fairly accurately by driving slowly and crisscrossing
the area.

Structures are of such variety that no
definite model can be given. Figure 23.9
shows some examples. As far as it is pos-

sible, we maintain our usual arrangement —height at left, width inside or above, grade or value on upper right, material on lower right. Again we mark only the significant features. For example, if the lookout tower is of standard construction, all we have to add to the symbol is the height. Driving through the industrial section of a large city, we have to limit ourselves to the most general notes.

Geostenography from a train is essentially the same; however, we can usually look only at one side. Here we carefully note the time of every notation and of every station passed. Reading mileposts

gives greater accuracy. From a *boat* paralleling the shore, we may make a continuous sketch of the coast as it appears when perpendicular to our course. If we sit on the starboard side, we have to sketch from right to left. We note frequently our estimate of the distance to the shore, as in Fig. 4.11.

Geostenography from a plane is most fruitful. We usually can see only to one side, but this will be more than enough to keep one busy. Here the greatest problem is the distances of features from the plane. This is expressed by drawing them at the proper distance from the horizon line. We

Fig. 23.9. In structures we show horizontal dimension inside or above, vertical dimension on the left, grade on upper right, material on lower right. Dimensions are in feet.

(*a*) Asphalt road 30 ft wide in good condition with a 20-ft cut on right and a 10-ft cut on left. The road runs steeply up from culvert to the cutbanks (small v's). At culvert the road is on 15-ft embankment.

(*b*) Intersection with a first-class four-lane divided highway. Our road continues in a 15-ft-wide dirt road. A second-class filling station on upper right.

(*c*) Steel bridge, 100 ft, over a 20-ft-wide river which flows to the left 30 ft below the bridge.

(*d*) Stone dam 25 ft high and 100 ft long with a sizable reservoir behind.

(*e*) Lookout tower, 100 ft high, of normal construction.

(*f*) Transmission line of three wires of normal construction.

(*g*) Microwave relay tower 100 ft high with a 15-ft-high and 50-ft-wide brick building below.

(*h*) Single-track railway crossing our road underneath a 50-ft bridge. Station consists of a III grade, single-story frame building, and a 200-ft-long shed.

(*i*) Large stone quarry 1,000 ft wide with 100-ft walls and a 60-ft-high crushing plant.

(*j*) Airport with 100-ft-long one-story station building and two normal-sized hangers. Runways are concrete.

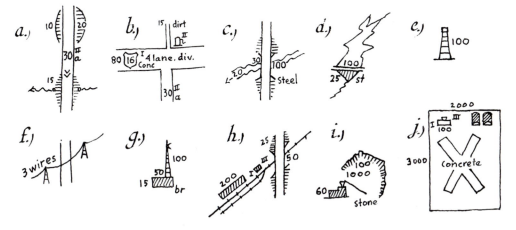

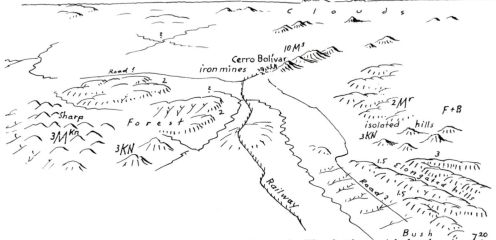

Fig. 23.10. Geostenogram from airplane over Venezuela. The sketch was inked, otherwise not altered. Compass direction was taken from a flight map.

draw the horizon line near the top of the page; the bottom of the page represents our flight line. Nearby features will recede quickly, while distant features can be drawn at greater intervals. A good eye for estimating distances helps greatly. One has to work fast; a page will be easily filled in a few minutes of flight. To mark the time on each page of every landing and take-off is most important for locating the features.

Notations upon Maps. Geostenographic symbols can well be used for marking up maps. Topographic sheets with good contour lines do not need much marking of land forms; texture may be added by simplified symbols. Most of our notes will indicate vegetation and culture. Small-scale or less-detailed maps, however, should be fully marked. The author had the task of annotating the 1:1,000,000 World Aeronautical Charts of Mexico from information obtained from actually flying and from air photos. The symbols described before were used, with one addition. Due to the extremely complex dissection of Mexico,

all mountainous regions were shaded with red pencil; the more dissected, the darker the red. This revealed a pattern in the Mexican plateau which hitherto had been only vaguely recognized.

Exercise 23.1

Select a rural road near your city and prepare a full geostenographic survey of about 15 miles. Do not mark maps or topographic sheets. Take notes there and back. Select about 3 miles of urban and suburban road and repeat the note-taking procedure. Prepare in the laboratory a panoramic section of the rural road with the road in center on 1 in. : 1 mile scale. Add your letter symbols with explanations. Name the various features. Add colors; inking is optional.

Economic Maps

Most of the problems of economic maps have been discussed already under the headings of statistical maps, diagrams, and cartograms. In view, however, of the emphasis on economic geography in our colleges, we present here a general overview of the problem.

Economic maps deal with the production, transportation, and marketing of goods. Economic life is extremely complex and its mapping has many problems. The cartography of economics is relatively new and offers a good chance to develop new devices. Our businessmen are aware of the help geographers can give and are willing to sponsor research and to underwrite colored maps. One type of economic map we have discussed already. The land-use map is a basic economic map, but there are many other kinds too. Here we are concerned more with small- and medium-scale designs.

Regional Economic Maps. To show all products of a region and their transportation in a single map has been attempted by many with only relative success. Economy is a somewhat abstract concept. Fields, factories, and mines are indeed visible, but their production is less so. To give a complete picture one should show not only the types, but also the value. We have to use diagrams, special symbols, and often colors.

Small-scale regional economic maps are standard features of school atlases. Agricultural products are shown by symbols, letters, and colors, and manufacturing areas are usually dark red. Some excellent regional economic maps are produced in Europe, such as the "Economisk Karta över Europa" of the Generalstabens Litografiska Anstalt in Stockholm. This shows agricultural acreage by small white squares on gray and the amount and type of manufacturing by colored squares. The "Wirtschaftsgeographische Karte der Schweiz" by Hans Boesch solves the problem with a striking array of colors, patterns, symbols, and letters in the most vivid

colors, and gives a good concept of Swiss economy. National, state, and school atlases have usually regional economic maps, some excellent, others bewildering in their complexity. The use of colors takes time and is expensive, yet the map is short-lived, as economic data are likely to be obsolete in a few years. Perhaps simpler maps reissued frequently would serve better.

Maps of Agriculture. Farming is by far the best-mapped economic activity. Agriculture is spread over the landscape and has rather clear-cut geographic relationships.

Fig. 23.11. Grouped dot maps are often used for regional economic maps. Here every unit represents 1 per cent of the world production of the respective commodity. (*From colored original in Ahlmann and Samuelsson, "Geografisk Atlas."*)

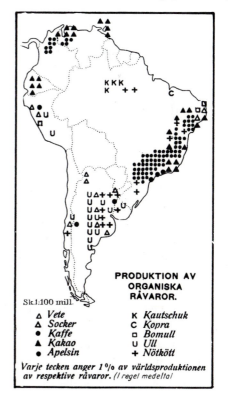

Sk.1:100 mill.

PRODUKTION AV
ORGANISKA
RÅVAROR.

△ *Vete*   к *Kautschuk*
△ *Socker*   c *Kopra*
● *Kaffe*   □ *Bomull*
▲ *Kakao*   u *Ull*
● *Apelsin*   + *Nötkött*

Varje tecken anger 1 % av världsproduktionen av respektive råvaror. (I regel medeltal)

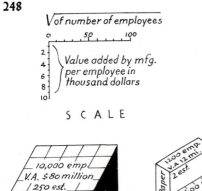

S C A L E

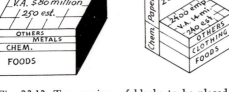

Fig. 23.12. Two versions of blocks to be placed upon maps of manufacturing. The diagram on the right discloses the fact that much of the value comes from two large establishments.

In the United States the most important producer of maps is the Department of Agriculture. The list of publications is very large and contains some of our basic maps such as F. J. Marschner's "Major Land Uses of the United States," which shows by colored dots the percentage of land in crops. The maps "Regionalized Types of Farming" and the "Natural Land Use Areas" present a multicolored carpet of our patterns of farming. The periodic "Graphic Summary" shows the various crops and animal products, with dots, choropleths, and graphs on page-size maps, as well as such items as farm machinery and value of real estate, tenancy, etc. Most farming states publish similar but more detailed maps of their own state.

For foreign countries we can find rich material in the national atlases. A summary treatment by Wm. Van Royen, "The Agricultural Resources of the World," has mostly dot maps.[2] As crops change con-

[2] Published by Prentice-Hall, Inc., Englewood Cliffs, N.J., 1954.

siderably from year to year, it is well worthwhile to be posted by the periodical *Foreign Agriculture* of the Department of Agriculture.

The cartographic problems of maps of agriculture are not very complex. Statistics are available by counties or farm districts and can be shown by dot maps (see Chap. 19). Very revealing are ratio maps such as production per acre, percentage of one crop against the other, or ratio of yield to investment. These can be handled by choropleths, isopleths, and superimposed diagrams.[3]

Maps of Manufacturing. Unlike agriculture, manufacturing is concentrated in and around cities and cannot be represented on a small-scale map by dots or isopleths. It varies also greatly in type and value, and the best presentation is by superimposed diagrams. Statistics by the Census Bureau are usually available for each district showing (1) kind of product, (2) value added by manufacture, (3) number of workers, and (4) number of establishments. While economically the value is more important, the geographer and sociologist are also interested in the workers and whether they are employed by large factories or small shops.

Figure 23.13 is an attempt to show all four factors in one diagram. For each city or industrial region we first draw an isometric base, the area of which is proportionate to the number of workers. This we divide into smaller units corresponding to the number of plants. The height of the block will be proportionate to the output per worker (value added, divided by num-

[3] J. C. Weaver, "Crop Combination Regions," American Geographical Society, 1954. This shows not only the present ratio of crops, but also changes in the past on large colored maps.

ber of workers). Thus the volume of the whole block will be proportionate to the total value added by manufacture. If few workers produce great value, as in the chemical industry, the block will be narrower and taller, while the food industries will produce more squat blocks. Type of product can be shown by dividing the block vertically or by simply naming the industries in decreasing order. The design of the block can be further elaborated. Where there are a few large factories and a number of small ones, we can show this by the division of the block, as in Fig. 23.12. Statistical data can be lettered at the top of the block.

No map with such elaborate diagrams has yet been produced. The most common are pie graphs dimensioned for the number of workers and sectioned for types of products. The least effective type of maps has even-sized symbols or letters giving no indication of the relative importance of the producer.

Mineral Production Map. Minerals are even more concentrated in certain localities and they are handled similarly to maps of manufacturing.[4]

Figure 23.13 shows metallic mineral production by block pillars according to the value of recovered metals referred to the locality of the mine, even if smelting was done elsewhere. The value of nonmetallic minerals is usually given at quarry or well. The block piles enable us to show values ranging from 1 to 10,000 times as great. No attempt has yet been made to shape the blocks with the number of workers, but it could be handled as it is in manufacturing. As with manufacturing, the least satisfactory type of mineral map is one showing every occurrence of a mineral with even-

[4] Statistics are available in the "Mineral Yearbook" of the U.S. Department of Interior. For foreign production, see William Van Royen and O. Bowles, "The Mineral Resources of the World," Prentice-Hall, Inc., Englewood Cliffs, N.J., 1952. See also the economic yearbooks of various publishers.

Fig. 23.13. Metallic mineral production is shown with subdivided and labeled block pillars. The largest producer is more than 2,000 times larger than the smallest shown.

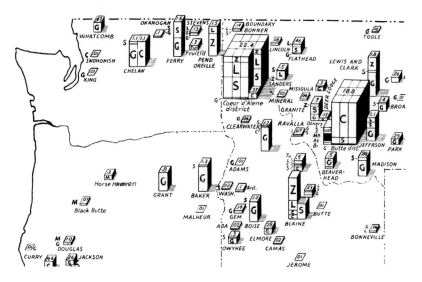

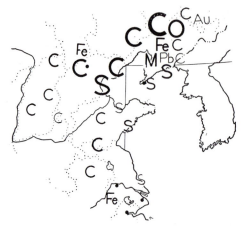

Fig. 23.14. Where exact figures are not available or are too variable, production can well be shown with proportionately sized capital letters. (*From George B. Cressey, "Land of the 500 Million," McGraw-Hill Book Company, Inc., 1955.*)

sized letters or symbols by which one is unable to discern the major producers. The highly generalized map of George B. Cressey in Fig. 23.14 is much more useful.

Fig. 23.15. Isephodic map showing cost of transportation. (*By J. W. Alexander, S. E. Brown, and R. E. Dahlberg, Econ. Georg.*)

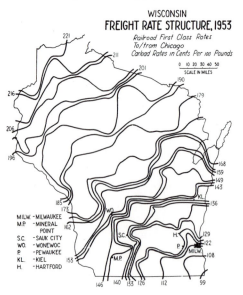

World Maps of Single Products. Showing production of wheat, rubber, oil, iron, and steel, such maps are contained in most economic atlases. The most successful ones show world production on one map with insets of the important producing areas on larger scale. Transportation and marketing are shown by traffic-flow lines. The maps are surrounded by all kinds of graphs showing trends of production, fluctuation of prices, types of consumption, and others. Obviously such atlases have a useful life for only a few years, and are kept simple.

Maps of Transportation. Shipping of goods is usually shown by traffic-flow lines. They can be quite diagrammatic, showing the actual route taken in a rather generalized way. Cartograms showing the movements of ore, coal, and grain in the Great Lakes area are published by the Corps of Engineers.

S. W. Boggs produced maps connecting places of equal transportation cost with lines called *isephodes* (see Fig. 23.15). These maps present the same difficulty as do isochronic maps. The cost may be modest to send a ton of coal by rail to Tapachula, Mexico, but it would be most expensive to send it to a nearby village in the mountains. The Institute for Ocean Transport Development in Bremen, Germany, publishes a remarkable map of international sea traffic.

Maps of Marketing. These still constitute a largely unsolved portion of the field of cartography. How far and which way people will travel to purchase and sell is a vital question in locating businesses. Considerable research has been done by sampling methods or questionnaires. The cartography is difficult because it has to take in consideration not only the density of the population of the area, but also the density of the kind of population interested in the particular business. There is a great deal of overlap of marketing areas of the various centers. Distribution may be shown with radiating lines or overlapping isopleths, depending upon the type of business. Often we use population-by-area or wealth-by-area cartograms (see Fig. 19.5). Research on marketing maps has been carried on by the Map Division of the Library of Congress.[5] The Production Marketing Administration of the Department of Agriculture is also actively interested in the problem.

[5] W. W. Ristow, "Marketing Maps of the United States," Map Division, Library of Congress, 1958.

24

Globes

As soon as the spherical form of the earth was recognized, the ancient Greeks experimented with globes. We know very little of the globe made by Crates about 150 B.C.; the design in Fig. 24.1 is largely conjecture. He knew the size of the earth from Eratosthenes and Hipparchus and the use of parallels and meridians. When he placed the known extent of the *ecumene* (the habitable world recognized by the ancient Greeks) on his globe, it covered only a small portion of it. This did not conform to the Greek concept of symmetry,

Fig. 24.1. We have only a vague description of the first globe known in history. It probably featured four continents separated by a north-south and an equatorial ocean.

so he included some continents for balance. This was the earliest anticipation of the Americas.

The first terrestrial globe which has survived was made by Martin Behaim in 1492 in Nuremberg. Columbus and Behaim were both in Lisbon in 1484 and there is little doubt that they exchanged ideas, as this globe closely approximates the concepts of Columbus when he sailed for the setting sun. Thousands of excellent globes were made during the Age of Discoveries. They served to explain the relation of America to Europe and Asia. These globes were highly ornate, a fusion of art and science characteristic of this period. Later globes are more simple, but also more accurate.

In this age of airplanes and rockets, globes are frequently in use. There is no better way to explain the global concepts of world relationships. Our first geography lessons in school are with globes. Day and night, summer and winter, poles and equator, parallels and meridians, local and mean time, date line—all these concepts and many others can hardly be understood without a globe. Later, when we learn about map projections, the tectonics of oceans and continents, climatic belts, ocean currents, we again turn to globes rather than to maps. We do the same to plan globe-girdling flights, ballistic missiles, radio beams, and artificial satellites.

Construction of Globes

Most globes consist of a spherical base upon which printed gores are pasted. The

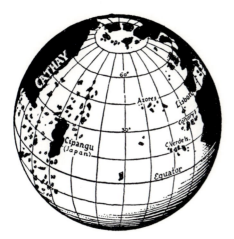

Fig. 24.2. Behaim's globe of 1492 is the first terrestrial globe which has survived. Behaim's geographic ideas come close to those which prompted Columbus to undertake his voyage.

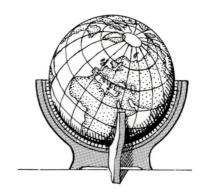

Fig. 24.3. Globes are often placed in a cradle from which they can be lifted and replaced in any position.

base can be of cardboard, metal, glass, rubber, plastic, or almost any light and strong material. Old globes were usually on a plaster of paris shell reinforced with strips of cloth or paper. Most globes are mounted on a stand with the axis tilted 23.5° from the vertical. This helps to explain the seasons. Today globes are often placed in a cradle, and we can turn Antarctica to be on top just as easily as the Arctic. Getting away from the usual orientation and looking at the world from a new angle gives us a better understanding of earth relationships. The mounting in Fig. 24.3 makes it easy to measure globe distances between any two points by placing them on the graduated arc of the cradle.

Globe Gores. Our presses are able to print only on flat paper. We can, however, print the surface of the globe on small enough gores, lune-shaped pieces of paper, so that by wetting and stretching them, they will conform to a spherical base. A system of 24 gores is shown in Fig. 24.4.

They are printed on two sheets, one for each hemisphere. The sheets are pasted on cardboard and the empty segments between the gores are cut out by machine. The gores are wetted and pressed into a hemispherical mold. To keep them together permanently, a second reinforcing cardboard hemisphere is pasted inside.

Fig. 24.4. Globes are often made from 24 gores arranged radially from the poles. The two hemispheres are joined together at the equator.

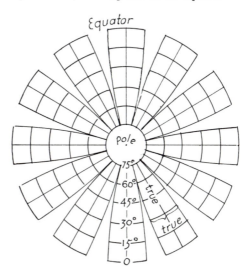

Metal globes are made similarly, but welded. The two hemispheres are held together by a graduated belt at the equator. Some globes are made of 12 gores, 30° wide, reaching from pole to pole. To paste this globe on a sphere requires highly skilled operators. By a somewhat similar process translucent plastic globes are manufactured by using a vacuum instead of pressure.

Inflatable globes were made of silk in Japan decades ago. Nowadays they are made of rubber, rubberized cloth, or plastic. They have the great advantage of easy storage when not in use.

Globes which open up like an umbrella with semicircular ribs were manufactured in England, but never became popular. Patents have been issued for paper or metal globe gores which dovetail together and can be disassembled when not in use. Various *near-globes*, in which the sphere is approximated by trapezoids, pentagons, or triangles and the planes assembled, were much in fashion in the forties. They are inexpensive and interesting to work with, but they fail in one respect—they do not give the impression of the solid planet of massive perfection which they aim to portray.

What Should a Globe Show?

Political globes are the most popular, with the various countries in gay colors. They look decorative, but they present unrealistic coloring over a realistic base. *Physical globes* are better. They are usually layer tinted and give a picture of the great mountain ranges, ocean depths, and the tectonics of our planet. Neither kind seems to be the best portrait of Mother Earth. The time is

ripe for new approaches. One such attempt is described at the end of this chapter.

The Analemma. Most globes have an 8-shaped figure (see Fig. 24.5) printed on them, which sometimes leads to some embarrassing question for the teacher. The analemma shows where the sun is directly overhead for each day of the year when the Local Mean Time is 12 noon on the meridian upon which the analemma is centered. Identical analemmas can be drawn on any meridian. The 8 shape is the result of two factors. The sun is overhead on June 21 on the Tropic of Cancer and on September 22 on the Tropic of Capricorn, which explains the North-South movement. The East-West motion is due to the Equation of Time. The earth revolves around the sun with varying velocity (Kepler's law), and the equatorial plane is variously inclined, as seen from the sun. For these two reasons, one day is not as long as another. The difference is only a few seconds, but they accumulate until on February 10, the sun is the highest at 12^h 15^m local time, and on November 2 at 11^h 43^m. As 4 min of time represents 1° of rotation, this time difference may be laid out as horizontal components of the 8-shaped curve.

With the help of the analemma, we can place the globe in the exact position as the earth on any day at any hour, and if we let the sun shine on it we duplicate the actual lighting. The analemma is also used for making sundials.[1]

Celestial Globes. The earliest globe which has survived is the Farnese globe, a small marble globe showing the constella-

[1] E. Raisz, The Analemma, *Jour. Geog.*, vol. 40, 1941, pp. 90–97.

tions, dating from about 25 A.D. In the Renaissance, celestial globes with magnificent mythological figures of constellations were popular, but they were works of art rather than of science. Celestial globes, however, are important for teaching mathematical geography and the solution of the problems of celestial triangles. Mechanical devices showing the sun, earth, moon, and planets and whirled around by clockwork were much in fashion in the eighteenth and nineteenth centuries. They were called "orrerys" after Charles Boyle, Earl of Orrery, and now they are prized antiques. The modern planetariums are descendants of the idea.

Some Notable Globes

We have already mentioned the Behaim globe of 1492, a small wooden globe, now exhibited in the Deutsches Museum in Nuremberg. One of the greatest globe makers was Vincenzo Coronelli, a Venetian priest. Among others, he made a 12-ft globe for the King of France; it was rotated by clockwork and one could enter it to see the light shining through little holes and looking like the stars in heaven. His name survives in the present Coronelli World League of Friends of the Globe.[2]

A glass "georama" in which one looks from the inside upon the spherical walls of a globe is in the Christian Science building in Boston. A photograph of this from one of the entrances shows the stereographic projection. A 28-ft globe covered with enamel sheets rotates out in the open at the Babson Institute in Wellesley, Massachusetts. Both of these globes are politically colored, and their value is chiefly in

[2] Vienna I. Gusshausstrasse 20.

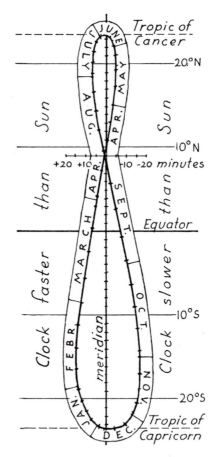

Fig. 24.5. The analemma shows for every day of the year the spot upon which the sun shines vertically when the local mean time is 12 noon on the meridian for which it was drawn.

their large size. A relief model of the United States, 64 ft in diameter, is also at Babson's and is actually a globe segment of great scientific value.

The author was called upon in 1956 to supervise the carving of a 6-ft relief globe which is intended to be the most realistic portrait of the earth ever made. No parallels and meridians and no names are visible. The globe is beautifully colored by

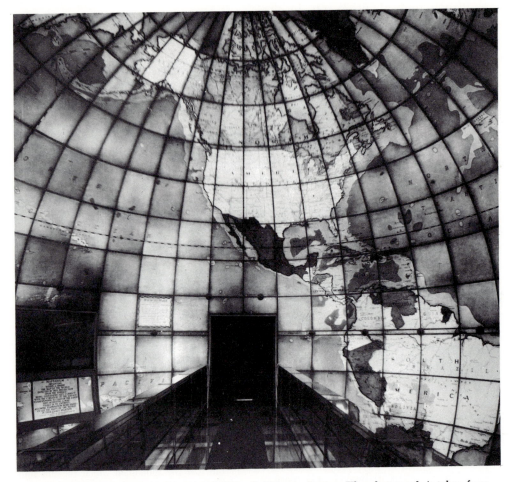

Fig. 24.6. Mapparium in the Christian Science Building in Boston. The photograph is taken from one of the equatorial entrances and thus shows the globe in stereographic projection. (*Christian Science Publishing Company.*)

Kenneth Fagg, showing dark-green forest, blue water, light-green grass, light-ochre deserts, dazzling white ice, etc. Indeed, except for the clouds, it looks as the earth would from a rocket ship thousands of miles up. Mountains, however, are exaggerated 15 to 60 times (more in the lowlands) because otherwise even the highest mountains would hardly be visible. The original was carved in plaster of paris. It was reproduced by some ingenious device in rubber, which is held in shape by an inner inflated balloon. This was necessary to carry it through doors. Several metal and plaster copies of this globe are also in use.[3]

[3] Color pictures are available from Rand McNally & Company, 405 Park Ave., New York. This company also has a 1-ft unpainted relief globe made of white plastic.

Fig. 24.7. Part of a 6-ft relief globe. The globe can be reproduced in inflated rubber, in plaster, and in metal. Several globes were hand painted in natural colors. (*Courtesy Rand McNally & Company.*)

25

Models

During the last war it became painfully obvious that in spite of all the training, most soldiers were not able to visualize a contour-line map. The Army had to go to great expense and trouble to provide relief models of at least the most important landing and target areas. Some were improvised from any kind of scrap material and were gibed at as "egg-crate" models, but they were eminently successful. Some were made of rubber or plastic by experts.

Terrain Models. These are made of any kind of material and are not only useful for soldiers, but a geographer, too, can learn much from them. The simplest kind is started on a board, on which a map is drawn and at every important point a peg of the proper elevation is driven in. Then the map can be built up with sand, cement, plaster, plastiline, paper mash, or any locally available material. Sand can be hardened by spraying it with lime, cement water, or water-glass. Grain or corn grit will keep its shape if mixed with thinned rubber cement. The Army recommends a mixture of 1 pt of sawdust, ¾ pt of plaster, ½ pt of library paste, and a few drops of glue. Thin the paste with water, add

plaster and sawdust, and knead it to the consistency of a tough dough; it will harden in about eight hours.

When hardened, the model can be painted with oil or poster colors, and a spray of liquid plastic will protect the paint even if it is exposed to rain. The ingenuity of our soldiers in making terrain models was inexhaustible. Very large-scale models were made by stretching canvas over a skeleton of laths. Smaller-scale models can be made rapidly by the "toothpick" method. For this, a copy of the map is placed on a wooden base; another copy is held in the hand. A graduated peg of toothpick will indicate the height to which the model should be built. Plastiline, a mixture of clay and mineral oil, is most commonly used. Modeling tools can be bought or fashioned from sticks and wire.

Cardboard Cutting. More exact models can be made by cutting cardboards. Each contour line of the map is copied on separate cardboard and cut with a knife or jig saw, starting with the lowest contour. The cutouts are pasted or nailed over each other. The model is smoothed over with plastiline; when a frame is placed around it, it is ready for casting.

Negative Method. For small-scale models of countries or continents, it is better to work the opposite way. We cut the topmost contour, discard the inside, and lay the outside upside down. Thus the highest peaks will appear as small holes. Proceeding from contour to contour, at the end we will have a negative of the model from

258

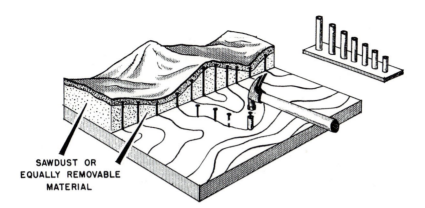

Fig. 25.1. Terrain model made with the help of nails driven into a board to proper height with the help of a set of tubes. The particular model is built up with sawdust and covered with plaster. The sawdust is removed to reduce weight. (*Courtesy of the U.S. Navy.*)

SAWDUST OR
EQUALLY REMOVABLE
MATERIAL

which positives are cast directly. The extra plaster, as shown in Fig. 25.2, is carved off and a new negative and positives are cast. In this method the many small peaks are not so easily lost as in the positive method.

Vertical Exaggeration

If we were to make a 4-ft relief model of the United States without vertical exaggeration, the Sierra Nevada would stand above the plains by about one twenty-fifth inch—hardly noticeable. This is far from our conception of the shape of our country, and we would not be able to sell a copy. We are very sensitive to elevation and exaggerate it in our minds. Thus, if we exaggerate the vertical dimension, we come nearer to our mental concept of the land. For instance, on the 4-ft model of the United States, we certainly want the highest peaks to stand about ½ to 1 in. high. This requires a vertical exaggeration of about 10 to 20 times. With this, however, Mount Rainier would stick up like a nail.

Fig. 25.2. Simple large-scale models are made by the positive method; the negative method is used for small-scale models with fine detail. In the positive method we add plastiline to fill the steps between contour lines, whereas in the negative method we carve off the extra plaster.

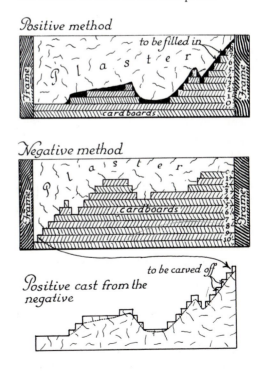

Positive method

Negative method

Positive cast from the negative

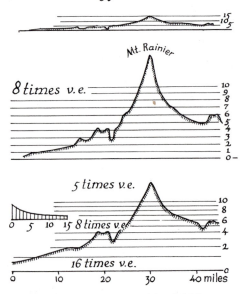

No vertical exaggeration

Mt. Rainier

8 times v.e.

5 times v.e.

8 times v.e.

16 times v.e.

Fig. 25.3. In small-scale models of high mountains, the vertical exaggeration has to be reduced at higher elevations.

To make a satisfactory model, the vertical exaggeration is not a purely arithmetic problem; rather, it is a psychovisual problem.

The amount of vertical exaggeration depends on the scale of the map, the relief of the terrain, and the amount of our generalization. A 1-in./mile map does not need any exaggeration at all unless the land is very flat. A mountainous land, broadly generalized, needs larger exaggeration than the same model with finely chiseled detail. The nearest empirical formula which can be given for hilly land is

$$VE = 2\sqrt{\text{miles per inch}}$$

twice the square root of the scale expressed in miles per inch. This, however, varies greatly. To give reasonable relief, it was necessary to exaggerate a 1-in./mile model

of the flat Gainesville area in Florida six times. In high mountainous regions, on the other hand, we use much less vertical exaggeration. On small-scale models, we usually reduce the vertical exaggeration with altitude, as in Fig. 25.3, because mountains tend to get steeper at higher elevations. No definite rule can be given for the reduction of vertical exaggeration. It will be about halfway between the arithmetic and logarithmic scale. Thus, a mountain four times as high as another will be modeled about three times as high. We would use somewhat greater reduction of vertical exaggeration where the range of altitudes is very great, as in the state of California.

If we cut out cardboards, we have to work with certain standard thicknesses, and adapting these to our planned vertical exaggeration is not always easy. For example, we have a 1:125,000 map of a rolling country with 50-ft contours and are working with thin cardboards $\frac{1}{20}$ in. thick. If we take one cardboard for each interval:

Horizontally: 1 in. to 10,417 ft
Vertically: 1 in. to $20 \times 50 = 1,000$ ft.

This would give us about ten times vertical exaggeration, which is too much, so we use only every second contour line, which means five times vertical exaggeration, which is permissible.

Plaster Casting. No material can be carved so easily and so finely as plaster of paris. It is light, fairly strong, and takes color well. For casting in plaster, we build a frame around the model, tighten all sides and corners with plastiline, and oil all surfaces. We calculate the volume and measure out the plaster. We take a dish and, figuring about 2 pt water for every pound of plaster, we slowly spread the plaster over the water until it is the consistency of rich cream. Although we do not stir, we should squeeze lumps apart. With a ladle, we pour it into the mold, starting in one corner and proceeding slowly so as to avoid formation of air bubbles. We may have to brush the plaster into the hollows. We add plaster until the frame is full and embed strips of cloth to reinforce the cast. Large models are left partly hollow; only the sides reaching the top of the mold are plastered. Filling the mold has to be completed in about 10 min, before the plaster sets. The plaster will almost completely harden in a few hours, but it is better not to open the frame and separate the cast until the next day, due to slight contraction in drying. If we are making a plaster cast from a plaster mold, we use green soap as a separator instead of oil; otherwise the process is the same.

Finishing. The final handling of the plaster surface is done best with dentist's tools used freehand or attached to an electric drill. We carve in the smaller tributaries, bluffs, and terraces. We check to be sure that every river flows downhill. We carve in railroad and road cuts. We often grain the surface—coarsely mottled forest, finely striped fields, and smooth pastures add much to the value of the model. If several copies are needed, we recast the carved model, making a negative; from that, we make as many positives as required. Lettering can be carved into the negative, right to left.

Painting. Plaster models are best painted in water color. We remove the oil from the surface with cleaning fluid. Plaster takes water color well and it remains bright for decades. Oil color or poster color usually

looks drab and peels off in a few years. Plastiline models, however, have to be painted with oil colors. Forest can be imitated with colored sand sprinkled over a gluey surface, cultivation by a checkerboard of colors, and bits of chipped eraser pasted on can represent cities. Lettering can be penned directly on the model or be pinned on. Relief shows best by the use of a strong side light. This effect can be imitated by side-spraying the model with airbrush, using dark gray.

Metal and Plastic Models. None of the new materials has the fine texture of plaster of paris, but the new materials possess greater strength, are lighter, and allow more rapid production.

Bround Reliefograph (see Fig. 25.4) shapes a metal sheet with an electric hammer. The contour map is pasted or photographed on an aluminum sheet. Then the hammer is carried along each contour, tapping it down to desired depth, thus producing a negative. The other side of the sheet can be used as a rough positive.

Fig. 25.4. The Bround Reliefograph has an electric hammer which works on a metal sheet, hammering every contour line to its proper level.

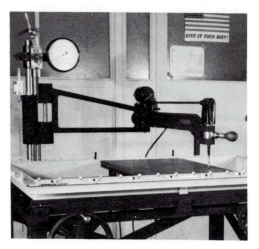

Metal Spray. A special alloy is sprayed into a plaster mold with a metal spray gun, producing a light and strong model. The layer of metal is so thin that it can be sprayed over a positive or into a negative mold.

Rotary Cutting Machines. The Karl Wenschow Company in Munich has developed a machine to cut models. An arm of this machine is carried along a contour line on a map, while a drill on the other arm cuts into a block of plaster of paris, removing all the plaster outside the contour. By repeating this with every contour from top to bottom, a positive model is produced. Wenschow uses these models in two ways. He illuminates the models from the side, and photographs of them form the plastic shading of his famous wall and atlas maps. He also makes printed Vinylite copies and sells them as models. Some years ago he had the map printed on thin paper and pasted it wet over a plaster model.

The Army Map Service has similar pantographs which cut the contours out of plastic sheets. These cutouts are pasted over each other to make the model.

Printed Vinylite Models. The map with all its detail is printed upon a white or transparent Vinylite sheet. A negative model is produced by any of the previous methods, and small holes are bored through it. The printed Vinylite sheet is heated until pliable and placed over the negative model in a vacuum frame. The vacuum sucks the Vinylite in, and within a few seconds a light, strong model is produced with the map on it. The Aero Service Corporation of Philadelphia and other companies reproduced several topographic sheets by this method. These models are now rarely missing from our classrooms.

Graphically Simulated Relief Models. These show that there is no limit to man's ingenuity in producing models. We take as many highly transparent plastic sheets as there are contours. Upon each successive sheet we print the part of the map which is between the corresponding contour line and the next higher interval. We do not cut any contours, but simply paste the sheets together with transparent plastic adhesive. We have now a solid block of transparent plastic in which a relief model seems to be embedded. This method is particularly good for mines and geological models.

Notable Models

Models have been made since very early times. Superb wax models have been made in Switzerland since the seventeenth century. A huge model of Flanders was made for Louis XV in 1726. In World War I, the Allied Command prepared the greatest relief model in existence: the entire Western Front on a scale of 1:20,000, with 1:2 vertical exaggeration, made of plaster with the maps wetted and pasted over the plaster.

At the Babson Institute in Wellesley Hills, Massachusetts, there is a 1:250,000 model of the United States, 64 ft wide. Vertical exaggeration is 12 times in the lowlands, and 6 times for altitudes over 5,000 ft. This is really part of a huge relief globe as the base is curved according to the earth's curvature. To see our country on such enormous scale, with all its mountains and rivers, is truly an inspiring sight.

Many of our readers are familiar with the huge relief model of California in the Ferry Building in San Francisco. The public square of Tuxtla Guttierez has an excellent concrete model of the state of Chiapas.

Guatemala City boasts a similar model of the republic.

Albert Heim's models of the Alps at Harvard and of Mount Everest at Clark University are visible witnesses of his genius. Among the American productions, E. E. Howell's models of hundreds of areas are found hanging on the walls of schools all over the country. The Kilauea Volcano and other models of G. C. Curtis at the Agassiz Museum of Harvard University are masterpieces. During World War II, a large organization under the direction of W. W. Atwood, Jr., produced hundreds of excellent models for strategic purposes.

Fig. 25.5. The Babson model of the United States is really part of a huge globe 170 ft in diameter. The United States is 64 ft wide.

Exercise 25.1

To construct a terrain model, select a small but interesting area from a topographic sheet. Photograph the map and enlarge it about five times. Make two copies. Attach one copy of the map to a board and keep the other at hand. Drive pegs into the board to proper height and fill in with selected material. Finish and paint as desired.

Exercise 25.2 *

Select an area on a topographic sheet and, without enlarging, make a positive by cutting cardboards. Cast a negative and a positive with plaster of paris.

---

* NOTE TO THE INSTRUCTOR: This is an alternate exercise.

26

Photography for Cartographers

Modern cartographic techniques are so interwoven with photography that students must have some idea of the general scope of this highly developed art. To have a photographic laboratory for course work not only helps in learning to handle the materials, but pays its way in actual use as well.

There is ample literature available on the subject and photo-supply stores are glad to give advice, supplemented by catalogues and price lists. Manufacturers have qualified field technicians available without charge for direct assistance to potential users of their materials. Here we assume that every geographer has some knowledge of photography, and the following comments are intended only as an incentive for further study. A geographer usually needs to be familiar with the following kinds of photographic work:

1. Contact prints, of the same size as the original
2. Photostats, reduced or enlarged on paper
3. Photographs, with finer detail and rich in halftones
4. Lantern slides

Contact Prints

Several types of sensitized paper and film are available to reproduce maps or any other graphics, of the same size as the original—such as blueprints and brown print paper, ammonia papers, and direct positive films. To be reproduced, maps must be drawn on transparent or translucent paper or plastic. A fair copy can be made from a map on thin paper, but thick paper must be made more transparent by oiling with glycerin or special compounds like Transparento to permit the light to pass. A drawing on heavier paper can also be reproduced by contact methods without being rendered transparent. To do this, papers or films which permit the making of "reflex" copies are used, or a photostat of the same size is made.

Frames. Maps and sensitized paper are placed in contact in a frame. If only one copy is needed we place both map and sensitized paper underneath, both facing the glass. But in this case, the thickness and opacity of the map paper will affect the sharpness of the print. A more faithful reproduction will result when the map placed face down in direct contact with the emulsion side of the sensitized material.

This chapter was originally written by Donald G. Bouma, cartographer, Modern School Supply Company, Goshen, Ind.

This then produces a mirrored image, that is, copy reading right to left (called "left reading"). In turn, any number of sharp left-to-right reading ("right reading") copies can then be made from this base copy. The best contact is achieved by a vacuum frame, but this is expensive. Very good results can be had with the frame shown in Fig. 26.1, provided the dimensions of the map are not over 24 in., or 36 in. at most. The map is placed under a glass plate in contact with the sensitized paper. A sheet of black paper or felt is then applied to absorb extra light during the exposure. All this is pressed against the glass plate by a sheet of sponge rubber held in place by a backboard.

Lighting. Contact papers and films are less sensitive to light than those used in cameras. Therefore they can be handled for a short time in subdued room lighting. For contact exposure we need very strong lights. A number of 500-watt bulbs, mercury vapor lamps, or arc lamps are placed on the two sides of the frame so that the light falls uniformly at a 45° angle. Bulbs should be placed far enough away so that they will not heat the paper excessively. Sunlight is excellent, but not always available.

Length of Exposure. This depends on the speed of the sensitized paper or film as well as the intensity of and the distance from the light source. Before risking a full-sized sheet of paper or film, it is best to lay a strip of the sensitized sheet in proper contact to serve as a test strip. We cover about nine-tenths of the strip with an opaque sheet, expose it, and lower the cover in 15-sec intervals to arrive at a suitable exposure time. The strip is then developed and the exposure time of the best

section is selected for this and similar copies. Most contact papers and films will take an exposure of 1 to 2 min of direct sunlight and will often need an even greater exposure time with some strong artificial light sources. This is several thousand times longer than the exposure time of ordinary snapshots.

Blueprints. When the blueprint paper is exposed in direct contact with a map, the resulting print will be negative, with white lines and blue interline areas. Blueprints are developed wet with an easily mixed solution, fixed, washed, and dried. To have positive copies, we first make a print by face-to-face exposure to act as a negative, and this is used for the negative.

Ammonia Prints. These include Ozalid, Diazo, DriPrint, Vapo, Helios, and others. Ammonia papers, films, and cloth are available in a variety of exposure speeds and in several colors. They provide a direct positive copy without requiring a negative. The coating consists of a dye which decomposes under light, but the unexposed dye darkens when exposed to ammonia fumes or alka-

Fig. 26.1. Frame for contact printing can be made in the workshop. (*By Donald G. Bouma.*)

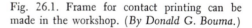
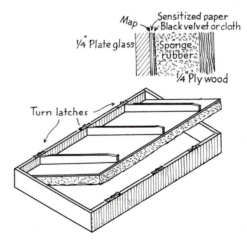

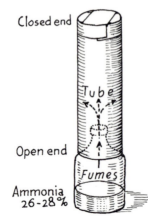

Closed end

Tube

Open end

Fumes

Ammonia
26-28%

Fig. 26.2. Home-made developing tube for ammonia prints. (*After Donald G. Bouma.*)

line solution. The print is dry-developed by
coming into contact with strong ammonia
fumes through rolling the copy into a paper
or metal tube closed on top and open at the
bottom. A fresh supply of 26 to 28 per cent
ammonia solution at room temperature
will give off enough fumes, when used as
in Fig. 26.2, to develop a print in about 2
min. The ammonia bottle should be kept
tightly closed when not in use to conserve
its strength. To avoid the presence of an-
noying odors, we use a well-ventilated room.
As the fumes weaken, the developing time
will increase slightly. No fixing is necessary.

A useful application of this process is in
proofing color-separation drawings. Each
color plate is reproduced on transparent
ammonia film of the proper color. These
color copies are placed over each other
upon the light table in proper register, and
this will give us an approximation of how
the printed map will look.

Modern cartography can make use of the
sensitized plastics like Eastman's Autopositive Film, Du Pont's Cronoflex, and
others. These may be handled in subdued

amber light without damage. When a map
drawn on a transparent sheet is exposed in
contact with this material by means of a
strong yellow light, it will make a positive
copy in perfect detail. Should a negative
be required instead, the film is simply exposed to strong yellow light without the
use of the map drawing. This "clears" the
sheet so that it has lost its sensitivity to
yellow light. A second exposure of brief
duration is then made with the map drawing in contact with the plastic using a
strong white light. This restores the sensitivity to the parts of the film not covered
by the line drawing and will produce a
negative copy when developed. By masking
portions of the film during these exposures,
it is possible to have parts of the map in
white on black while the rest is in black on
white. After the film is processed and dried,
additional information may be added to
the matte surfaces on either side.

If a manuscript is drawn enlarged, it
must be reduced to publication scale by
means of a camera with regular ortho film
as the direct positive films cannot be used
in camera.

Photostats

One of the most common processes in
map making is the enlargement or reduction of a base map to fit a certain scale.
This can be done in various ways (see
Chap. 3), but generally the quickest way
is the use of a photostat. Most colleges
and large map-making establishments have
their own photostat laboratories. The
equipment consists of a big enlarging-reducing camera which can be moved back
and forth on a track in reference to the
frame. Different lenses may be used for
enlarging and for reducing. The frame can

be simpler than the one described previously. It usually consists of a glass plate in a hinged frame which can be pressed against a large board covered with black velvet. The map to be photographed is placed between the two. Lighting is oblique from the two sides.

Photostats differ from photographs only in that we use low-speed sensitized paper instead of film. Paper is cheaper and faster to dry, but it is not as dimensionally stable. The paper comes in rolls and has a tendency to expand differently lengthwise than crosswise Papers of various speeds and contrast can be bought.

The camera has a large ground glass upon which the image appears. The camera is moved forward or back until the image is of the desired size. Refocusing and readjustment for size may be necessary. An indicator showing the distance of the camera from the drawing can easily be made for each ratio and for each lens. Exposure time is several seconds even with strong lights. Processing is similar to that of photographs, and a negative is produced. From this negative, positives are made by repeating the whole process. For enlargement or reduction of more than four times, the process may have to be repeated. For instance, the negative is reduced to one-quarter length; this is further reduced to positives one-sixteenth of the original, instead of pulling the camera to a great distance.

Photographs

The best investment a geographer-cartographer can make is a good camera—or better still, three cameras: one for 35-mm color rolls, the second for black-and-white pictures, and the third a large portrait camera. The smaller cameras will be used in field work, the portrait camera for map work. Many good books are available on the art of photography. The following are only a few bits of advice born of long experience:

1. Buy a camera with a lens f:2.8 or f:3.5 to make color snapshots possible in cloudy weather or late in the day. The second camera need not be over f:4.5, but the use of fast orthochromatic film is recommended for black-and-white work.
2. Carry a tripod, light meter, and telephoto attachment. The tripod will help in taking time pictures in dim light. For exposure time, set the camera first by estimation and then check it with the light meter. One should be able to set the camera under normal conditions without a light meter. The telephoto attachment is used for distant buildings or structures; it can be used also for close-ups, as when photographing portions of a map on a larger scale. A haze-eliminating filter which does

Fig. 26.3. A small enlarging-reducing camera which can be used for photostats or for direct drawing upon plastic or thin paper. (*Courtesy Lacey-Luci Products Company.*)

not reduce exposure time can be permanently attached to the lens of the color camera. Other filters are rarely needed for geographic recording.

3. Carry both color and black-and-white film, using two cameras. The black-and-white pictures are faster to take, cheaper, and can be developed at home immediately. They are also better for reproduction in books or articles. Color pictures are more expressive and make better slides for lectures. Color slides must be sent to processing laboratories in order to be developed, and it may take one or two weeks before the pictures are returned as 2- by 2-in. or larger mounted slides.

4. Artistic composition adds much to the beauty of a picture, but it should not be the primary concern of a cartographer. It is more important that the picture have geographic significance. Most people like to take pictures of the unusual rather than the usual. The picture of a featureless Brazilian campos or the monotonous cattle country of Florida may not be much to look at, but for a geographer it means the record of an important type of terrain.

The large portrait camera is used for photographing large maps, either for compilation or for reproduction. Some large map collections are equipped with their own photographic laboratory and reproductions can be ordered, but this is not always the case. In photographing maps for reproduction, one has to decide whether they can be reproduced in line cut or by halftone process. For line cuts we use shorter exposure and shorter developing films, and in making the prints "hard," strong-contrast film is used so that the copy will have pure blacks and whites and no grays. The map will then be reproduced by line cuts or by offset plates without a screen. Maps with plastic shading, varying colors, and faded or dirty old maps have to be reproduced through a halftone screen (see Chap. 14). They have to be photographed like portraits with relatively long exposure and using soft paper for the prints.

The large enlarging-reducing camera used for photostats can handle film just as well as paper and can be used for photographing maps. The recommendation for acquiring a smaller portable portrait camera is offered here, because with the larger camera one must bring the map to the camera instead of carrying the camera to the map. The portrait camera will be useful also in photographing pictures, air photos, and all kinds of objects where more detail is needed.

The simplest way to photograph maps is to fasten them to a gray cardboard and take them outdoors. The light meter will give the exact time for exposure for a snapshot. Exact focusing is important, and the camera must point perpendicularly to the center of the map. No shadow should fall on the map. As most hand cameras take in a view of 50 to 60°, a 24-in.-wide map has to be photographed from about 3 ft in height to fill the viewfinder.

Processing. The exposed photographic film is opened in the darkroom under red light, which does not affect the film. Processing consists of developing, fixing, washing, drying, trimming, and mounting. The minimum equipment of the darkroom consists of a table large enough for four trays of a size to hold the films, running water and sink, a drying rack or string, a cutter, frames for positive copies, and a shelf for chemicals, films, papers, etc.

Developers. These are usually sold in double boxes which separate the chemicals for two solutions. When mixed with water and stored separately, they will keep for a long time; but once mixed together, as required for the developing process, they last only a day or so. Therefore we mix the two only in the amount needed at a given time.

We avoid touching the exposed surface of the film with damp fingers or gloves until after it is immersed in the developer. Immersion should be as instantaneous as possible; therefore it is advisable to insert the film in the solution face down, pushing it into the developer with the fingers. When completely immersed, the film may be turned over to watch the progress of development. When development is completed, the film is transferred to the next tray, containing the weak solution of acetic acid, which halts the development process in about 10 sec. The stop bath is not absolutely necessary but preserves the life of the fixing solution (popularly called the "hypo"). It is then placed in the third tray containing the hypo for from 2 to 5 min, after which the light can be turned on. Here the image is retained while the rest of the emulsion is washed into the solution. The hypo could be reused later if only a few prints were processed in it. The fourth bath may be a tray or sink. It is supplied with plenty of fresh running water. Hypo salts are very persistent; with plenty of agitation the wash bath will remove them in about 10 min. The film is dried on racks or hung on clips. Wiping with a soft sponge and fanning with warm (not hot) air will hasten the drying process.

A few cautions should be noted here: All chemicals contain active agents and should be mixed outside the darkroom in a well-ventilated place. We should allow a few seconds for solutions to drip off the film before moving to the next tray to avoid breakdown of the solutions. One should not agitate the solutions too much while processing. To be properly developed

and fixed, all parts of the film must always be under the surface of the solution.

Lantern Slides

With the amazingly fine grain of modern films, simple maps can be shown by 2- by 2-in. (35-mm) slides. The older type of 3¼- by 4¼-in. slides is now rarely used. Color pictures are coming back from the processing companies in the form of lantern slides, and most camera owners have a projector— at least for small audiences. Some of the common problems with lantern slides are listed here:

1. *Readability.* The most common complaint of audiences is that the maps are on too small scale to be readable across the room. Chapter 6 describes how large lettering must be in order to be readable on the screen. If we prepare the maps ourselves, the lettering can be made the proper size, but if we use ready-made maps this may not be feasible. One cannot show, for instance, a topographic sheet or an atlas page in a single slide and have all the lettering readable. It would be a mistake, however, to forego the use of such slides if we wish to show general relationships. For instance, on a slide showing Indonesia, we cannot make every name readable, but with colors and shading we can present the pattern of settlements, land use, climate, etc. When lettering is important, we must show the significant parts on larger scale by using a telephoto or portrait attachment.

2. *Overlay map slides.* We can make one multipurpose base map and superimpose upon it, using transparent plastic, such items as climate, vegetation, cultivation, and population, one after the other or in combination, and make a slide of each. Another way is to make several contact prints of the base and color it by hand or apply cellotones to each special map. Transparent cellotones are particularly good for tinting and coloring patches on the map.

3. *Labeling slides*. We mark the date, place, object, running number, and index letters designating the field trip or category. One, two, or three bars on the side may be used to mark the excellence of the picture. The red paint or crayon mark on the upper right edge of the slide is a means of preventing the embarrassment of the operator trying to put an upside-down canoe on even keel, as there are eight different ways a slide can be put in the projector.
4. *Two screens and two projectors*. It is often advisable to use one projector to project the map and the other the pictures—thus allowing the lecturer to point out the location of each picture on the map. This method is better than using a wall map and flashlight.
5. *Selecting slides*. Too often lecturers show too many slides. For the lecturer it may be heartbreaking to weed out beautiful but insignificant slides, but it is better to show fewer slides and say more about them. The rough rule is a slide per minute—if there is not enough to say about a slide for a minute, then it is hardly worth showing. In travelogues two pictures per minute may be permissible.

Moving Pictures. For a geographer, moving pictures are less significant than for the average traveler. The pictures do not reproduce well and are not quite as easy to use as slides in connection with a lecture. But they fascinate the audience and they serve very well for showing motions of people, animals, boats, carts, etc. Sweeping the camera slowly around the horizon is also effective. In general, however, a geographer does better if he concentrates on good stills and uses a movie camera as an auxiliary.

The Impact of Photography on Cartography

Photography has had a profound effect on the development of modern cartography. When photoengraving was introduced almost a century ago, it first enabled map makers to draw maps on paper on a larger scale instead of cutting them into copper or drawing them on lithographic stone in the same size and in mirrored image. Photography helped to separate the cartographer's job from that of the engraver.

Next, the introduction of the halftone screen made easy the photographic reproduction of plastic shading and thus the laborious method of hachuring was gradually eliminated. The halftone screen, in combination with color filters, made the reproduction of fully painted maps possible. The new methods of modern cartography, such as negative scribing, photomechanical etching, phototypography, etc., are all based on photography. They are described in the next chapter. Their full effect is still to be felt.

Letters and symbols are often drawn or composed on large scale and reduced photographically to the desired size. They can be printed on negative or positive, on transparent or opaque film or paper. With adhesive backing the cut-out letters and symbols can be applied to the positive or negative copy, producing black or white lettering. Special machines are constructed to set up the type and photograph it to size with great economy of time. The process of phototypography is described in the next chapter.

All the new technical advances make maps more precise, easily readable, yet less expensive, and they have helped to put maps into the hands of the great masses of people. On the other hand, some of the artistic quality and charm of the older maps has undoubtedly been lost.

27

Modern Techniques

If a cartographer versed in prewar methods were to visit some of the modern establishments, he would hardly understand what people are talking about. Such expressions as "negative scribing," "Mylar," "photoscribe etching," "drystrip," and dozens of others would be new to him. Indeed, a modern plant looks more like a photomechanical laboratory than a studio.

These new processes concern mostly the reproduction and not the original layout of the map, but they influence the designer in his composition. The whole character of our maps has changed since the introduction of modern methods. Our present maps have finer lines, more clear-cut symbols and lettering, but they lack something of the personality of the older ones.

The methods discussed here are used by the great government agencies and by large private companies. The best sources of information are the Aeronautical Chart and Information Center technical reports, Army Map Service bulletins, the Hydro-

graphic Office, and publications of the U.S. Geological Survey and the U.S. Coast and Geodetic Survey. The International Cartographic Conferences, held in Stockholm in 1957 and in Evanston in 1958, published reports from the world's leading cartographers. Most agencies give a long and detailed course of instruction for their new operators. It is not expected that a student of geographical cartography know them in detail, but he should know enough to plan accordingly to speak the language of modern cartography.

Scribing

The difference between drawing and scribing is that in the former we add some material such as ink, graphite, or paint, or an emulsion to the base sheet, while in the latter we remove some.

Negative scribing on glass was introduced to the U.S. Coast and Geodetic Survey with the instruments of S. Sachs in 1940 and on coated plastics in 1946. Since that time this method has become almost universal in large map-making offices, which claim 25 to 50 per cent reduction in time with improvement in quality. They use transparent plastics covered with a photographically opaque *scribe coating*. Into this coating, lines, patterns, and symbols are cut with special *gravers*. Thus light can pass through the cut-away lines or areas, as in the case of a photo negative. The cutting is usually done mirror-wise, right to left reading.

Negative scribing has many advantages:

1. The lines are more even than those drawn with pens because the width of lines is accurately controlled by the gravers. More people can learn to scribe well sooner than they can learn to draft well with pen and ink.
2. Scribing saves the time of inking, for it can be done directly after a penciled manuscript guide copy.
3. It saves the photographing to size, as most scribing is done on publication size. It also takes less time, space, and materials to work in smaller size.
4. Scribed negatives are usually reproduced by simple contact with the sensitized metal offset plate, which elimates much of the camera work, with its inherent sources of error. This is particularly advantageous in color work.
5. Scribed negatives are adaptable to various techniques, such as using templates. They can be transformed into positives by solvent etching and other processes. These will be described later.
6. Revisions are easier to make. The portions to be changed are simply recoated and rescribed.
7. Less space is needed for storing, and the sheets are very durable and resistant to heat, moisture, and mildew.

Negative scribing has some disadvantages too:

1. Working in negative is a difficult mental process. It does not have the direct appeal of positives, and mistakes are not as easily discovered. Working in a mirror image is also difficult, yet scribers seem to learn this easily enough.
2. Lettering is not easily applied.
3. Negative scribing is not adaptable to half-tone reproduction or three-color process work.

Negative scribing is particularly good where many maps have to be produced having the same style and lettering and symbol system, such as topographic sheets, charts, atlas sheets, and road maps. For a geographer, whose work is reproduced in books and periodicals, every map is a problem of its own, so for him negative scribing is of less value.

Lettering. Letters could be scribed directly from templates but most establishments find the following methods more economical:

1. Two plates are used in combination (see Chap. 14). The lettering and symbols are applied in positive to a transparent plastic overlay, usually by stick-up (see Chap. 6) but occasionally by hand. A negative is made of this and lighted through upon the sensitized metal printing plate in succession with the scribe plate, which has all the line work. The lettering plate can also carry cellotones of patterns and symbols.
2. We make a contact positive of the scribing plate which contains all lines and solidly colored areas, and upon this directly apply the letters, symbols, and cellotints. From this, a negative has to be made for the metal offset plate. Sometimes we have the positive of the scribe plate on one side of the film and lettering on the other; this makes it easy to remove the lines and black surfaces where they interfere with lettering.
3. We prepare the scribe plate and we make "windows" by removing the scribing coat from places where we want lettering or symbols and from areas where we want cellotones. We make negatives of all the lettering, symbols, and cellotones. Then we strip off the parts of the film and apply them to the proper windows. The negatives of letters, symbols, and cellotones can be stored on thin film and can be stuck up in reverse, the same way as positive stick-up.
4. We can use the "photoscribe" process, which employs phototypesetting machines. These will be discussed later.

The Base Sheet. For scribing, the base sheet must be transparent, sufficiently hard not to be scratched easily by the scribing tools, and highly stable under changes in temperature and humidity. It must also be strong, flexible, and thin enough to allow lighting from either side in contact printing.

Glass is thick, inflexible, and breakable, yet it has an ideal surface because of its great stability and is still used, particularly in Switzerland. Most establishments, however, use either of two plastics—vinyl chloride acetate (Vinylite) or a polyester like Mylar. Mylar is less affected by changes in temperature while Vinylite resists changes in humidity somewhat better. For strength and resistance to chemicals, Mylar is superior. Thus the average thickness of Mylar base sheets is 0.0075 in., while the most popular Vinylite scribing sheets are 0.01 in. thick. Sheets as large as 76 by 52 in. are on the market. With the rapid advance in the plastics industry, the above characteristics may soon be out of date. The British Astrafoil and the German Astralon are vinyl compounds which are extensively used in Europe.

Coatings. A good coating has to be translucent enough for the eye to follow a design over the light table, yet it has to be perfectly opaque for photography. This can be achieved by red- or orange-colored coatings whose long-wavelength lights are photographically the same as black. The coating has to be soft enough to be easily cut, but hard enough to withstand handling. It has to absorb sensitizing solutions easily. There are several coatings on the market which can be whirled, sprayed, or roller-coated upon the base, but most map makers buy ready-coated sheets.

Watercote. This is a sensitizer which comes in 40 colors. It is wiped or whirled upon the matte side of a wet scribe coat. Any design, positive or negative, can be contact-printed upon the sensitized scribe sheet. The sheet is then developed with ammonia water, washed, and dried. The lines are now visible and the scribe coat is ready for the graver. Colored Watercote sheets are used also for colorproofing and for the photoscribe process. Both will be discussed later.

Gravers. A good graver has to be sharp enough to cut the coating but not so pointed on the tip that it would cut the base sheet easily.

The *Pen-type, Freehand Gravers* (see Fig. 27.1) is a needle attached to a holder. It is used for very thin lines not over 1/200 in. Pen points with very small yet rounded points are on the market. These are less apt to damage the base, even if the pen is held at an angle. Phonograph needles can be made into good pen-type gravers.

The *rigid graver* (see Fig. 27.2) has a conical point with a flat tip held vertically and is used for single, even-width lines not much more than 1/100 in. thick. The graver is attached to a carriage with two ball-pointed legs. For wider lines, a wedge-type point (see Fig. 27.3) is used, and the carriage turned so that the wedge should

Fig. 27.1. Pen engraver is used for scribing very fine lines. (*From Inter-Agency Committee on Negative Scribing Report*, 1957.)

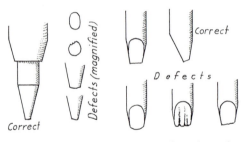

Points

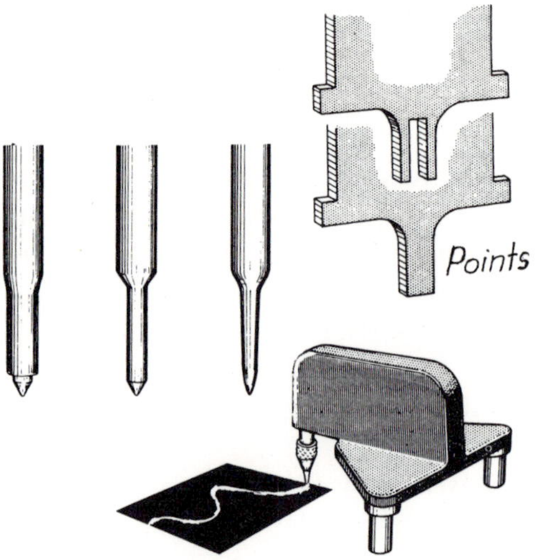

Fig. 27.2. Rigid graver is used for scribing even lines. (*From Inter-Agency Committee on Negative Scribing.*)

Correct

Defects (magnified)

Correct

Defects

Fig. 27.3. On the left are a conical cutting point and some of its common defects; to the right are a chisel cutter and some of its defects. (*By F. B. Meech.*)

Fig. 27.4. Swivel graver is used for double lines and very wide lines.

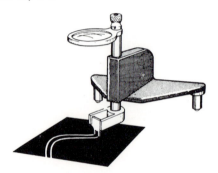

be always perpendicular to the direction of the line.

The swivel graver (see Fig. 27.4) resembles a pivot pen. The scribing point is on the end of a free turning arm. It is used for double lines or very wide lines.

The dot graver (see Fig. 27.5) has a wedge-shaped point and is attached to a carriage having a mechanism which will rotate the point when one pushes a button, somewhat on the principle of the push drill. The carriage usually holds points of various sizes. Electrically driven dot gravers are also available.

The building graver (see Fig. 27.6) produces a square or rectangular symbol for buildings. It has wedge-type points of various widths. These are pushed down into the coating; the graver is then pulled back only as far as a preset screw allows it to go.

The dash graver is similar, but more complex, as it allows the scribing of evenly spaced dashes of equal width and length. Dashed lines are more easily produced, however, by cutting a continuous line and covering it at intervals.

Fig. 27.5. Strongly magnified view of the point of a dot graver.

The turret graver has a rotating disk with eight needles of various shapes and sizes, each of which can easily be locked into scribing position. It saves time when constant change of needles is necessary, as in scribing rivers.

New tools are constantly introduced. The electromechanical dotter, which engraves dots of various sizes and density by an electrically driven device, is very promising. The Stabilene scriber provides the pen holder with legs (see Fig. 27.7).

Sharpening is one of the most delicate jobs in maintaining the graver points. Some agencies prefer sapphire-tipped or tungsten carbide points which do not need sharpening for a long time. The common steel points, however, have to be sharpened at intervals. Special rigs are marketed which hold the points at proper angle to sharpen wedge and conical points, yet the operation needs a fine touch and a good magnifier.

After scribing, the plate must be brushed or rubbed so that no loose slivers of coating remain and the lines are carefully checked for imperfections. Pinholes are covered up; imperfections are coated and regraved.

Guide Image (*or Guide Copy*). This is what scribers call the manuscript or photograph of the manuscript which has to be scribed. The original manuscript may be of larger size but in the case of standardized maps it is usually on the same scale as the guide image. Usually a negative is made of this. The scribe coat is sensitized with Watercote or similar sensitizer and allowed to dry in a semidark room. The negative is placed over it in a frame (see Chap. 26) and exposed to intense light. The exposed scribe sheet is washed with ammoniated water, rinsed with clear water, and dried. The lines will show up clearly on the scribe

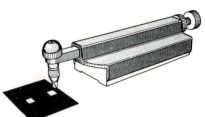

Fig. 27.6. Building gravers can be used also for wide dashed lines. (*From Inter-Agency Committee on Negative Scribing.*)

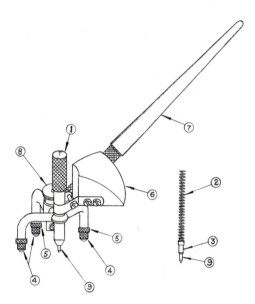

Fig. 27.7. The Stabilene scriber. (1) Pressure-adjusting cap, (2) spring, (3) scribing-point holder, (4) ball-bearing feet, (5) lock nut, (6) adjustable magnifier, (7) scriber handle, (8) handle lock screw, and (9) scribing point. (*By Keuffel & Esser Company.*)

coat, which is now ready for scribing. This is done usually in left-reading (mirror-like) fashion so when the scribe sheet is laid face down upon the printing (metal) plate, the image will be a normal right-reading copy. The offset printing process (see Chap. 14) will now produce a right-reading map.

Color Separation. Most of the maps produced by scribing are in flat colors such as topographic sheets or navigation charts. The original manuscript may have all the colors, although the contour lines or plastic shading are often on different drawings. From these as many negative guide images are made as there are colors, and on each guide image the lines and areas unnecessary for this color are opaqued. The negatives are contact-printed on the sensitized scribe coat and are scribed.

Large areas of even color, such as water areas on the blue plate, are windowed. Either the scribe coat is scraped off, or all the land areas are covered with a resistant ink, and the scribe coat over the window area is washed away with a solvent.

Lettering, symbols, and patterns are applied as described before. Figure 27.11 shows how the topographic sheets of the United States are scribed and printed by using a combination plate.

Altitude Tints. These are usually produced by using a combination of screens printed in yellow, blue, brown, and gray of varying density. For instance, a half-density yellow and quarter-density blue screen will produce a pale green, while a full-yellow and quarter-brown screen will produce the light brown of middle heights (see Chap. 13). In the alternate-band method, every second zone of elevation is opaqued on one sheet, the alternate zones are opaqued on the other sheet. The necessary number of screen sheets are prepared, and, by combining them with the alternate-band sheets for each elevation and rendering opaque all the other zones, the proper color tints are produced. For the details of the process a "Flow Diagram of Chart Reproduction Processes" by the Aeronautical Chart and Information Center in St. Louis can be obtained for display and elucidation.

Proofing is usually done on opaque white plastic. This is sensitized, for instance, with yellow Watercote. The scribed negative for the yellow part of the map is contact-printed on the opaque sheet; after developing and drying, the sheet is coated with red Watercote and the scribed negative for the red color in contact-printed on it. All the colors are proofed this way, and a fair image of the final map is obtained.

Deep-etch Plates. All the foregoing is written in the supposition that the map will be printed by the usual surface offset process, described in Chap. 14. In this process, the press plate is contact-printed from negatives. In the deep-etch process, the press plate is contact-printed from positives. Positives have several advantages: They read normally, to the right, and one sees at once the final appearance of the map. They are easier to superimpose on other positives for checking. Corrections are easier with pen and ink. Stick-up lettering and symbols do not need windows. On the other hand, they do not have the quality of line of the scribe negatives.

The press plate is sensitized, as before, but now the lines become soluble and the space between is resistant. After washing out the lines, we etch them deeply with acid and the plate is now somewhat like the copper engraved plate of old times. If we roll greasy ink into the lines and clean and wet the interline areas it will print as any other offset. Deep-etched plates have a longer life but are more expensive than surface offset plates, and are used only when large quantities are printed. All the scribing processes will have to be modified to produce positives.

Exercise 27.1

For practice in negative scribing, take a part of a marine chart about 4 in. square and place over it a coated scribe sheet. Place both over the light table and using the available tools prepare a scribed negative. Make a contact negative of the chart and cut out some of the names. Make windows in the scribe sheet by scraping off the coating and paste on the names in reverse. Make a contact print, and color the sea with shades of light blue.

The Photoscribe Process. This is a method to produce scribe-coat negatives photomechanically. In principle it is similar to the deep-etch process, using the same sensitizer solution on the scribe coat. Over this is placed a positive guide copy and a contact print is made. After exposure the sheet is developed with Deep-Etch Developer and squeezed. Next the photoscribe etching solution is poured over the sheet. Etching is completed in a few minutes. The sheet is squeezed, washed, dried, and ready to use—the same as a scribe negative. Photoscribe process is used when a finished positive copy is available, particularly if a quantity of lettering and symbols must be reproduced. It is quick and does away with mechanical scribing, but the lines are less clear-cut.

The Dystrip Technique.[1] This is used with advantage for color separation of maps with altitude tints, where large areas such as forest marshes, water areas, etc., are evenly colored or patterned.

The original drawing has to be on transparent plastic in positive with the contour lines, or the outlines of all colored areas (windows), clearly outlined. The color-separation sheets are $\frac{1}{100}$-in.-thick vinyl

[1] "Dystrip Technique of Color Separation," Aeronautical Chart and Information Service Tech. Rep. 70, 1955, St. Louis, Mo.

or Mylar sheets; upon this a bluish-colored transparent coating is applied, consisting of a rubberized substratum and a carbon tissue type strip coat. A sensitizer of special formula is brushed on the sheet under amber light and the sheet is dried. Presensitized strip coats are now on the market and are used widely. Contact exposure with the original drawing is made immediately after drying. The exposed interline areas become "tanned," insoluble in water. The unexposed lines and areas (black on the original) are washed away in warm water (110°) and we have a somewhat transparent negative. The sheet is blotted to remove excess water and is bathed in 10 per cent glycerin solution to make the strip coat pliable and dry.

Next comes the stripping over the light table. Any area can be windowed easily according to the color-separation plan, as the lines on the original drawing which outlined the windows are now washed away. We use a needle, knife, and the oval-shaped "pry knife"; thus we produce the required open windows for any color (see Fig. 27.8), keeping in mind that we

Fig. 27.8. Stripping of areas to be windowed in the Dystrip process. (*From Aeronautical Report No. 70.*)

Fig. 27.9. Phototypesetting machine prints all the desired names on 70-mm film and with the help of an electrocoordinatograph positions them on the place according to the position indicator template. (*Staphograph of the Society of Optics and Mechanics, drawings after C. W. Westrater, Army Map Service.*)

work in the negative with the picture reversed. Thus we produce a negative which could be used for the regular surface offset printing.

This process, however, is used mostly for applying lettering symbols and patterns to be reproduced by deep-etch offset printing; therefore Dystrip sheet has to be converted into a positive. We dye the sheet in a red-colored ketone-type solution of special formula, which will impregnate the vinyl base in the windows. Then we place the sheet in warm water, which after a while will soften the tanned coat. This will be removed with a scrubbing brush until any trace of the original coating is gone, but the ketone-impregnated areas will be resistant. Now the design will appear positive, and can be used for deep-etch reproduction. Most agencies do not dye the strip but use it as a window negative after opaquing unwanted lines and areas.

Phototype Setting. Most of the lettering in major map-making establishments is made by phototypesetting machines. Several types are on the market. Most of them have a separate disk or a negative matrix for each type of lettering containing all letters and numbers. Each letter can be brought into position to be photographed by using a keyboard selector. The size of the letter can be regulated by changing the distance between the disk or matrix and the lens. Each letter in sequence is exposed upon a film which is automatically developed, fixed, and dried. The machines make a negative copy which can be used as stick-up on windows or a scribed negative, or a positive copy for the direct stick-up on the positive of the map. Thousands of letters can be set in one hour. It is economical to phototype together all names in the same type and size and to cut up the film and paste it on the map or scribe sheet.

In recent years some machines were developed which actually position names and letters on the map with the help of a guide template (see Fig. 27.9). They can print the letters on their proper places on film negatives or positives. There are different types of machines, such as the Intertype, the Hadego (Dutch), the British Phototype-setter and Photonymograph, the French Nomafot and Staphograph. The Staphograph first photographs the already composed names on a 70-mm roll film in negative. Each name is guided to its place by an electrocoordinatograph over a map-size film, according to a position indicator template. After it has been exposed, the large film will become the positive lettering plate.

The Rotograph. This instrument engraves lettering in a scribing coat with the help of a small electromotor (see Fig. 27.10). It works like an ordinary template lettering machine but, instead of a hollow pen, a graver, rotating at 3,000 rpm, will engrave the lettering. The depth of engraving is regulated by a screw and the instrument rides on legs. The machine can produce vertical or slanted lettering at various widths and heights. The same manufacturer (Gebr. Haff, Pfronten-Ried, Germany) also produces a machine which similarly engraves symbols, such as circles, squares, and small trees.

Indeed, new developments in scribing and related techniques change at such a rapid rate that the methods here described can be out of date in a few years. A cartographer needs to keep abreast of the latest developments through periodicals, reports, and advertisements of the leading manufacturers not only in America, but also in Europe.

Every instructor of cartography does well if he asks the Aeronautical Chart and In-

Fig. 27.10. The Rotograph engraves lettering into a scribe coat with the help of an electromotor. It works like a template lettering set. (*By Gebr. Haff, Pfronten-Ried, Bavaria.*)

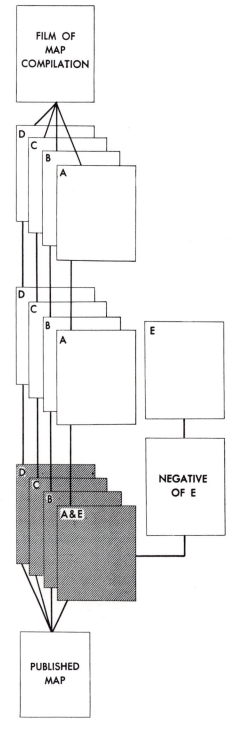

Map Feature	Published Color
A Culture_____black	
B Hypsography_____brown	
C Hydrography_____blue	
D Road classification_____red	
E Lettering_____black	

 Press plate

Guide image

The film of the map compilation is used to process its image in left-reading position on scribe-coated sheets; one for each color that is to appear on the published map.

Scribing

Following the guide image, the delineator scribes the detail required for each color, producing a series of negatives. Lettering printed on adhesive-backed, transparent film is applied to an uncoated base sheet. A negative of this overlay is then made for further processing. Individual or composite proofs may be made as required for checking, editing, or other purposes.

Press plates

The scribed negatives and the lettering negative are contact printed to the offset press plates, producing right-reading printing images. The culture features and lettering are combined on a single press plate.

Published map

The multicolor map is printed from the several color-separated press plates.

formation Center for their "instruction pocket." This contains a flow diagram showing the 80 stages through which a chart is processed before becoming ready for distribution. It illustrates the extreme complexity of modern cartography. The pocket contains other valuable material, and studying it is good preparation for a cartographer who wishes to work for large establishments.

Silk-screen Printing

This method lends itself to printing maps in color when relatively few copies are required. The entire producing and printing can be done at home with little expense, but considerable time. Materials can be bought in art stores and introductory kits are available for classroom use.

The process is simply a high-grade stencil operation for printing on paper, cloth, metal, or glass. It is often used by sign painters. A finely woven silk is used with 125 strands per inch. It is washed with soap to remove dirt and sizing and is stretched over a frame while wet. It will dry as tight as a drum. The frame should be a few inches larger than the map on each side.

The map design is applied to the silk screen either by the hand or by photographic methods. There are two methods of application by hand. In the first method the design is laid upon the light table and a transparent plastic sheet is placed over it for protection. The framed silk is placed over the plastic and held rigidly in place.

The design is copied on the silk with pen or brush, using a special ink called *touche* (*Tusche* in German). First the areas which are of one color (for example, green) on the map are covered with *touche*. After it is completely dry, liquid glue is applied over the entire silk screen, including the areas to which *touche* had not been applied, and it is left to dry. Reapplication of glue is recommended to ensure complete masking. Next the *touche* areas are cleared by soaking both sides of the silk with turpentine; this removes the *touche* even if some glue is on it, but does not affect the pure-glue areas. We wait until the *touche* becomes softened and remove it gently with a cotton pad.

The second hand method requires a commercially prepared material consisting of a film of emulsion with a waxed-paper or plastic backing. All areas to be printed are cut away from the coating with a sharp knife. The stencil is placed under the frame with the coated side in direct contact with the silk. A special adhering liquid is rubbed over the silk with a cotton rag until all areas under the coating become clogged. Then the waxed-paper backing is peeled off, and the screen is ready for printing. This method has advantages: It is quick, inexpensive, and requires less skill to produce crisp lines and small symbols.

In printing, the frame of the silk screen is usually hinged to the working table. The paper is exactly positioned by register guides on the corners. The frame is brought down and locked into position so that the silk is suspended just above the paper but does not touch it. Then printing ink is pressed through the screen with a squeegee, pushing the silk down on the paper for a moment, after which the silk snaps away from the paper. The paper is now removed, and a new print is made. The silk may need occasional flushing with turpentine to prevent clogging. Water-soluble inks, however, can be used with the emulsion-coated stencils, and they can be cleaned more easily and rapidly without fear of damaging the design. After the printing is finished, any remaining ink is washed out, and the masked areas of the design are removed with a solvent or destenciling liquid. The screen is now ready to be used again for another color or a new design.

In the *photoscreen process,* the silk is sensitized. The map is contact-printed from the positive. The exposure hardens the interline areas, and the printing surfaces are washed out. From here on, the printing process is the same as before. The photoscreen process is particularly useful if the map has much lettering and small symbols.

The Du Pont Company developed a screen process film which works the same way as silk screen except that a vinyl plastic material coated with photographic emulsion is used in place of the silk.

Maps on Television

With over 60 million television sets used in the United States, a cartographer has to be concerned about how his maps will appear on the screen. Many of our geography departments offer lectures on the educational networks and we are often asked to prepare maps or give advice on their use.

Most lecturers use existing maps. Obviously the fine detail of maps will be lost on the vibrating screen. Wall maps with bold design serve the best. The colors are lost, but show as various shades of gray. Hachuring and plastic shading show up on maps very much better than altitude tints. The televisor can use strong telephoto lenses and throw on the screen maps which are only a few inches wide. Thus the lecturer does well if he concentrates on close-ups of the region about which he is currently talking, with occasional reference to the location on a general map.

Globes are most effective on television. A strongly shaded physical globe is the best. The camera can be focused on any part of it. To show first the whole globe and to bring it nearer and nearer to the camera until it focuses on the spot in question is a most effective display. A spinning mounted globe is often not enough; we may also want one which can be turned in a cradle.

Particularly effective are black and blue "blackboard" maps and globes. The lecturer can draw on them with chalk and thus, eliminating all other detail, he can forcefully concentrate on the subject in question. Even more effective is the use of the bare blackboard itself. The outlines of the land can be traced beforehand lightly with a special nonphotographic dark-red chalk and thus the lecturer may impress his audience with his ability to draw freehand maps.

Many experiments were made on maps for television purposes. As we have to dispense with colors, we make the most of the range between white and black. All small lettering can also be dispensed with. The maps do not have to be very large; 2 to 3 ft will be sufficient for the lecturer to point out what he wishes. A most effective way is to take medium-gray paper. On this we whiten the sea and lakes with chalk or light-gray paint. We may cut out the land areas and mount them on light-colored cardboard which will give a raised appearance to the land. Large pure-white areas should be avoided because they vibrate on the screen. Rivers can be drawn with white chalk or paint. Mountains are best shown by bold plastic shading with charcoal or black wash. Highlighting the northwest-facing slopes with white chalk adds very much to their plasticity. Startling effects can be made by holding the chalk sideways. Roads and railways can be shown by black lines and cities by crisscross lines. One may actually cut out small blocks from balsa wood in proportion to the size of the cities.

To prepare maps for television is a fascinating task for a cartographer and offers a fertile field for his imagination.[2]

Cartography of the Space Age

The impact of man's penetration into space has already been felt in cartography. The Mercury capsule orbiting our first astronaut had a 1:52,000,000 world map rotating on a cylinder, so as to show the proper part of the Earth below. A 70-mm film strip in full color simulated the appearance of a 1,640-mile strip of the Earth, for which special studies were made of the outline of land forms, rock structure, and vegetation cover.[3] Much of this information came from rockets which photographed the Earth in natural color from a height of hundreds of miles. Although the cloud cover dominates these pictures, they reveal the true colors of our planet. For instance, contrary to common concepts, most deserts appear gray rather than the conventional reddish brown. The Tiros satellites send televised pictures of the Earth, showing cloud cover, ice and snow conditions, sea, and land remarkably well. Even our concepts of the Earth's size, shape, and structure have been improved by analyzing the differences between the calculated and observed orbits of satellites. To show the Earth's great radiation belts has also created interesting cartographic problems. An atlas of the moon is in preparation by the U.S. Geological Survey; it will show photographs and detailed maps with contour lines, without having a sea level for a datum plane.

Even more than by actual space travel, cartography is influenced by automation, computers, data plotters, and other new developments of the Space Age. Machines are already in existence which can draw a detailed weather map in 30 sec if the data are fed in a specified way. Or, reversing the process, there is equipment which can store information from maps and photographs on punch cards or magnetic tape. Facsimile transmission of maps is already much in use by newspapers, weather stations, and the military. Even the color separation of maps by scanning devices is a definite possibility.[4]

An interesting example of a scanning device is the *Stereomat*, under development by the U.S. Geological Survey. Much of the photogrammetrist's time is spent in positioning stereo pairs. The Stereomat does this automatically by correlating the amount of coincidence between two photographs. The *Orthophotoscope* corrects the tilt and relief displacement of photographs by exposing a sensitized film through a narrow slit in a movable screen and varying

[2] Mildred Danklefsen, Televising Geography, *Jour. Geog.*, 1953, pp. 253–258. Henry J. Warman, Telecasting Techniques in Geography, *Jour. Geog.*, vol. 55, 1956, pp. 217–226. Allen Philbrick, University of Michigan, East Lansing, prepared many excellent maps.

[3] John E. Dornbach, Charting the Project Mercury, paper presented at the American Association of Geography meeting of 1961 in East Lansing, Mich.

[4] W. R. Tobler, Automation and Cartography, *Geog. Rev.*, vol. 49, 1959, pp. 527–534.

the elevation of the film according to the terrain.[5]

The GIMRADA (Geology, Intelligence and Mapping, Research and Development Agency of the Corps of Engineers, Fort Belvoir, Va.) has an interesting "Orientation Brochure 1962" describing many new methods and devices, such as the Multicolor Electrostatic Printer which prints remarkably clear maps from 70-mm microfilm by photoelectric methods. The widespread use of the Geodimeter which uses light rays and the Tellurometer which uses short-wave impulses for measuring distances has already been described in Chap. 15. The Elevation Meter uses an electronic pendulum mounted in an automobile and can give elevations with reasonable accuracy.

Cartography has come a long way since Mercator and Cassini. The present emphasis is on automation and mechanization handled by highly trained technicians with little interest in geography. There is a danger here of losing touch with the real picture of the Earth. Let us always remember Proteus, who, wrestling Hercules, gained strength every time he touched the Earth and perished when he lost contact with the ground.

[5] G. O. Whitmore, N. M. Thompson, and J. L. Speert, Modern Instruments for Surveying and Mapping, *Science*, vol. 130, 1959, pp. 1059–1066.

Appendix I:

Tables

The following tables are included in the text:

TABLE A.1. SQUARES, SQUARE ROOTS, CUBES, AND CUBE ROOTS OF NOS. 1 TO 100 *

No.	Square	Cube	Square root	Cube root	No.	Square	Cube	Square root	Cube root
1	1.000	1.000	1.000	1.000	51	2,601	132,651	7.141	3.708
2	4	8	1.414	1.259	52	2,704	140,608	7.211	3.732
3	9	27	1.732	1.442	53	2,809	148,877	7.280	3.756
4	16	64	2.000	1.587	54	2,916	157,464	7.348	3.779
5	25	125	2.236	1.710	55	3,025	166,375	7.416	3.803
6	36	216	2.449	1.817	56	3,136	175,616	7.483	3.825
7	49	343	2.645	1.913	57	3,249	185,193	7.549	3.848
8	64	512	2.828	2.000	58	3,364	195,112	7.615	3.870
9	81	729	3.000	2.080	59	3,481	205,379	7.681	3.893
10	100	1,000	3.162	2.154	60	3,600	216,000	7.746	3.914
11	121	1,331	3.316	2.224	61	3,721	226,981	7.810	3.936
12	144	1,728	3.464	2.289	62	3,844	238,328	7.874	3.957
13	169	2,197	3.605	2.351	63	3,969	250,047	7.937	3.979
14	196	2,744	3.741	2.410	64	4,096	262,144	8.000	4.000
15	225	3,375	3.873	2.466	65	4,225	274,625	8.062	4.020
16	256	4,096	4.000	2.519	66	4,356	287,496	8.124	4.041
17	289	4,913	4.123	2.571	67	4,489	300,763	8.185	4.061
18	324	5,832	4.242	2.620	68	4,624	314,432	8.246	4.081
19	361	6,859	4.358	2.668	69	4,761	328,509	8.306	4.101
20	400	8,000	4.472	2.714	70	4,900	343,000	8.366	4.121
21	441	9,261	4.582	2.758	71	5,041	357,911	8.426	4.140
22	484	10,648	4.690	2.802	72	5,184	373,248	8.485	4.160
23	529	12,167	4.795	2.843	73	5,329	389,017	8.544	4.179
24	576	13,824	4.899	2.884	74	5,476	405,224	8.602	4.198
25	624	15,625	5.000	2.924	75	5,625	421,875	8.660	4.217
26	676	17,576	5.099	2.962	76	5,776	438,976	8.717	4.235
27	729	19,683	5.196	3.000	77	5,929	456,533	8.775	4.254
28	784	21,952	5.291	3.036	78	6,084	474,552	8.831	4.272
29	841	24,389	5.385	3.072	79	6,241	493,039	8.888	4.290
30	900	27,000	5.477	3.107	80	6,400	512,000	8.944	4.308
31	961	29,791	5.567	3.141	81	6,561	531,441	9.000	4.326
32	1,024	32,768	5.656	3.174	82	6,724	551,368	9.055	4.344
33	1,089	35,937	5.744	3.207	83	6,889	571,787	9.110	4.362
34	1,156	39,304	5.831	3.239	84	7,056	592,704	9.165	4.379
35	1,225	42,875	5.916	3.271	85	7,225	614,125	9.219	4.396
36	1,296	46,656	6.000	3.301	86	7,396	636,056	9.273	4.414
37	1,369	50,653	6.082	3.332	87	7,569	658,503	9.327	4.431
38	1,444	54,872	6.164	3.362	88	7,744	681,472	9.380	4.448
39	1,521	59,319	6.245	3.391	89	7,921	704,969	9.434	4.464
40	1,600	64,000	6.324	3.420	90	8,100	729,000	9.486	4.481
41	1,681	68,921	6.403	3.448	91	8,281	753,571	9.539	4.497
42	1,764	74,088	6.480	3.476	92	8,464	778,688	9.591	4.514
43	1,849	79,507	6.557	3.503	93	8,649	804,357	9.643	4.530
44	1,936	85,184	6.633	3.530	94	8,836	830,584	9.695	4.546
45	2,025	91,125	6.708	3.556	95	9,025	857,375	9.746	4.562
46	2,116	97,336	6.782	3.583	96	9,216	884,736	9.798	4.578
47	2,209	103,823	6.855	3.608	97	9,409	912,673	9.848	4.594
48	2,304	110,592	6.928	3.634	98	9,604	941,192	9.899	4.610
49	2,401	117,649	7.000	3.659	99	9,801	970,299	9.949	4.626
50	2,500	125,000	7.071	3.684	100	10,000	1,000,000	10.000	4.641

* This table and the next will be particularly useful in drawing two- and three-dimensional graphs.

TABLE A.2. COMMON FRACTIONS REDUCED TO DECIMALS

8ths	16ths	32ds	64ths	8ths	16ths	32ds	64ths	8ths	16ths	32ds	64ths
			0.015625				0.359375		11	22	0.6875
		1	0.03125	3	6	12	0.375				0.703125
			0.046875							23	0.71875
	1	2	0.0625				0.390625				0.734375
			0.078125			13	0.40625	6	12	24	0.75
		3	0.09375				0.421875				
			0.109375		7	14	0.4375				0.765625
1	2	4	0.125				0.453125			25	0.78125
			0.140625			15	0.46875				0.796875
		5	0.15625				0.484375		13	26	0.8125
			0.171875	4	8	16	0.5				0.828125
	3	6	0.1875				0.515625			27	0.84375
			0.203125			17	0.53125				0.859385
		7	0.21875				0.546875	7	14	28	0.875
			0.234375		9	18	0.5625				
2	4	8	0.25				0.578125				0.890625
			0.265625			19	0.59375			29	0.90625
		9	0.28125				0.609375				0.921875
			0.296875	5	10	20	0.625		15	30	0.9375
	5	10	0.3125				0.640625				0.953125
			0.328125			21	0.65625			31	0.96875
		11	0.34375				0.671875				0.984375
								8	16	32	1

TABLE A.3. UNITS OF LENGTH AND AREA *

Inches	Millimeters	Feet	Meters (m)	Miles	Kilometers
	1 = 25.4001		1 = 0.304801		1 = 1.609347
	2 = 50.8001		2 = 0.609601		2 = 3.218694
	3 = 76.2002		3 = 0.914402		3 = 4.828042
	4 = 101.6002		4 = 1.219202		4 = 6.437389
	5 = 127.0003		5 = 1.524003		5 = 8.046736
	6 = 152.4003		6 = 1.828804		6 = 9.656083
	7 = 177.8004		7 = 2.133604		7 = 11.265431
	8 = 203.2204		8 = 2.438405		8 = 12.874778
	9 = 228.6005		9 = 2.743205		9 = 14.484125
0.03937 = 1		3.28083 = 1		0.621370 = 1	
0.07874 = 2		6.56167 = 2		1.242740 = 2	
0.11811 = 3		9.84250 = 3		1.864110 = 3	
0.15748 = 4		13.12333 = 4		2.485480 = 4	
0.19685 = 5		16.40417 = 5		3.106850 = 5	
0.23622 = 6		19.68500 = 6		3.728220 = 6	
0.27559 = 7		22.96583 = 7		4.349590 = 7	
0.31496 = 9		26.24667 = 8		4.970960 = 8	
0.35433 = 9		29.52750 = 9		5.592330 = 9	

Square inches	Square centimeters	Square feet	Square meters	Acres	Hectares	Square miles	Square kilometers
	1 = 6.452		1 = 0.09290		1 = 0.4047		1 = 2.5900
	2 = 12.903		2 = 0.18581		2 = 0.8094		2 = 5.1800
	3 = 19.355		3 = 0.27871		3 = 1.2141		3 = 7.7700
	4 = 25.807		4 = 0.37161		4 = 1.6187		4 = 10.3600
	5 = 32.258		5 = 0.46452		5 = 2.0234		5 = 12.9500
	6 = 38.710		6 = 0.55742		6 = 2.4281		6 = 15.5400
	7 = 45.161		7 = 0.65032		7 = 2.8328		7 = 18.1300
	8 = 51.613		8 = 0.74323		8 = 3.2375		8 = 20.7200
	9 = 58.065		9 = 0.83613		9 = 3.6422		9 = 23.3100
0.15500 = 1		10.764 = 1		2.471 = 1		0.3861 = 1	
0.31000 = 2		21.528 = 2		4.942 = 2		0.7722 = 2	
0.46500 = 3		32.292 = 3		7.413 = 3		1.1583 = 3	
0.62000 = 4		43.055 = 4		9.884 = 4		1.5444 = 4	
0.77500 = 5		53.819 = 5		12.355 = 5		1.9305 = 5	
0.93000 = 6		64.583 = 6		14.826 = 6		2.3166 = 6	
1.08500 = 7		75.347 = 7		17.297 = 7		2.7027 = 7	
1.24000 = 8		86.111 = 8		19.768 = 8		3.0888 = 8	
1.39500 = 9		96.875 = 9		22.239 = 9		3.4749 = 9	

* For instance, a peak 2,856 meters high will be

$$
\begin{array}{r}
6,561.67 \\
2,624.67 \\
164.04 \\
19.68 \\
\hline
9,370.06 \text{ feet high}
\end{array}
$$

* (*Courtesy of the World Almanac.*)

Table A.4. Natural Sines, Cosines, Tangents, and Cotangents *

°	sin	d	tan	d	cot	d	cos	d	°
		+		+		−		−	
0	0.0000		0.0000		∞		1.0000		90
		175		175				2	
1	0.0175		0.0175		57.290		0.9998		89
		174		174				4	
2	0.0349		0.0349		28.636		0.9994		88
		174		175				8	
3	0.0523		0.0524		19.081		0.9986		87
		175		175				10	
4	0.0698		0.0699		14.301		0.9976		86
		174		176				14	
5	0.0872		0.0875		11.430		0.9962		85
		173		176				17	
6	0.1045		0.1051		9.514		0.9945		84
		174		177				20	
7	0.1219		0.1228		8.144		0.9925		83
		173		177				22	
8	0.1392		0.1405		7.115		0.9903		82
		172		179				26	
9	0.1564		0.1584		6.314		0.9877		81
		172		179				29	
10	0.1736		0.1763		5.671		0.9848		80
		172		181		526		32	
11	0.1908		0.1944		5.145		0.9816		79
		171		182		440		35	
12	0.2079		0.2126		4.705		0.9781		78
		171		183		374		37	
13	0.2250		0.2309		4.331		0.9744		77
		169		184		320		41	
14	0.2419		0.2493		4.011		0.9703		76
		169		186		279		44	
15	0.2588		0.2679		3.732		0.9659		75
		168		188		245		46	
16	0.2756		0.2867		3.487		0.9613		74
		168		190		216		50	
17	0.2924		0.3057		3.271		0.9563		73
		166		192		193		52	
18	0.3090		0.3249		3.078		0.9511		72
		166		194		174		56	
19	0.3256		0.3443		2.904		0.9455		71
		164		197		157		58	
20	0.3420		0.3640		2.747		0.9397		70
		164		199		142		61	
21	0.3584		0.3839		2.605		0.9336		69
		162		201		130		64	
22	0.3746		0.4040		2.475		0.9272		68
		161		205		119		67	
23	0.3907		0.4245		2.356		0.9205		67
		160		207		110		70	
24	0.4067		0.4452		2.246		0.9135		66
		159		211		101		72	
25	0.4226		0.4663		2.145		0.9063		65
		158		214		95		75	
26	0.4384		0.4877		2.050		0.8988		64
		156		218		87		78	
27	0.4540		0.5095		1.963		0.8910		63
		155		222		82		81	
28	0.4695		0.5317		1.881		0.8829		62
		153		226		77		83	
29	0.4848		0.5543		1.804		0.8746		61
		152		231		72		86	
30	0.5000		0.5774		1.732		0.8660		60
		150		235		68		88	
31	0.5150		0.6009		1.664		0.8572		59
		149		240		64		92	
32	0.5299		0.6249		1.600		0.8480		58
		147		245		60		93	
33	0.5446		0.6494		1.540		0.8387		57
		146		251		57		97	
34	0.5592		0.6745		1.483		0.8290		56
		144		257		55		98	
35	0.5736		0.7002		1.428		0.8192		55
		142		263		52		102	
36	0.5878		0.7265		1.376		0.8090		54
		140		271		49		104	
37	0.6018		0.7536		1.327		0.7986		53
		139		277		47		106	
38	0.6157		0.7813		1.280		0.7880		52
		136		285		45		109	
39	0.6293		0.8098		1.235		0.7771		51
		135		293		43		111	
40	0.6428		0.8391		1.192		0.7660		50
		133		302		42		113	
41	0.6561		0.8693		1.150		0.7547		49
		130		311		39		116	
42	0.6691		0.9004		1.111		0.7431		48
		129		321		39		117	
43	0.6820		0.9325		1.072		0.7314		47
		127		332		36		121	
44	0.6947		0.9657		1.036		0.7193		46
		124		343		36		122	
45	0.7071		1.0000		1.000		0.7071		45
		−		−		+		+	
°	cos	d	cot	d	tan	d	sin	d	°

* This table can be used in finding the true lengths of degrees of longitude and for construction of map projections in general. The functions at the bottom refer to the degrees on the right; thus sin α = cos $(90° - \alpha)$, making sin, 60° cos 30° = 0.8660, and in the same way tan 60° = cot 30° = 1.732. If fractions of degrees are wanted, a proportionate amount of d can be used. For very exact large-scale maps, however, a book of logarithms is necessary.

TABLE A.5. CONVERSION OF DEGREES TO MILS *

Degrees	Mils	Degrees	Mils	Degrees	Mils	Minutes	Mils	Minutes	Mils
1	17.8	31	551.1	61	1,084.4	1	0.3	31	9.2
2	35.6	32	568.9	62	1,102.2	2	0.6	32	9.5
3	53.3	33	586.7	63	1,120.0	3	0.9	33	9.8
4	71.1	34	604.4	64	1,137.8	4	1.2	34	10.1
5	88.9	35	622.2	65	1,155.6	5	1.5	35	10.4
6	106.7	36	640.0	66	1,173.3	6	1.8	36	10.7
7	124.4	37	657.8	67	1,191.1	7	2.1	37	11.0
8	142.2	38	675.6	68	1,208.9	8	2.4	38	11.3
9	160.0	39	693.3	69	1,226.7	9	2.7	39	11.6
10	177.8	40	711.1	70	1,244.5	10	3.0	40	11.9
11	195.6	41	728.9	71	1,262.2	11	3.3	41	12.1
12	213.3	42	746.7	72	1,280.0	12	3.6	42	12.4
13	231.1	43	764.4	73	1,297.8	13	3.9	43	12.7
14	248.9	44	782.2	74	1,315.6	14	4.1	44	13.0
15	266.7	45	800.0	75	1,333.3	15	4.4	45	13.3
16	284.4	46	817.8	76	1,351.1	16	4.7	46	13.6
17	302.2	47	835.6	77	1,368.9	17	5.0	47	13.9
18	320.0	48	853.3	78	1,386.7	18	5.3	48	14.2
19	337.8	49	871.1	79	1,404.5	19	5.6	49	14.5
20	355.6	50	888.9	80	1,422.2	20	5.9	50	14.8
21	373.3	51	906.7	81	1,440.0	21	6.2	51	15.1
22	391.1	52	924.4	82	1,457.8	22	6.5	52	15.4
23	408.9	53	942.2	83	1,475.6	23	6.8	53	15.7
24	426.7	54	960.0	84	1,493.3	24	7.1	54	16.0
25	444.5	55	977.8	85	1,511.1	25	7.4	55	16.3
26	462.2	56	995.6	86	1,528.9	26	7.7	56	16.6
27	480.0	57	1,013.3	87	1,546.7	27	8.0	57	16.9
28	497.8	58	1,031.1	88	1,564.5	28	8.3	58	17.2
29	515.6	59	1,048.9	89	1,582.8	29	8.6	59	17.5
30	533.3	60	1,066.7	90	1,600.0	30	8.9	60	17.8

* Conversion factor—1 degree = 17.77778 mils; 1 minute = 0.29630 mil.
  Examples:

1. Convert 64°29′ to mils.
    64° = 1,137.8 mils
    29′ =   8.6
   64°29′ = 1,146.4 mils

2. Convert 87.95° to mils.
    87° = 1,546.7 mils
    .90° = 16.0
  .05° = 0.889 = 0.9
   87.95 = 1,563.6 mils

(From U.S. War Department, *Tech. Man.* 5–236.)

TABLE A.6. METHODS OF EXPRESSING GRADIENTS *

Angle, degrees	Feet per 100 feet horizontal or per cent	Feet to the mile horizontal	1 vertical on or in — horizontal
¼	0.44	23.0	229
½	0.87	46.1	115
¾	1.31	69.1	76
1	1.74	92.2	57
1¼	2.18	115.1	46
1½	2.62	138.3	38
1¾	3.06	161.2	33
2	3.49	184.4	29
2½	4.37	230.5	23
3	5.24	276.7	19
3½	6.12	322.9	16
4	6.99	369.2	14
4½	7.37	415.5	13
5	8.75	461.9	11.4
6	10.51	555	9.5
7	12.28	....	8.1
8	14.05	....	7.1
9	15.84	....	6.3
10	17.63	....	5.7
15	....	....	3.7
20	....	....	2.7
25	....	....	2.1
30	....	....	1.7

* The different methods of expressing gradients have their values given for the usual range and to the customary degree of accuracy of their use. (From U.S. War Department, *Tech. Man.* 5–236.)

TABLE A.7. MAP SCALES IN ENGLISH MEASURE AND METRIC UNITS *

Scale, 1 to—	Miles per inch	Kilometers per inch	Feet per inch	Inches per mile	Inches per kilometer
500,000	7.891	12.700	41,666	0.1267	0.07874
250,000	3.945	6.350	20,833	0.2534	0.15748
125,000	1.972	3.175	10,416	0.5068	0.31496
100,000	1.578	2.540	8,333	0.6336	0.39370
63,360	1.000	1.609	5,280	1.000	0.6213
62,500	0.986	1.587	5,208	1.0137	0.6299
40,000	0.6313	1.016	3,333	1.584	0.9842
31,680	0.500	0.804	2,640	2.000	1.2427
25,000	0.3945	0.635	2,083	2.534	1.5748
20,000	0.3156	0.508	1,666	3.168	1.9685
10,000	0.15782	0.254	833.3	6.336	3.937
2,500	0.0394	0.0635	208.3	25.34	15.748
1,200	0.0189	0.0305	100	52.8	38.808

* (From U.S. War Department, *Tech. Man.* 5–236.)

Appendix 2:

Glossary

The purpose of this glossary is to help students with unfamiliar words. The terms listed often have additional meanings, but here only the cartographic use is considered. However, no attempt has been made to define all cartographic terms. Individual map projections and the various kinds of diagrams and special devices have been defined in the text and are not redefined here. More information may be found in the Index and the appropriate parts of the text. For additional details, the reader is referred to the following works:

Inter-agency Comm. on Negative Scribing: Preliminary List, 1956, p. 33.

Mitchell, Hugh C.: "Definitions of Terms Used in Geodetic and Other Surveys," U.S. Coast and Geodetic Survey Spec. Pub. 242, 1948, p. 87.

U.S. Army Map Service: "Glossary of Cartographic Terms," Tech. Man. 28, 1948, p. 95.

U.S. Coast and Geodetic Survey, Aeronautical Chart Division: "Geographical Glossary," 1955, p. 12.

U.S. Geological Survey: Definitions, Topographic Instructions, chap. 6A4, 1957, p. 98.

———: Glossary of Names for Topographic Forms, Topographic Instructions, chap. 6A3.

292

Aeronautical chart   Chart used for air navigation.

Alidade   The part of a surveying instrument which consists of a sighting device with reading or recording accessories. *Telescopic alidade* is used on a plane table.

Altitude tints   A sequence of colors usually varying from green to brown, marking zones of elevations between successive contour lines. Called also *layer tints, gradient tints, hypsometric coloring*.

Ammonia process (Diazo, Ozalid)   A positive printing process in which sheets are coated with a compound which decomposes on exposure to light. The unexposed lines will darken in ammonia fumes.

Astrolabe   Instrument used for measuring the altitudes of celestial objects. "Altitude" in this connection means the vertical angle between the horizon and the celestial object.

Azimuth   In navigation it is the horizontal angle reckoned clockwise from the North 0 to 360°. *Magnetic azimuth* is reckoned from magnetic north. In geodesy and astronomy azimuth is usually reckoned clockwise from the South.

Azimuthal projections   The projection of a part of the globe upon a plane from a given eye point. Some azimuthal projections are derivatives of the geometric concept.

Barometer   An instrument for measuring atmospheric pressure. *Aneroid barometer* uses a vacuum box. Barometers in airplanes are called *altimeters*.

Bathymetric   Relating to the measurement of ocean or other water depths.

Calibration   The determination of subdivisions in a measuring instrument by interpolation.

Cartogram   A highly abstracted map on which the actual outlines or locations are distorted to express a geographic concept.

Cartography   The art and science of making maps, charts, globes, and relief models.

Chart   (1) Map used for navigation in air or in water. (2) A large special-purpose map or diagram.

Chronometer   A timekeeping instrument of great precision.

Combination plate   Halftone and line work combined on one press plate by exposure to two separate negatives. Also called *surprint plate*.

Compass   (1) An instrument to draw circles. (2) A device for finding directions by means of a magnetic needle which swings on a pivot. One end of the needle points to *magnetic north* of the locality.

Conformal map projection   A projection on which the shape of any small area remains unchanged.

Conic projections   A group of projections which are derived from the concept of projecting the parallels and meridians of a globe upon a tangent or secant cone and then developing the cone into a plane.

Contact printing   A process which produces a print the same size as the original by lighting through the design, which is in direct contact with a sensitized sheet.

Contour   An imaginary line on the ground, all points of which are at the same elevation above a specified datum surface. The difference in elevation between successive contours is called the *contour interval*.

Control   A system of control stations established by geodetic methods. *Horizontal control* is established by triangulation and *vertical control* by leveling or by measuring vertical angles.

Cylindrical projections   A group of projections with horizontal parallels and evenly set vertical meridians.

Datum   A reference element, such as a line or plane, to which the positions of other elements are related.

Declination, magnetic   The horizontal angle between the true north and the magnetic north. In aeronautics it is called *variation*.

Diapositive   A photographic positive print on glass or film often used in photogrammetry.

Ephemeris   A table containing the positions and related data for a celestial body for given dates.

Equal-area (equivalent) projections   A group of projections upon which any area, large or small, is the same as on a globe of corresponding scale.

Equation of time   Difference in time between the mean solar noon and the actual culmination of the sun.

Error of closure   The distance by which the end point of the computed legs of a closed traverse fails to agree with the starting point.

Fathometer   An echo-sounding instrument used for measuring depth of water.

Fiducial marks   Marks on sides of air photographs which are used to find the principal point.

Fix   The position on a map of a point of observation obtained by surveying.

Flattening of the earth   Ratio of the difference between the equatorial and polar radii of the earth to the equatorial radius. This is the measurement of the ellipticity of a meridian: $F = \dfrac{a-b}{a} = \dfrac{1}{298}$.

Gelatin   An organic colloid which swells in cold water and dissolves in hot water. It is used in most photographic emulsions.

Geodesy   The science which treats mathematically the shape and size of the earth and the position of points, lines, and areas on it.

Geoid   The shape of the earth considered as a mean sea-level surface extended continuously under the continents.

Geostenography   A method of rapid notations of geographic data in a notebook or on maps.

Globe   A spherical body. In cartography it refers to a small sphere representing the earth.

Globe gore   A lune-shaped map to be fitted to a globe.

Gradient   An angle of slope. A gradient of a river is a profile along its course.

Great circle   A circle on a sphere produced by any plane which passes through the center of the sphere.

Grid   A network of two sets of regularly spaced straight lines intersecting usually at right angles.

Guide copy   A map which is sufficiently complete to be given to the engraver or scriber for the preparation of printing plates. Also called *guide image*.

Hachuring   A method of relief representation on maps by short lines which run parallel to the dip of the slope. The steeper the slope the heavier the line.

Halftone   (1) A shade between black and white. (2) The process of reproducing halftones by a halftone screen is called the *halftone process*.

Horizon (*apparent*)   A circle along which the cone, with its apex at the observer's eye, is tangent to the sphere. In geodesy and astronomy the *true horizon* is a great circle on the celestial sphere whose plane is perpendicular to the direction of the plumb line at the point of observation.

Horizontal   (1) Parallel to the water level or perpendicular to the plumb line. (2) Parallel to the surface of a spheroid at a certain point.

Hydrography   The representation of water features on maps.

Hypsography   Parts of a map showing relief.

Intermittent river or lake   A river or lake which is dry for three months or more, on the average.

Intersection   Obtaining the position of points by observing angles from two or more inter-measured stations.

Isometric diagram   A drawing of a three-dimensional body related to three axes. The dimensions parallel to the axes are true to scale. One of the axes is usually vertical.

Isopleths   Lines drawn on maps connecting points of equal value. They are arranged according to regular isopleth intervals. Also called *isarithms, isolines*. Examples: *isotherms, isobars, isogons, isohyets, etc.*

Landform map   Small- or medium-scale map showing the nature of the relief by semi-pictorial symbols.

Latitude   An arc distance from the equator measured in degrees.

Layer tints   See *altitude tints*.

Legend   An explanation of symbols on a map.

Leveling (*spirit*)   The determination of elevation of points with respect to each other by means of a spirit level.

Light table   A glass-topped table with lights underneath the glass which is used for copying.

Lithography   Printing from the surface of limestone (or kerneled metal sheets) on which the features are drawn with greasy ink or crayon. The stone, if wetted and rolled over with greasy printing ink, will take the ink only on the previously greased surfaces and repel it elsewhere.

Longitude   An arc distance from a prime meridian measured in degrees.

Loxodrome   A line on the map which crosses the successive meridians at constant angle. Also called a *rhumb line*.

Map   A selective, symbolized, and generalized picture on a much reduced scale of some spatial distribution of a large area, usually the earth's surface.

Map projection   Any regular set of parallels and meridians upon which a map can be drawn. (For individual map projections, see text.)

Mean sea level   The average height of the sea for all stages of the tide.

Mean solar time (*mean time*)   Time measured by the daily motion of a fictitious body, the "mean sun" which is supposed to revolve uniformly in the plane of the equator, completing one revolution in one day.

Meridian   The trace of plane on the earth's surface which passes through the poles. Meridians are in North-South direction.

Meter   Unit of length derived from 10-millionth part of the arc distance between the equator and poles. It is 39.37 in. long.

Mil   An angle derived from subtending a foot in a 1,000-ft distance. This was modified to the *military mil*—1:6,400 part of the circle. A degree equals 17.8 mils.

Moiré   The formation of regular light and dark patches by interference of two halftone screens over each other.

Mosaic Several air photos mounted together to form a continuous picture of a larger area.

Neatline The inner border of a map.

Offset printing A method of printing by the lithographic principle in which a map is applied to a kerneled metal sheet with greasy ink. This is attached to a roll, wetted and inked. This roll prints on a rubber cylinder and the rubber cylinder transfers the design to paper on a third roll.

Origin of grid systems A point from which the grid lines are laid out, usually in the center of the grid zone. *False origin* is a point west and south of the grid zone from which grids are actually numbered to avoid negative values.

Pantograph An instrument for copying maps on larger or smaller scale. Most pantographs are made of rods forming a parallelogram jointed on the four corners.

Parallax The difference in the direction of an object as seen by an observer and as seen from some standard point of reference. The apparent displacement of an object as seen in an airplane photograph because of difference in height.

Parallels Lines on the earth's surface cut by planes parallel to the equator. In cartography they are regarded as circles. Points on a parallel have the same latitude. Parallels are in an East-West direction.

Peneplain A nearly level surface derived from prolonged erosion of an uplifted region to near sea level.

Perspective Projection of an object upon a *plane of projection* from an eye point or light source. In true perspective the more distant the feature, the smaller it will appear.

Photogrammetry The science and art of preparing maps from photos.

Photolithography A process consisting in making a negative of the map and contact-printing it on an albumen-sensitized metal printing plate.

Photostat A design photographed directly on sensitized paper.

Physical maps and globes Term used by commercial map publishers for maps with altitude tints. This group is in contrast to *political maps*, in which the various countries are differentiated by color.

Plane surveying Surveying on such small areas that the earth's surface can be considered as a plane.

Plane table A board used for direct surveying of small areas by sighting points with an alidade.

Planimeter Instrument for measuring the area of a plane surface.

Planimetric map A map not showing relief, in contrast to a *topographic map*.

Plat A map showing boundaries and subdivisions of a tract of land determined by surveying.

Prime meridian The meridian from which all other meridians are reckoned. At present the standard zero-degree meridian passes through the Greenwich Observatory in London.

Principal point In an air photograph, the point where the axis of the camera pierces the ground or its photographic image.

Profile A vertical section along a straight or curved line.

Proportional divider An X-shaped divider, formed by jointing a pair of two-pointed arms, which is used for enlargement and reduction.

Protractor An instrument for laying out angles.

Pull-up A tracing of a map, or part of it, on transparent paper or plastic. This is often done by contact printing. *Light-blue pulls* are used in color separation.

Refraction The change in the direction of a ray of light due to its passing from one medium into another in which the speed of light is different.

Register marks Pinpoints or crosses, on four sides of the map, by which the color-separation drawings are adjusted to each other.

Relative relief Height of hills and mountains over the adjacent valleys, basins, or plains.

Representative fraction Map scale giving the ratio between any small distance on the map and the corresponding distance on the ground, as 1:62,500.

Rhumb line A line which crosses the successive meridians at a constant angle. Also called *loxodrome*.

Scale The relationship between a distance on the map and the corresponding distance on ground.

Scribing Engraving lines, symbols, and windows in a scribe coating, usually for the preparation of a negative for map reproduction.

Series maps A set of maps which are sections of a larger area, such as topographic sheets.

Shades Relative darkness of gray tones.

Shadient relief Plastic shading combined with gradient (altitude) tints.

Spherical excess The amount by which the sum of the three angles of a triangle on a sphere exceeds 180°.

Spheroid A mathematical figure closely approaching the geoid in form and size and used as surface reference for geodetic surveys. The Clarke spheroid which is used in North America is an ellipsoid, with flattening of $\frac{1}{295}$.

Spot height A point whose elevation is noted on a map.

Squeegee A T-shaped tool to squeeze out superfluous ink or any other liquid from a surface.

Stadia survey The determination of distance by reading the intercept between the images of two horizontal hairlines on a graduated rod.

Stereoplotting A method which permits drawing contour lines and other detail by using two overlapping air photos under a stereoscopic instrument.

Stereoscopic vision The mental process which fuses the images of the two eyes into a three-dimensional impression.

Symbols Designs on maps used to represent various features.

Theodolite A precision-surveying instrument for reading horizontal and vertical angles.

Three-color process A method of color reproduction using filters and halftone screens.

Tick Short lines perpendicular to the neatline marking grid systems.

Tints Color gradations on maps, like altitude tints. Also patterns applied to a map, such as cellotints or Ben-day tints.

Tone The relative darkness of grays on air photos.

Topographic map A general map of large or medium scale showing all important features, including relief.

Trachographic relief drawing A method of using small, curved hill-shaped lines to indicate relief on small-scale maps. The height of the curves indicates relative relief; the width, the average slope.

Transit A theodolite in which the telescope can be reversed (plunged) by rotating it about its horizontal axis.

Transverse projections Map projections turned at right angles to their usual orientation.

Traverse A connected series of straight lines on the earth's surface, the lengths and bearings of which are determined.

Triangulation A method of surveying which consists in precisely measuring the length of a baseline and computing the location of a chain of stations by measuring angles.

Vanishing point The point in perspective where parallel lines meet.

Vertical exaggeration Enlarging the vertical component on a profile, relief model, or block diagram to make it more apparent.

Vinylite A synthetic resin of great dimensional stability, used in drawing maps.

Appendix 3:

Bibliography[1]

General Cartography

Balchin, W. C. V., and A. W. Richards: "Practical and Experimental Geography," Methuen & Co., Ltd., London, 1952, p. 136.

Bauer, A. H.: "Cartography," Vocational and Professional Mon. 60, Bellman, Boston, 1945, p. 31.

Birch, T. W.: "Maps, Topographical and Statistical," Oxford University Press, London, 1949, p. 240.

Bormann, Werner: "Allgemeine Kartenkunde," Astra, Lahr/Schwarzwald, 1954, p. 175.

"Columbia-Lippincott Gazetteer of the World," Columbia University Press, New York, 1952, p. 2148.

d'Agapayeff, A., and E. C. R. Hadfield: "Maps," Oxford University Press, New York, 1943, p. 140.

Debenham, F.: "Map Making," M. S. Mill Co., New York, 1936. 239 pp.

Deetz, C. H.: "Cartography," U.S. Coast and Geodetic Survey Spec. Pub. 205, 1943. 124 pp.

Deetz, Charles H., and O. S. Adams: "Elements of Map Projections . . . ," "U.S. Coast and Geodetic Survey Spec. Pub. 68.

Eckert, M.: "Die Kartenwissenschaft," 2 vols., Walter De Gruyter & Co., Berlin, 1921; 1925.

"Encyclopedia Americana": Cartography, Cartograms, Chart, Diagram, Map, 1962.

General Drafting Company: "Of Maps and Mapping," Convent Station, N.J., 1959.

"Glossary of Cartographic Terms," Army Map Service Tech. Man. 28, 1948, p. 95.

Greenhood, David: "Down to Earth," Holiday House, Inc., New York, 1944. 262 pp.

Imhof, E.: "Gelände und Karte," E. Rentsch, Erlenbach-Zürich, 1950, p. 400.

International Geographical Union, Geographische Institut d. Universität, Zürich. Reports of the Cartography Sections of all congresses. Secretary: Hans Boesch.

Jervis, W. W.: "The World in Maps," Oxford University Press, New York, 1938.

Kosack, H. P., and K.-H. Meine: "Die Kartographie 1945–1954," Astra, Lahr/Schwarzwald, 1955, p. 216.

"Lists of Names," Royal Geographical Society, Permanent Committee on Geographical Names, London.

Mackay, J. Ross: Geographical Cartography, *Canadian Geog.*, vol. 4, 1954, pp. 1–14.

Monkhouse, F. J., and H. R. Wilkinson: "Maps and Diagrams," Methuen & Co., Ltd., London, 1952, p. 330.

Musham, H. A.: "The Technique of the Terrain," Reinhold Publishing Corporation, New York, 1944, p. 228.

Philbrick, A. K.: Principles of Geo-cartography, *Annals Assn. Am. Geog.*, vol. 43, 1953, pp. 201–215.

Raisz, Erwin: "General Cartography," 2d ed., McGraw-Hill Book Company, Inc., New York, 1948, p. 270.

———: "Mapping the World," Abelard-Schuman, Inc., Publishers, New York, 1956, p. 112.

[1] For abbreviations see list of abbreviations in the Preface.

Robinson, Arthur H.: "The Look of Maps," University of Wisconsin Press, Madison, Wis., 1952.

———: Geographical Cartography, "American Geography: Inventory and Prospect," Association of American Geography, 1954, pp. 553–557.

———: "Elements of Cartography," John Wiley & Sons, Inc., 1960, p. 254.

Salishchev, K. A., and A. V. Getsimin: "Kartografiya," Geograficheskoi Literatur, Moscow, 1955, p. 407.

Sloane, R. C., and J. M. Montz: "Elements of Topographic Drawing," McGraw-Hill Book Company, Inc., New York, 1943, 251 pp.

Sylvester, Dorothy: "Map and Landscape," George Philip & Son, Ltd., London, 1952, p. 295.

U.S. Army Map Service: "Applied Cartography," Training Aid No. 1, 1951, p. 185. (Mimeographed.)

———: "Geographical Names for Military Maps," Memo No. 453, 1945, p. 126.

U.S. Department of the Army: "Topographic Symbols," Field Man. 21–31, 1952, p. 112.

U.S. Department of the Army and Air Force: "General Drafting," Tech. Man. 5–230, 1955, p. 240.

U.S. Navy Department, Bureau of Aeronautics: "Air Navigation, Part I, Introduction to Earth," U.S. Navy Flight Preparation Training Series, McGraw-Hill Book Company, Inc., New York, 1943.

"Webster's Geographical Dictionary," G. & C. Merriam Company, Springfield, Mass., 1949, p. 1300.

Wright, J. K.: Map Makers Are Human, *Geog. Rev.*, vol. 32, 1942, pp. 527–544.

Historical Cartography

Almagiá, R.: "Monumenta Cartografica Vaticani," 3 vols., 1941; 1948; 1952.

Bagrow, Leo: "Geschichte der Kartographie," Safari, Berlin, 1951, p. 478.

Brown, Lloyd A.: "Map Making . . . ," Little Brown & Company, Boston, 1960, p. 217.

———: "The Story of Maps," Little, Brown & Company, Boston, 1949, p. 395.

Cottler, J., and H. Jaffe: "Map Makers," Little, Brown & Company, Boston, 1938.

Crone, G. R.: "Maps and Their Makers," Hutchinson's University Library, 1953, p. 125.

Curnow, I. J.: "The World Mapped," Sifton Pread & Co., Ltd., London, 1930.

Dickinson, R. E., and O. J. R. Howarth: "The Making of Geography," Oxford University Press, New York, 1933.

Durand, Dana B.: "The Vienna-Klosterneuburg Map Corpus," E. J. Brill, Leiden, 1952, p. 520.

Fite, E. D., and A. Freeman: "A Book of Old Maps," Harvard University Press, Cambridge, Mass., 1926.

Fordham, H. G.: "Maps, Their History, Characteristics and Uses," Cambridge University Press, New York, 1927.

Friis, Herman R.: Highlights in the First 100 Years of Surveying and Mapping and Geographic Exploration of the U.S. by the Federal Government, 1775–1880, *Surveying and Mapping*, 1958, pp. 186–206.

Fuechsel, C. F.: "Geographic Exploration and Topographical Mapping of the U.S. Government," National Archives, 1952, p. 52. (Mimeographed.)

Heidel, W. A.: "The Frame of the Ancient Greek Maps," American Geographic Society, New York, 1937.

Humphreys, A. L.: "Old Decorative Maps and Charts," Minton, Balch & Co., New York, 1926.

Karpinski, L. C.: "Bibliography of the Printed Maps of Michigan," Michigan Historical Commission, Lansing, Mich., 1931.

LeGear, Clara E.: "U.S. Atlases," Library of Congress, Map Division, 1950, p. 444.

———: "A List of Geographical Atlases in the Library of Congress," vol. 5, Library of Congress, Map Division, 1958.

Lynam, E.: "The Mapmaker's Art," The Batchworth Press, 1953, p. 140.

———: "British Maps and Map Makers," William Collins Sons & Co., Ltd., New York, 1944. 48 pp.

MacFadden, Clifford H.: "A Bibliography of Pacific Area Maps," American Council,

Institute of Pacific Relations, San Francisco, New York, Honolulu, 1941.

Miller, Konrad: "Mappae-mundi, Die Ältesten Weltkarten," 6 vols., J. Roth, Stuttgart, 1898.

Nordenskiöld, A. E.: "Facsimile Atlas to the Early History of Cartography," Stockholm, 1889.

Paullin, C. O., and J. K. Wright: "Atlas of the Historical Geography of the U.S.," Carnegie Institution of Washington, Washington, and American Geographic Society, New York, 1932.

Phillips, P. L.: "A List of Geographical Atlases in the Library of Congress with Bibliographical Notes," 4 vols., Library of Congress, Map Division, 1909–1920.

Ristow, W. W., and Clara E. LeGear: "A Guide to Historical Cartography," Library of Congress, Map Division, 1954, p. 14.

Skelton, R. A.: "Decorative Printed Maps of the Fifteenth to Eighteenth Centuries," Staples Press, London, p. 192.

Stevenson, E. L.: "Portolan Charts," The Hispanic Society of America, New York, 1911.

Tooley, R. V.: "Maps and Map Makers," B. T. Batsford, London, 1949. 136 pp.

Wagner, H. R.: "The Cartography of the Northwest Coast of America to the Year 1800," 2 vols., University of California Press, Berkeley, Calif., 1937.

Wheat, C. I.: Mapping the American West, 1540–1857, *Proc. Am. Antiquarian Soc.*, vol. 64, 1954, pp. 19–194.

Wieder, F. C.: "Monumenta Cartographica—Reproduction of Unique and Rare Maps, Plans and Views in the Actual Size of the Originals," 5 vols., M. Nijhoff, The Hague, 1925–1933.

Winsor, Justin: "Narrative and Critical History of America," 8 vols., Houghton Mifflin Company, Boston, 1884–1889.

Map Projections

Birdseye, C. H.: "Formulas and Tables for the Construction of Polyconic Projections," U.S. Geological Survey Bull. 809, 1929.

Deetz, Charles H.: "The Lambert Conformal Projection," U.S. Coast and Geodetic Survey Spec. Pub. 47, 1918.

———— and O. S. Adams: "Elements of Map Projection," U.S. Coast and Geodetic Survey Spec. Pub. 68.

Driencourt et Laborde: "Traité des projections des cartes géographiques," Paris, 1932.

Fisher, Irving, and O. M. Miller: "World Maps and Globes," Essential Books, New York, 1944.

Kellaway, G. P.: "Map Projections," 2d ed., E. P. Dutton & Co., Inc., New York, 1949.

Steers, J. A.: "An Introduction to the Study of Map Projections," University of London Press, Ltd., London, England, 1927.

Thomas, P. O.: "Conformal Projection in Geodesy and Cartography," U.S. Coast and Geodetic Survey Spec. Pub. 251, 1952, p. 142.

U.S. Army Map Service: "Universal Transverse Mercator Grid," Tech. Man. 19, p. 66.

————: "Grids and Grid References," Tech. Man. 36, 1950, p. 100.

U.S. Coast and Geodetic Survey: "State Coordinate Systems," Spec. Pub. 235, 1945, p. 62.

————: "Equal-area Projections for World Statistical Maps," Spec. Pub. 245, 1949, p. 45.

————: "Tables for Polyconic Projection of Maps," Spec. Pub. 5, 1930.

U.S. National Bureau of Standards: "Construction and Application of Conformal Maps," 1952, p. 280.

Wagner, Karl-heinz: "Kartographische Netzenentwürfe," Bibliographisches Institut, Leipzig, 1949, p. 300.

Wellman, Chamberlin: "The Round Earth on Flat Paper," National Geographical Society, Washington, D.C., 1947.

Map and Air-photo Reading

Abrams, Talbert: "Essentials of Aerial Surveying and Photo Interpretation," McGraw-Hill Book Company, Inc., New York, 1944, p. 289.

Air Forces Advanced Flying School: "Aerial Photography and Photograph Reading," Brooks Field, Tex., 1942.

Anderson, M. L.: "Steps in Map Reading," Rand McNally & Company, Chicago, 1949, p. 156.

Dury, C. H.: "Map Interpretation," Pitman Publishing Corporation, 1952, p. 207.

Garnett, A.: "The Geographical Interpretation of Topographic Maps," George C. Harrap & Co., Ltd., London, 1953, p. 334.

International Civil Aviation Organization: "Aeronautical Charts," Montreal, 1942, p. 50.

Lobeck, A. K., and W. J. Tellington: "Military Maps and Air Photographs," McGraw-Hill Book Company, Inc., New York, 1944.

McCurdy, P. G.: "Manual of Coastal Delineation from Aerial Photographs," U.S. Department of Navy, Hydrographic Office Pub. 592, 1947, p. 144.

Putnam, W. C.: "Map Interpretation with Military Applications," McGraw-Hill Book Company, Inc., New York, 1943.

U.S. Department of Defense: "Advanced Map and Aerial Photograph Reading," Field Man. 21-26.

Surveying and Techniques

American Society of Photogrammetry: "Manual of Photogrammetry," 2d ed., George Banta Publishing Company, Menasha, Wis., 1952, p. 863.

Bagley, Lt. Col. J. W.: "Aerophotography and Aerosurveying," McGraw-Hill Book Company, Inc., New York, 1941.

Bosse, H.: "Kartentechnik," 2 vols., Astra, Lahr/Schwarzwald, 1954, pp. 174, 232.

Bygott, J.: "Mapwork and Practical Cartography," London University Technical Press, 1947, p. 260.

Church, E. A., and A. E. Quinn: "Elements of Photogrammetry," Syracuse University Press, Syracuse, N.Y., 1948.

Fuschsel, C. F.: Scribing Today, *Surveying and Mapping*, vol. 14, 1954, pp. 331-338.

Jenks, G. F.: Pointilism as a Cartographic Technique, *Annals Assn. Am. Geog.*, vol. 43, 1953, pp. 174-175.

McCurdy, P. G.: "Manual of Aerial Photogrammetry," U.S. Department of the Navy, Hydrographic Office, 1940.

Means, Ruth: "Shaded Relief," U.S. Coast and Geodetic Survey, Aeronautical Chart Division Tech. Man. RM-895, 1958, p. 108.

Mitchell, H. C.: "Definitions of Surveying Terms," U.S. Coast and Geodetic Survey Spec. Pub. 242, 1948, p. 87.

"Nachrichten aus dem Karten-und Vermessungwesen," Verlag d. Institut für angewandte Geodäsie, Frankfurt, A. M., 1958.

Sharp, H. O.: "Geodetic Control Surveys," John Wiley & Sons, Inc., New York, 1943.

Smith, H. T. U.: "Aerial Photographs and Their Application," Appleton-Century-Crofts, Inc., New York, 1943.

U.S. Army Map Service: "Color," Bull. 13, 1949, p. 47.

U.S. Coast and Geodetic Survey: "Horizontal Control Data," Spec. Pub. 277, 1941.

U.S. Department of the Army: "Elements of Surveying," Tech. Man. 5-232, 1953, p. 233.

————: "A Guide to the Compilation and Revision of Maps," Tech. Man. 5-240, 1955, p. 165.

U.S. Department of Defense: "Surveying," Tech. Man. 5-235, 1940.

————: "Surveying Tables," Tech. Man. 5-236.

Whitmore, G. D.: "Advanced Surveying and Mapping," Scranton, Pa., 1949, p. 619.

Special Cartography

Boggs, S. W., and D. C. Lewis: "Classification and Cataloguing of Maps and Atlases," Special Library Association, New York, 1945, p. 175.

Expenshade, E. B.: Problems in Map Editing, *Sci. Monthly*, vol. 65, 1947, pp. 217-226.

Gerlach, Arch C.: "An Adaptation of the Library of Congress Classification for Use in Geography and Map Libraries," International Geographical Union, 1956, Washington, D.C., p. 30.

LeGear, C. E.: "Maps, Their Care, Repair and Preservation in Libraries," Library of Congress, Map Division, 1949, p. 46.

Lobeck, A. K.: "Block Diagrams," John Wiley & Sons, Inc., New York, 1924.

Minogue, A. E.: "The Repair and Preservation of Records," U.S. Natl. Arch. Bull. 5, 1943.

Olson, E. C., and A. Whitmarsh: "Foreign Maps," Harper & Brothers, New York, 1944.

Ristow, W. W.: Journalistic Cartography, *Surveying and Mapping*, vol. 7, 1957.

———: "Aviation Cartography," Library of Congress, Map Division, 1956, p. 114. (Mimeographed.)

Roepke, H. G.: Care and Development of a Wall Map Collection, *Prof. Geog.*, vol. 10, May, 1958, pp. 11–15.

Thiele, W.: "Official Map Publications," American Library Association, Chicago, 1938.

"Typography and Design," U.S. Government Printing Office, 1951.

Periodicals and Publications

Annals of the Association of American Geographers: 1785 Massachusetts Avenue, N.W., Washington, D.C. (Quarterly.)

Current Geographical Publications: Additions to the research catalogue of the American Geographic Society, New York 32. (Monthly.)

Economic Geography: Clark University, Worcester, Mass. (Quarterly.)

Geographical Journal: Royal Geographical Society, Kensington Gore, London, S.W. 7, England. (Quarterly.)

Geographical Review: American Geographic Society, Broadway at 156th Street, New York 32. (Quarterly.)

Globen: Generalstabens Litografiska Austalt, Vasagatan 16, Stockholm, Sweden. (Bimonthly.)

Der Globusfreund: Coronelli Weltbund d. Globusfreunde, Wien IV, Gusshausstrasse 20, Austria.

Imago Mundi: Mouton & Co., 's Gravenhage, Netherlands. (Annual.)

Kartographische Nachrichten: C. Bertelsmann Verlag, Gütersloh, West Germany. (Monthly.)

Military Engineer: 808 Mills Building, Washington, D.C. (Bimonthly.)

National Geographic Magazine: National Geographical Society, Washington 6, D.C. (Monthly.)

Photogrammetric Engineering: American Society of Photogrammetric Engineers, P.O. Box 18, Beniamin Franklin Station, Washington 4, D.C. (Monthly.)

The Professional Geographer: Journal of the Association of American Geography, 1785 Massachusetts Avenue, N.W., Washington, D.C. (Bimonthly.)

Special Libraries Association, Geographical and Map Division Bulletin: M. A. Lucius, 31 East 10th Street, New York 3. (Monthly.)

Surveying and Mapping: American Congress on Surveying and Mapping, P.O. Box 470, Benjamin Franklin Station, Washington 4, D.C. (Quarterly.)

U.S. Aeronautical Chart and Information Service Technical Reports: St. Louis 18, Mo.

U.S. Army Map Service Bulletins: Washington 25, D.C.

U.S. Board on Geographic Names: "Decisions," Washington 25, D.C.

World Cartography: United Nations, New York. (Irregular.)

Appendix 4:[1]

Sample

Examination

Questions

The ability and industry of students can best be judged from the exercises. The questions suggested for selection here are designed solely for testing their theoretical knowledge and may count less than the exercises toward their final grade. The answers to the questions should be precise and short, taking about three to ten lines and requiring an average of four minutes for completion.

1. Compare the functions of the surveyor, the cartographer, and the geographer.
2. List the steps in producing a map from compilation to distribution.
3. What are the functions of the different types of cartographers?
4. What were the contributions of the ancient Greeks to cartography?
5. What were Ptolemy's contributions to cartography?
6. Characterize the cartography of the Middle Ages.

[1] Questions marked (2) count double in grading.

302

7. Describe the portolan charts of the fourteenth century.
8. What events, inventions, and attitudes produced the renaissance of cartography? (2)
9. Characterize the reformation of cartography. (2)
10. Describe the cartography of the nineteenth century.
11. Describe American cartography up to the mid nineteenth century. (2)
12. What influences have shaped American cartography in the last hundred years?
13. Give a short classification of maps. (2)
14. Describe five different types of pens used by cartographers.
15. Describe five different types of papers or drawing sheets used for map making.
16. In what regions could the air photo given by the instructor have been taken? State your reasons. (2)
17. What are the chief considerations in recognizing features on an air photograph?
18. Why is the tone of features on air photos often misleading? Give examples.
19. List various rural land types in the Eastern United States and the corresponding textures on air photos.
20. How and why can stereovision be applied to airplane photography?
21. Define (a) a parallactic displacement; (b) principal point; (c) stereoscope.
22. Give at least five basic differences between an air photograph and a map.
23. What are the five fundamental principles of cartography? In addition, mention two other principles which are very important, although some maps are without them.
24. What is the numerical scale if 1.1 in. on the map represents 10 miles?
25. What are the usual methods of reducing or enlarging the scales of maps?

26. What are the requirements of a good symbol?
27. What are the usual five categories of symbols and what color is used for each?
28. What are the advantages and limitations of compass traversing?
29. What is the principle of stadia measurements? Make a drawing.
30. How do we correct for inclined readings on stadia? Make a drawing.
31. Compare field sketching with photography.
32. Letter the names of five different styles of lettering, such as ROMAN.
33. What styles of letters are used in the U.S. Geological Survey topographic sheets for various features?
34. When are capital letters used? lower-case letters?
35. Make a sketch showing the readability of letters from various distances.
36. How do printers express the height of letters in points? How much is a pica?
37. How large should be the smallest lettering on a 12-in. map which is being prepared for lantern slides to be shown in a 60-ft room?
38. What are the rules for placing the names of cities, rivers, and lakes on small-scale maps? (2)
39. Draw a freehand sketch of Florida, spreading the name over it.
40. Letter your name carefully in Gothic capitals.
41. What is stick-up lettering? What are its advantages and disadvantages?
42. What are the usual contour intervals on maps of various scales? Give formula.
43. What is the formula for vertical exaggeration of profiles in a hilly country? When would you use more exaggeration? less?
44. Give five different ways to express slopes.
45. What is the origin of our conventional altitude tints?
46. What is the usual direction of the source of light in plastic shading? Why?
47. What are the usual combinations of various relief methods?
48. What is the Kitiro method of showing relief? Draw sketch.
49. Draw a landform map showing a submaturely dissected plateau with about 30 per cent flat upland.
50. Draw a landform map of a small but high mountainous island. (2)
51. Express, with landform symbols, a canyon land. (2)
52. Draw a landform map of a volcanic region. (2)
53. Draw a landform map of a karst region. (2)
54. Draw a landform map of a low gneissic (elongated) area. (2)
55. What is a slope category map and when is it used?
56. Compare a relative relief map with a map showing average slopes.
57. What is the hypsographic curve? the hypsographoid curve?
58. What is the general land-slope curve?
59. List 10 items of "marginal data" on topographic sheets.
60. What is the International Numbering System? Give example.
61. What are for us the most important publications of the U.S. Geological Survey?
62. Which are for us the most important publications of the Department of Agriculture?
63. Which offices in the United States publish sea and air navigational charts?
64. What map sets would you order for a small-budget college? (2)
65. Describe the various methods of map mounting.
66. What is the difference between the design and the layout of a map?
67. What are the chief considerations in our over-all plan for a map? (2)
68. What are the advantages and liabilities of maps on end papers (cover linings)?
69. What are the chief considerations in drawing a guide map to copy for our base?
70. How much larger than publication size should a map be drawn?

71. What kind of insets do maps usually have and how are they handled?
72. What is the order of inking in the various features on a map?
73. Draw 10 different types of lines used on maps and label them.
74. What is the "curve of gray spectrum"?
75. Give four different ways to apply shades to manuscript maps intended for printing.
76. What is vignetting and how can it be accomplished?
77. What produces color? Which are the primary colors? the subtractive and complementary colors?
78. How is the Munsell color chart constructed?
79. Describe the process of copper engraving used on old maps.
80. What is the principle of lithography?
81. What is the principle of the offset process? Sketch.
82. Describe the steps in offset printing. (2)
83. List five different types of map manuscripts from the printing point of view.
84. How are halftones reproduced? What are their limitations?
85. How would you reproduce a map with the relief in halftone, the sea solid black, with white parallels and meridians? (2)
86. What are combination plates and how are they reproduced?
87. State five ways for color separation of flat colored maps.
88. What is the three-color process for printing paintings?
89. Describe the wax-engraving method.
90. Describe the steps in the photoengraving of maps for books.
91. Describe the spirit duplicating process.
92. Describe the stencil duplicating process.
93. Describe the multilith or offset duplicating process.
94. Describe the shape of the earth, and give its approximate dimensions.
95. What ellipsoid is used by the U.S.G.S.? Are there any others and how much do they differ?
96. What results of the earth's rotation affect cartography?
97. What effects of the earth's revolution are shown on maps and globes?
98. Define parallels, meridians, latitude, and longitude.
99. How does the length of 1° of latitude vary from equator to pole?
100. How does the length of 1° of longitude vary from equator to pole?
101. What prime meridians are and were used on maps?
102. How do navigators measure latitude and longitude at sea?
103. How are latitude and longitude usually measured on land?
104. What are orthodromes and loxodromes? How are they used for sailing?
105. Discuss the various hemispheres in use.
106. What is the formula for the radius of the horizon at sea visible from various altitudes?
107. Describe the metric system.
108. What are the origin and length of a nautical mile?
109. Make a sketch showing the basic parts of a transit. (2)
110. How does a vernier work?
111. How do modern theodolites usually differ from a transit?
112. What is the great advantage of triangulation?
113. How are bases for triangulation measured?
114. What is a Bilby tower? How is it used?
115. What is done in the office computation of triangulation?
116. Why do triangulation and leveling parties rarely work together?
117. What is the principle of leveling? Sketch.
118. How does a Multiplex stereoplotting machine work?
119. What is radial line plotting? (2)
120. How is a "controlled" mosaic made?
121. What is trimetrogon photography? Sketch.
122. How is shoran used for locating an airplane?
123. What is a map projection?
124. When is a projection "equal-area"?

125. What are the necessary criteria for a conformal map projection?
126. Why is it important to know which lines are "true" in the various projections?
127. What are the advantages of projections with horizontal parallels?
128. What is the principle of the equirectangular projector? What is it used for?
129. What is the principle of the Mercator projection? What is it used for and why?
130. What is the principle of the sinusoidal projection? the use? Sketch.
131. Describe the Mollweide projection.
132. What are the advantages and disadvantages of interrupted projections for world maps?
133. What is the principle of the Eckert projections?
134. What is the formula for the radius of the standard parallel of the conic projection? Sketch.
135. How is the conic projection with two standard parallels constructed?
136. Describe the Albers conic projection. What is it used for?
137. Describe the Lambert Conformal Conic projection with two standard parallels. What is it used for?
138. Describe and sketch the construction and uses of the Bonne projection. (2)
139. Describe and sketch the construction and uses of the polyconic projection. (2)
140. What are the advantages and disadvantages of the conic projections in general?
141. What are the common properties of all azimuthal projections? (2)
142. What are the principle and use of the gnomonic projection?
143. What are the advantages and disadvantages of the orthographic projection?
144. What is the "armadillo" projection? What is its main advantage?
145. Describe the stereographic projection.
146. What is the azimuthal equidistant projection used for? What does the antipodal point to the center look like?
147. Describe the azimuthal equal-area projection.
148. Describe the Aitoff and Hammer projections.
149. Describe the Briesemeister projection of the world.
150. Describe the transverse Mercator projection.
151. Name five equal-area projections and three conformal projections.
152. What are the advantages and disadvantages of square grid systems? (2)
153. Describe the Universal Transverse Mercator grid. (2)
154. What are state grids? What is their use?
155. Describe the British system of square grids. Draw sketch. (2)
156. Describe the GEOREF system of the U.S. Air Force.
157. What are chorochromatic maps used for?
158. How is the value of a single dot on dot maps calculated?
159. Describe the "dasymetric" method. (2)
160. What are the advantages and disadvantages of choropleth maps?
161. What are the virtues and deficiencies of isopleth maps?
162. When are bar graphs used? line graphs?
163. Describe logarithmic line graphs.
164. Make a sketch showing the rapid construction of a generalized "central" curve for scattered points on two coordinates.
165. What are star diagrams used for? What is the radial scale if areas are compared?
166. Soil is composed of 60 per cent sand, 30 per cent clay, and 10 per cent gravel. Draw a triangular percentage graph.
167. What are the advantages of a block-pillar system?
168. When are superimposed diagrams used on statistical maps?
169. What methods are used for density-of-population maps?
170. How is a value-by-area cartogram constructed? Sketch. (2)
171. How can a traffic-flow map be drawn if the range of traffic is 1 to 200 and the widest line is ¼ in.?
172. Discuss the median point and pivotal point in centrograms.

173. How are isometric block diagrams constructed?
174. Draw a block diagram in one-point perspective. How do you lay out peaks and rivers?
175. What is a Dufour diagram? Sketch. (2)
176. What is the best way to show the races of mankind on a map?
177. What is the chief problem in planning maps to show languages and religions? What are the solutions?
178. What categories are used on the British land-use maps?
179. What categories do functional city maps show?
180. How are land forms shown on geostenograms? Give examples.
181. How is vegetation shown on geostenograms? Give examples.
182. How are buildings shown on geostenograms? Give examples.
183. What four factors do maps of manufacturing show? Draw a block showing all four.
184. What is the chief problem in planning mineral-production maps? What methods can be used?
185. What are the chief uses of globes?
186. How are globe gores prepared for 24 sections?
187. What does an analemma show? Sketch. (2)
188. How would you rapidly prepare a terrain model of a hilly region, 1:24,000? (2)
189. When is the negative method of cardboard cutting used for models? Why?
190. What vertical exaggeration would you use on a 1:62,500 model of Central Florida? What for Mount Rainier?

191. Describe the process of casting relief models in plaster of paris.
192. How does the Bround Reliefograph work? How is it used for printed plastic models? (2)
193. What are graphically simulated relief models?
194. Describe the making of blueprints.
195. Describe the making of ammonia prints.
196. Describe the making of photostats.
197. Give some rules of good photography for a geographer. (2)
198. Describe the developing and fixing of photographic films.
199. How did photography influence cartography in the last one hundred years? (2)
200. What are the advantages and liabilities of negative scribing?
201. How is lettering applied to scribe sheets?
202. What materials are used for base sheets for scribing?
203. What is Watercote and how is it used?
204. Describe five different gravers for scribing.
205. How are color-separation sheets made in scribing? (2)
206. What is the "alternate band method" for layer tints?
207. How are scribe sheets for colored maps proofed?
208. Describe the Deep-Etch process. (2)
209. Describe the photoscribe process.
210. Describe the Dystrip technique for color separation. (2)
211. Describe the phototypesetting machine.
212. Describe the process of silk-screen printing. (2)
213. How are maps for television made? (2)

Laboratory Syllabus

The best way to learn how to do something is by doing it. To learn cartography without exercises is unthinkable. The amount and selection of exercises depends on the time available and whether cartography is taught in one, two, or more semesters. Not always will there be enough time to finish all the suggested exercises, but it is better to distribute them all among the students and discuss them collectively than to omit any. Many students will have their own map problems which could be substituted where they best fit in, but the suggested exercises in the first part of the book are fundamental. All maps should be displayed on a large board with the instructor's remarks added, and the best maps should be selected for permanent exhibit.

The following syllabus is a correlation of lectures and laboratory work. Even with the best effort lectures could not be kept fully ahead of the exercises. For instance, the fundamentals of map design come only in Chap. 12, yet in connection with the field work we have already made a map. The instructor may suggest reading ahead, but this will rarely be needed. We learn most by our own mistakes, and the chapter on map design will be more meaningful after the student has already struggled with its problems. Note that the specifications for exercises are not overly detailed. We want to give the student all possible freedom in designing his work.

The exercises belong to two classes: (1) long-term maps lasting through several laboratory meetings and most likely to be finished as homework or after hours; (2) short-term works closely following the lectures and interspersed with the long-term maps.

SYLLABUS FOR CORRELATING LECTURES AND LABORATORY WORK

Chapter subject	*Short-term works*	*Long-term works*
Introduction	Tools and airplane photos to be presented	
1. Tools and Equipment	Practice with tools, p. 19	
2. Air-photo Reading	Marking air photo in field Scale exercises Map from air photo, p. 29	
3. The Principles of Map Making		
4. Field Methods		
5. The Principles of Lettering		
6. The Practice of Lettering	Practice in hand and machine lettering, pp. 64 and 65	Field work (3 afternoons.)
7. Relief Methods	Fill in contours, p. 69 Draw profiles, p. 71 Imaginary landscape, p. 78	Large-scale map of field work, Chap. 4
8. Land Forms and Land Slopes	Slope-category maps (optional), p. 91	
9. Government Maps		
10. Private Maps		
11. Map Collections and Compilation		
12. Map Design and Layout		
13. Lines, Shades, and Colors	Use duplicating machines and contact printing	
14. Map Reproduction		Medium-scale map, p. 124
15. The Earth		
16. Surveying	Practice with transit, p. 154	
17. Map Projections	Constructing projections, pp. 173, 178, and 190	
18. Azimuthal Projections, Grid Systems		
19. Thematic (Statistical) Maps		
20. Diagrams		
21. Cartograms		
22. Science Maps		Economic map, p. 214
23. Land-use and Economic Maps	Geostenography and panoramic section, p. 246	(Geology students may make Dufour diagram instead, p. 229)
24. Globes		
25. Models	Terrain model, p. 263	
26. Photography for Cartographers		
27. Modern Techniques	Practice in negative scribing, p. 297	

A set of examination questions is in the Appendix

Index

Pages references in **boldface** type indicate illustrations or tables, which are indexed only when not shown on the same or facing page as the text reference.

207770

DATE DUE